Royko

ALSO BY F. RICHARD CICCONE

Daley: Power and Presidential Politics

Chicago and the American Century

Royko

A Life in Print

F. RICHARD CICCONE

 PublicAffairs New York

Book design by Mark McGarry, Texas Type & Book Works.
Set in Monotype Dante.

Library of Congress Cataloging-in-Publication data
Ciccone, F. Richard.
Royko: a life in print / by F. Richard Ciccone.
p. cm.
ISBN 1-58648-172-X (PBK.)
1. Royko, Mike, 1932–1997. 2. Journalists—United States—Biography.
3. Chicago (Ill.)—Biography. I. Title.
PN4874.R744 C53 2001
070.92—dc21
[B]
00-054779

10 9 8 7 6 5 4 3 2 1

For Cristin and Richard

Acknowledgments

This book could not have been undertaken and written without Judy Royko, who shared not only her memories but her home, allowing me to spend the better part of three months in Royko's third-floor office with his many boxes of memorabilia and giving me access to read almost all his 8,000 columns; and I will be always grateful to Sam and Kate, who graciously let me prowl among them. Judy Royko was also the person who provided encouragement when I found myself overwhelmed by the details of her husband's long career and abbreviated life.

David Royko was extremely generous with his time and made available the more than 100 letters that his father wrote in a literary courtship of his mother. He, his wife, Karen, and his brother, Rob, were also gracious in discussing their life with their father.

Robert Royko shared his recollections of his brother and the nearly two dozen audiotapes he had made with family and friends. Royko's sister, Eleanor Royko-Cronin, provided family background.

Eleanor's memories of her parents and grandparents were rich with detail that provided some ancestral clues of Mike Royko's future eccentricities and talent.

Lois Wille was invaluable, having known Royko for all his professional life. Lois remembered the history of her and Royko's early days at the *Daily News* and a perspective of Chicago journalism in which she was a pivotal figure for thirty-five years. Her encouragement was as valuable as her recollections.

Many old friends and colleagues contributed greatly. Royko's closest friend, John Schackitano, talked openly about some of Royko's happiest and unhappiest days; Dan Hurley and Tim Wiegel, his golf partners of more than a decade, were also a great help. Phil Krone shared and understood Royko's fascination and knowledge of Chicago politicians, and Krone also helped shape them.

Royko's legmen, Terry Shaffer, Ellen Warren, Paul O'Connor, Hanke Gratteau, Wade Nelson, Nancy Ryan, Paul Sullivan, Susan Kuczka, Janan Hanna, and Pam Cytrynbaum, put his work and work habits in perspective.

Studs Terkel shared his memories of Royko from his earliest successes to his final triumphs.

Pam Warrick was gracious in describing her relationship with Royko. Sam Sianis and Rick Kogan knew the after-hours Royko best; Don Karaiskos provided background about Royko's air force days; Brad Burnside talked about Royko lunches; Ray Coffey, Dorothy Collin, Bill Garrett, Paul Zimbrakos, James Strong, Howard Tyner, Dorothy Stork, Ed Gilbreth, Ed Rooney, James Squires, Charley Madigan, Peter Gorner, and dozens of other Chicago newsmen talked not only about Royko, but the city and its newspapers in the years when Royko wrote in them. Ira Berkow shared letters he exchanged with Royko.

During his life, Royko was profiled by dozens of newspapers and magazines. Some of the ones used for this book were the *Washington Journalism Review, Esquire, Chicago, Time, Newsweek,* the *Reader, MORE,*

Chicago Independent, Chicago Scene, the *Washington Post,* the *Seattle Times,* the *Fort Lauderdale Sun-Sentinel,* the *St. Louis Post-Dispatch,* and the *Wall Street Journal.* Neal Grauer's *Wits and Sages* included a lengthy interview with Royko.

Royko also gave a lengthy interview to WBEZ public radio in Chicago on the 30th anniversary of his column and a similar interview in 1996 to Tower Productions for the 150th anniversary of the *Chicago Tribune.* Both were valuable resources. Doug Moe's *World of Mike Royko* (University of Wisconsin, 1999) was a fine map.

I would like to thank the *Chicago Sun-Times* for permission to excerpt columns that ran in the *Chicago Daily News* and the *Chicago Sun-Times,* and the *Chicago Tribune* for permissions to run columns that appeared in the *Tribune.*

The person who had the most to say about Mike Royko was, of course, Mike Royko and he said almost all of it in his more than 8,000 columns, which were the most valuable resource of all for this project.

A special note of thanks to Howard Tyner for arranging a meeting with his old friend Peter Osnos at PublicAffairs. My debt to Peter is much greater since this book is his idea, and I can only hope this book reaches his expectations and justifies his faith. Kate Darnton's contribution was the hardest, but she painstakingly and caringly edited, cut, moved, trimmed, and repaired something on just about every page. I have known many great editors in four decades as a newspaper man, but I have never found one any better.

Always, there was the love and patience of Joan.

Introduction

The day after Jimmy Carter's inauguration, I was sitting in the Class Reunion, a Washington saloon that had become a favorite hangout for the city's newspapermen. The bar's popularity among newsmen attracted political operatives, and in the next four years it would become a watering hole for many of the new President's top aides such as Hamilton Jordan and Jody Powell.

The *Chicago Tribune*'s Washington bureau was at 1717 H Street, just across the street from the Class Reunion, and it was often easier to find some of the *Tribune* staff in that bar than in the office. They were joined by those reporters from the *Los Angeles Times* and the *Boston Globe* who felt a certain psychological and professional kinship with their *Tribune* counterparts; they were all competing to climb up the ladder to parity with the *New York Times* and *Washington Post*.

Mike Royko was alone at the bar. Someone at my table invited him to join us. Introductions began and someone said, "Of course, you know Dick Ciccone."

Royko snarled, "He's a greaseball from the *Tribune*. Probably connected to the mob."

Royko knew exactly who I was. He knew I was a political writer for the *Tribune* and he knew I had spent most of the year covering the presidential campaign. The reason he knew was because he read the opposition papers carefully. He saw my byline, and he had also spotted me more than a few times at one or another of the Chicago taverns where he was always a center of attraction. Except for a condescending nod or grunt I had never had a conversation with Mike Royko. I knew he was fiercely competitive and disliked anyone who didn't work for his beloved *Chicago Daily News*. He hated *Trib* writers most of all. I also knew Royko's reputation for verbal savagery. I knew more than anyone else at the table—all of these guys were based in Washington—that Royko could use ethnic slurs as humor and self-deprecation as well as insult.

I was not fazed by his opening barb. I grew up in a western Pennsylvania steel-mill town that was as ethnically mixed as Royko's Milwaukee Avenue neighborhood. Calling people Polacks, Hunkies, dagos, spicks, or Chinks was as normal as using their names. It was the choice of adjectives surrounding such ethnic appellations that determined whether the usage was intended to be affectionate or hostile.

Royko decided to be hostile. "Yeah, I know him. He's a phony wop. I may not stay at this table. I don't like to drink with wops."

No one laughed. David Nyhan of the *Boston Globe*, who towered over Royko, interjected. "Hey, Ciccone's my friend and if you don't want to sit here, fine, leave."

Royko took a sip and grinned. "Dick knows I was just kidding. In Chicago we all talk that way to each other."

I grinned back, perversely proud that Mike Royko had picked on me. After all, in Chicago in 1977 and in any place in America where newspapermen gathered, Royko was known as the best in the business. He was the star.

As the drinking continued, the tablemates shifted. Some returned to work to finish a story; others who, like Royko and me, had been in town just to cover the inauguration started leaving for home.

After several more martinis, Royko asked me when I was heading back to Chicago.

"I've got an eight o'clock flight out of National," I said.

"You'll be home by ten, right?" he said.

"Sure, when are you leaving?" I asked.

"I'm taking a train tomorrow. I never fly. I won't be home until Saturday."

"That's crazy. Why don't you fly home with me?" I suggested.

"I'm afraid to fly. I was on a plane in the service that almost crashed."

"Just have a few more martinis and you won't be afraid of anything," I said.

A few more drinks and I had convinced Royko to come to the airport. He resisted at every step. When we got out of the taxi at National he announced that he was getting a cab back to the bar. I persuaded him to go inside the terminal. At the ticket counter, he protested that he didn't have enough money. I told him they'd be happy to take a credit card. He bought the ticket, and we proceeded to the gate waiting area.

"I can't do this," Royko said. He began shouting, "I've got a bomb, I've got a bomb." Two airline security people came running up. I convinced them Royko was hopelessly drunk. Terrorist bombing was not a real fear in 1977. The security people gave us a smirk and allowed us to board the plane.

As the plane engines revved up for takeoff, Royko unfastened his seat belt and bolted for the door. A stewardess insisted he sit down. We were at the front of the plane, only a few rows back from the curtain separating coach from first class, and everyone could hear the ruckus he was causing. As the plane climbed, one of the first-class passengers stuck his head through the curtain and said, "That's Mike Royko, let him come up here with me." The interloper was Charles Stauffacher, chairman of Field Enterprises, which operated the *Chicago Sun-Times* and *Daily News*. Royko dragged his new best friend —me—along and we settled in first class. The wise stewardess quickly slipped two glasses of vodka into our hands.

Also sitting in first class was the heavyweight champion of the world, Muhammad Ali, who was traveling with his wife and infant daughter. Ali was lifting the child high above his head. She was giggling. Royko walked over and introduced himself to Ali, who replied with a sullen look. He clearly had no idea who Royko was. "I'm the only guy who defended you when you changed your name from Cassius Clay," Royko proclaimed. Another sullen look.

Royko tried another tack. "You're not that tough, you know. You ever hear of Tony Zale? Tony Zale was the 'Man of Steel' from Gary. Never got knocked out. Tony Zale would have kicked the shit out of you."

Ali handed the baby to his wife. I thought, "He is going to get up and with one punch hit both of us and we will be dead."

Instead, Ali reached behind his back for an airline pillow, turned his face from us, and closed his eyes. The stewardess brought us another drink. We eventually landed. Royko left in Stauffacher's limousine. I caught a cab.

On the following Monday, I realized exactly how many people read Mike Royko. His column was devoted to the joys of flying. I was mentioned in a single line, "A friend, Dick Ciccone, suggested I get drunk."

The phone started ringing. I heard from college friends I hadn't seen in fifteen years and marine buddies I hadn't talked to in ten years. I got calls from relatives in small California towns and Pennsylvania valleys. I was a minicelebrity in the *Tribune* newsroom.

I saw Royko occasionally after that. He was always warm and

friendly. He stunned me once by telling me, "After me, I think your column is the most influential in town." That was probably the greatest compliment I ever received. He was sober, too.

In 1981 I was appointed metropolitan editor of the *Tribune*, and my first assignment—since I was the only *Tribune* management person who had remotely friendly relations with Royko—was to recruit Royko to the *Tribune*. I called him, and he said, "I just signed a new deal with Field. I'm happy here. But let's have a drink."

Three years later, Royko finally joined the *Tribune* where I, as managing editor, was responsible for having his column edited. Although he worked directly for the editor (if he worked for anyone), he would clear all his vacation time, days off, and other logistic problems through me. He resigned to me on several occasions and cursed me out on several others. For years, I chose which of his columns to submit for prize contests. I golfed with him a half-dozen times and sometimes shared his despondency over smoking and drinking, which were our common enemies. I could never share his deep depressions over the slightest insult or his own fears that one of his columns was not perfect.

Royko suffered from debilitating fits of anxiety and self-doubt. Like his family and his friends and his admirers, I was baffled that this most talented and most celebrated newspaperman had such difficulty accepting his own fame. Newsrooms everywhere are full of people who have achieved some moderate form of success and yet fret that tomorrow is the day when they will be found out. It is probably the same in any profession where success is judged on a subjective basis. But Royko had reigned for more than thirty years as the best columnist in the country. He won every journalism prize worth winning. It seemed he should be able to enjoy his fame. It seemed he should be comfortable knowing that whatever he wrote it would be—on most days—better than what anyone else was writing.

But he couldn't. His friend Studs Terkel called it "the demon." It was the demon that drove Royko's insecurity, that forced him to labor over each word, each phrase, and then wonder if it was any good. In his head, Royko knew it was good. But at night, when he couldn't sleep, the demon appeared and reminded him that in a few hours he would have to get up and do it all over again. And day after day, he got up and did it better than anyone else possibly could.

He did it better than anyone.

chapter 1

Sunset Ridge Country Club is one of the exclusive playgrounds on the North Shore of Chicago. It is in Northfield, nestled just north of Glenview and west of Winnetka, which straddles the ravines at the edge of Lake Michigan. The Sunset Ridge golf course was the site of the 1972 Western Open Championship won by the rather obscure Jim Jamieson. On most days the club's parking lot is filled with cars driven by women coming for lunch in the white clapboard clubhouse; or moms dropping off children for swimming or golf lessons; or lawyers, chief executives, and stockbrokers who sneak away from their downtown glass-enclosed worlds for a quick eighteen holes.

On a cool gray Chicago sort of day in March 1996, the small driving range at the western end of the tree-lined parking lot was not crowded. Mike Royko was hitting some balls. At age sixty-three he was fretting about losing distance on his fairway woods. He methodically checked his grip, tried different stances, twisted at the hips, and rolled his wrists, running through his standard routine. For four decades, Royko had adhered slavishly to the principles of Ben Hogan, whom Royko worshiped both for the work ethic that made Hogan the world's greatest golfer in the 1940s and 1950s and for his total immersion in the study of the game. These were also the traits of Mike Royko: relentless work and total dedication.

Royko did not merely play golf or listen to classical music or collect movies or catch bass or pitch softball. He attacked these hobbies as intellectual opponents to be vanquished. He studied, reported, consulted, and involved himself in every aspect of whatever it was he was interested in to the exclusion of family, friends, and civility.

For forty years Royko, when not absorbed with one of the many other things that fascinated him, was absorbed in golf. He would lecture friends about how to improve their game, pass out Xeroxed pages from Hogan's book, and demonstrate Hogan's version of the "grip" the way Moses must have enumerated the commandments. He would launch into critiques of various courses he had played and fantasize the design of the perfect golf course. Constructing the ultimate golf course was one of the options he considered for his retirement years, along with playing blues at a piano bar, operating a Florida fishing-shack, owning a magazine, buying a newspaper, and writing a novel.

Why Mike Royko, the son of a bartender, who became the most famous newspaperman in Chicago's fabled newspaper history, was standing on a golf course that had long been the preserve of the very rich, the very elite, and the very Protestant, and which epitomized all Royko envied and ridiculed for nearly forty years is as difficult to explain as Royko's moods, as ironic as his satire, and as complicated as the egomaniac twists and frightened loneliness of his genius. Royko was indeed a true genius, and the newspaper business did not produce many of them.

Newspapers dominated the information and entertainment needs of America for a century. They were the supreme provider of news and enlightenment. They groomed literary figures who changed the way Americans thought. They were an indispensable part of most people's everyday lives. For the final third of the twentieth century, no American newspaperman was as prolific, entertaining, poignant, vicious, reflective, sentimental, probing, courageous, contrarian, and provocative as Mike Royko.

Royko was the writer that others copied. He was the reporter who could not be beaten, possessing the wit that defied imitation and providing inspiration to almost everyone still trying to practice journalism by the rules of fair play, decency, and that elusive search for accuracy. He would despise today's world of digital and electronic reporting of rumor or hearsay shamelessly devoid of attribution. He hated anonymous reporting. He detested invasions of privacy that injured innocent family members. He abhorred factual errors. Of course, at one time or another in his forty-year career he committed all those journalistic sins, but he apologized for them. He was rigorous. He checked everything, made every telephone call, gave everyone a fair chance to rebut charges or defend their actions. Usually such responses gave him more ammunition, but he always gave them the chance.

In thirty-three years of writing columns five times a week, Royko became, more than any person during the twentieth century, the personification of Chicago, save perhaps his foe and foil, Mayor Richard J. Daley. His column was read in more than 600 newspapers all over America. He symbolized Chicago's heritage as the reckless frontier town turned into a megalopolis where millions of immigrants struggled in the blood and muck of stockyards and the sweat and fire of steel mills to buy a bungalow and educate their families, survive a depression, and win a few world wars. Although he gruffly denied that he spoke for anyone, Royko was the voice of the Chicago that worked hard, paid taxes, and slapped their kids when they needed it.

He was the voice of the little man, the illegal Polish immigrant doctor tossed out the country, the black civil rights marchers of the 1960s. At times he was also the voice of the Lake Forest privileged, the politicians, the Gold Coast millionaires, the cops, and the bureaucrats. He defended the Asians, the Hispanics, the Italians, the Greeks, the Swedes, the Hungarians, the Croatians, and the Serbs. He made fun of them all, of their drinking and cooking, and their muddled romances and their foolish whining about racism and prejudice when they didn't bother to vote often enough to throw the bigots out of office.

He wrote passionately about the civil rights movement of the 1960s, driving white suburban readers of the *Chicago Daily News* to cancel subscriptions. He wrote hundreds of columns about the city's abhorrent treatment of blacks, Hispanics, Native Americans, and other minorities. When most of white America thought its heavyweight champion, Cassius Clay, was an American disgrace, Royko noted that almost every successful entertainer in the country had changed his name. Muhammad Ali was all right with him. He savaged Jesse Jackson for twenty years for various acts of hypocrisy and self-promotion, but he defended and praised Jackson's 1984 bid for the Democratic presidential nomination. No other newspaper commentator or media figure of any stripe in the last half of the century had such an abundant record of defending the downtrodden.

If, between his practice swings, Mike Royko gave a few thoughts to retirement—he thought about retiring constantly those days—it would have only magnified the most ironic aspect of that cold first day of March 1996.

Just twenty miles south of Sunset Ridge, a block north of the Chicago River on Michigan Avenue, more than 1,000 people were marching around the Chicago Tribune Tower demanding that Royko retire. Actually they were naively hoping that the *Chicago Tribune* would fire Royko because of a column he had written Tuesday, February 27. The column stated that the only thing of value Mexico produced in the twentieth century was tequila. He said a lot of other sharp things, too. The piece was satire, a shrill mocking of Pat Buchanan's xenophobic speeches about Hispanics. Royko noted that Buchanan was suspected of privately calling Hispanics "beaners."

The protesters considered the column demeaning and insensitive to Mexican Americans. They burned and tore up copies of the *Tribune*, waved huge Mexican flags, and carried dozens of signs, including ones that read: "*Tribune* must act now to fire Royko," "Retire Mike. Shame on you!" "Stop insulting us." Chanting, "Kick him out" and "Royko go to hell," the protesters demanded a front-page apology from the *Tribune*.

How could these protesters attack the man who, in winning the Heywood Braun award, was described as the "ombudsman for the underdog"? How could they attack the man whose columns "could attack evils without showing the bitterness of his heart...for his sardonic, bold and courageous writing, always stressing the little man and what our society does to him"?

In his February 27 column Royko had written, "If Mexico is sincere about wanting to improve itself, it would stop pushing drugs and border hopping. Instead, it would invite us to invade and seize the entire country and turn it into the world's greatest golf resort. Let us be open about this. There is no reason for Mexico to be such a mess except that it is run by Mexicans, who have clearly established that they don't know what the heck they are doing."

The telephone began ringing at Tribune Tower that Tuesday night. The callers were furious. The *Tribune* corporate hierarchy responded with a statement on Thursday supporting Royko, and defending the column as political satire: "Anyone who has read Royko over the past 30 years knows he is not reluctant to speak sharply and sarcastically. We recognize some people were offended by Royko's column and regret that they have misinterpreted the intent."

Tribune publisher Jack Fuller agreed to meet with Hispanic leaders. On Thursday, Royko also responded to angry callers:

> The paper with the offending column had been on the newsstands for fourteen hours before the first protest came in. It had been delivered to homes about eight hours before the first phone call.
>
> Not until Hispanic broadcasters lifted words and lines out of context did the salsa suddenly hit the fan. But if people were upset by what they heard—or even read—then I will oblige them with an apology. A resignation? Very soon, but not quite yet. I have my own schedule for retirement.

Royko went on in his usual fashion of apology: He continued the attack. He apologized to Mexican drug lords and said that he now was against any kind of immigration laws so anyone who wanted could swarm across the border. Naturally, Royko's apology did not deter the planned protest.

A Mexican language radio station that had taken up the politically correct mantle demanded its listeners boycott the *Tribune*. The Hispanic members of the City Council, and, surprisingly, Alderman Edward Burke, one of the brightest and best-read Chicago pols, who surely understood satire when he read it, joined the protest.

Royko's column on the morning of the protest began:

> Unfortunately, a previous engagement will prevent me from attending the protest demonstration that is to be held outside of my workplace some time today. But I don't want the people who will be there demanding my firing or resignation to think that I am snubbing them.
>
> Nothing could be further from the truth. For the last two days, I have done little but listen to hundreds of phone callers, both live and on my overflowing voice mail machine.

So I thought it would be a good idea to print some of the calls here as they came in, except for a few colorful words and phrases that shouldn't be viewed by children.

Also, there will be no names since most of the callers were surprisingly shy about identifying themselves.

As usual, Royko found the crudest and silliest callers to quote. But he did not omit, as he easily could have, quoting a caller who brought up Royko's embarrassing and highly publicized arrest a year earlier: "The article you wrote yesterday, I disagree about it. I remember last year when you were a drunken driver. Were you drinking tequila or what?"

And he was uncharacteristically upset with another caller whose message said: "Mike, you (deleted), don't speak (deleted) with the Mexicans, you mother (deleted). I am Mexican, you stupid (deleted). Hey, (deleted) your momma."

In almost four decades, and more than seven million words, Royko had rarely lashed out in a personal diatribe against a reader whose mail and call he had received. But not this one.

I didn't intend to comment on this matter. But this last comment requires a response. My mother, after a life of hard work to keep our family together, died quite young of cancer. So if the last caller would be so kind as to contact me and identify himself and allow us to arrange a place to meet, I assure him that I will rip his dirty tongue out of his mouth and stuff it where it belongs. You, sir, are human garbage.

Other than that, hey, everybody, have a nice protest.

It was not the first time Royko's columns had been the subject of reader anger. White readers threatened to picket the *Chicago Daily News* during the 1960s when Royko defended black civil rights actions and longhaired antiwar protesters. Some of his Northwest Side neighbors threw a brick through his window that landed near his sleeping son. And on countless nights in countless bars, Royko had to withstand countless critics. He often shrugged off these critics by offering them the quarter or thirty-five cents they had paid for the newspaper.

This time it was different.

A few days after the protest, the *Tribune* wrote an editorial that sounded like it still supported Royko but not quite as enthusiastically.

Newspapers are not in the business of offending people who have done nothing wrong. Sometimes we manage to do so anyway, even though we don't set out to. But when one causes pain to people who don't deserve to suffer it, the only proper human response is regret. That's the proper journalistic response as well.

The Tribune is sorry that many Mexican-Americans and others were deeply insulted by the Royko columns. We are particularly sorry that the first public statement the company made in the aftermath added to the injury. That statement was not political satire. It was not journalism. It was a corporate statement, and it should have been on the mark.

We are not sorry that the *Tribune* publishes Mike Royko, who for three decades has stood courageously for the underdog and the rights of individuals against the power of the majority. A newspaper has to be willing to speak forthrightly in its news columns, to speak bluntly on its editorial page, and to let many voices speak pointedly, even knowing that occasionally the result will be discomfort.

In earlier times, newspapers wouldn't have acknowledged a protest. In the 1950s and 1960s, newspapers were arrogant. They did not worry about readers. They did not create focus groups to determine what the editorial page should say. When Royko became a newspaperman, American newspapers were still lingering from the effects of private ownership. Newspapers said what their owners wanted to say. The reader could do what he liked.

By the 1990s, corporate ownership and corporate sensitivity had tried to change the image of newspapers. For an institution as old and as steeped in tradition as the *Chicago Tribune* this was difficult, if not impossible. The looming Romanesque-Gothic tower that Colonel Robert R. McCormick had built in 1925 at the head of Michigan Avenue had long stood as a citadel of Republican conservatism for the entire Midwest. By the time Royko joined the *Tribune* in 1984, much of that Republican legacy had dissipated, but the image of the *Tribune* as a powerful member of the establishment persisted.

Ironically, Mike Royko grew up hating the *Chicago Tribune* for its many rigid conservative postures. He only went to work there as the lesser of journalism evils. Yet the once haughty *Tribune*, where a column perceived as insulting to immigrants would have been relished inside the newsroom, was now the arena of dissent that was most upsetting to Royko. The monolithic white male, Republican *Tribune* of his youth was now staffed almost equally with women, and was filled with many African Americans, Hispanics, and Asians. Most of all, it was young. Many of the reporters and editors who scuttled sideways when Royko passed them in the hallway without a nod or a mumble of recognition were not only quite oblivious of his long and honored record, but were practicing a different kind of journalism, one that he openly attacked and ridiculed. Political correctness was as disgusting to him as the racist attitudes he had found in newsrooms forty years earlier.

The reaction to the Mexican-American column, outside and inside Tribune Tower, was not the first time Royko was made aware that not all his colleagues were still filled with awe and admiration for him. In 1991, in the aftermath of the Rodney King–related rioting in Los Angeles, Royko had obtained a Los Angeles Police Department transcription of a conversation between the police dispatcher and a patrol car. He wrote a column as a parody of the 1950s *Dragnet* radio and television programs which featured the sparse dialogue of the fictional Los Angeles detective Joe Friday. In part, he quoted the police transcript and its references to blacks as "monkeys."

9:40 A.M. Drove down Slausen toward home of complainant. Monkeys in the trees, monkeys in the trees, hi ho dario, monkeys in the trees. Too bad we don't have time for some monkey-slapping. My partner says he wishes he could drive down Slausen with a flame thrower. We could have a barbecue.

9:45 A.M. Arrived at home of complainant. Asked her, "Ma'am, did you see the perpetrators?" She said she got a glimpse of two men running through her back yard. Asked her: "Be's 'dey Naugahyde?" She said: "What?" I said: "You know, be's 'dey Negrohide?" She asked if I meant African-Americans. I asked my partner if African-American meant the same as monkeys. He said he believed so. I said: "Yes, ma'am." She said: "No, they did not appear to be African-Americans." I said, too bad, I was in the mood for some monkey-slapping. Could have gone out on a monkey hunt and questioned a Buckwheat or a Willie Lunch Meat. But she described them as being swarthy. That was a good clue. Meant that we might get some Mexercise. Or maybe they were Indians, the towel-head kind, not the feather kind. Thanked her and left. No sense in staying. Didn't have a huge set of kazoopers.

Royko finished the column, as he usually did, about 6:30 P.M. and left the office. The column was edited on the city desk and processed as usual. About 8 P.M. a young African-American editor was searching the computer and came upon the column, which she read. She became incensed that Royko would include such degrading language in his column. She rounded up several colleagues and urged them to read it. All of them were young; some were minorities. They complained to the ranking editor on the floor, Carl Sotir, the News Editor. Sotir read the column. He agreed it was offensive. He said he would try to reach me, as I was the managing editor at the time. I was not home. Sotir then called Howard Tyner, the deputy managing editor, and explained the complaints about the column. He read several of the paragraphs to Tyner who then ordered the column killed. It was the first time in Royko's seven years at the *Tribune* that anyone even suggested his column be killed. When he was notified at home he was furious.

I returned home from a concert about 10:30 P.M. to a ringing telephone. It was Royko. "I am resigning from that fucking newspaper and never writing another word for that fucking rag."

It was not the first time an angry Royko had resigned to me or to many of the other editors he had worked for over the years "What's wrong now?" I asked.

"Those fucking assholes killed my column."

"Let me call the paper and find out what the hell's going on," I replied.

"It's too late, it's almost deadline. They killed it and I quit." Royko replied, hanging up the phone.

I reached Bill Parker, the assistant news editor, at almost 11 P.M., and Parker explained what had happened and that Tyner had made the decision. Parker then read me the column.

"Put it back in the paper," I said.

"It's getting late, we'll probably miss deadline," replied Parker. Allowing the presses to start late inevitably caused a chain-reaction that resulted in newspapers arriving late at commuter stations and causing dreaded declines in the daily sales, the ultimate sin for a news editor in charge of getting the edition out on time.

"Put the column back in the paper," I said. A few minutes later I called Royko at home. "The column is no longer killed. It's in the paper," I said.

"I still quit. I'm not coming in." Royko said.

The next morning Royko trudged through the office at his usual 10:30 arrival time, and said nothing.

The column with its references to African Americans as monkeys did not stir one complaint from the *Tribune*'s 600,000 readers. The only unrest was inside the newsroom. The death and resurrection of the column became media news. Howard Kurtz of the *Washington Post* wrote about it in his media column and so did *Editor & Publisher*. Royko eventually calmed down and said that he did not blame Tyner for initially killing the column, because Tyner had not had the chance to read the entire piece. The problem was with the naive young staffers more interested in political correctness than good journalism. Thankfully, this crisis disappeared quickly.

The crisis over the Mexican-American column did not. Some of Royko's closest friends thought he had not quite pulled off the satire. They told him the column seemed cruel.

Lois Wille, a two-time Pulitzer winner and the journalist Royko most admired for nearly forty years, knew him longer than anyone and was the single person whose judgment he most trusted. When Royko called Wille at her Virginia retirement home seeking consolation for the criticism over the column, she suggested maybe the phrasing could have been better. "You're wrong," he answered.

But Wille defended Royko publicly. "Mike Royko has not changed. Newspapers have changed. At the *Chicago Daily News* he did things that offended a lot of people. The only person he had to answer to was the editor. There was no hierarchy at the paper that commented or got nervous about what Mike wrote. Marshall Field [owner of the *Daily News* and the *Chicago Sun-Times*] would never have thought of responding to protests."

Studs Terkel, Pulitzer Prize–winning author and Chicago icon who formed a mutual admiration society with Royko, said, "His aim was lethal. Sometimes he went after the vulture and got the sparrow. In the last couple of years, he was hitting the wrong target unnecessarily." Terkel, who never drove a car and became an aficionado of Chicago taxi drivers, used to berate Royko for his columns deriding Middle Eastern immigrants who eagerly sought cab-driving jobs.

"He never let up," Terkel sighed.

Hanke Gratteau, one of Royko's sixteen "legmen," said, "He was mad that I thought the Mexican column missed the mark. His genius was that he could go right to the edge with something and never fall over. I thought on that one, he fell over."

Others did not tell him the column was bad. They looked for excuses. Rick Kogan, whose father, Herman, was one of Chicago's greatest all-around writer-editors and an idol of Royko's in his early newspaper days, said, "He would never have written the column in such a way that anyone could have missed the satire had he been working in the office. He wrote it from home and didn't have the usual people around to get feedback." Kogan, a *Tribune* feature writer, was one of the handful of people who were welcome visitors at Royko's corner office in the *Tribune*. But Royko didn't have many advisers, and while he sometimes tried out column material on visitors, he hardly ever asked his assistants or anyone to actually edit his work.

"At least half the columns he wrote when I worked for him, I had no idea what he was writing until I read it in the paper, " said *Tribune* columnist Ellen Warren. "Most of the time I left at five and he was still staring at the typewriter."

In the weeks following the Mexican-American furor, the *Tribune* printed several letters allowing offended Mexican Americans to vent their displeasure. The newspaper printed one defending Royko. It was written by Jon Leonard:

After reading the column carefully, I explained to these students that Royko's intention was to poke fun at Buchanan's extremist stances, as well as at voter's aptness to fall prey to such demagoguery. I then spent the next 45 minutes giving them a lesson on political satire, explaining its history going back to Jonathan Swift's "Modest Proposal," in which he proposed eating the children of England as a way to cut down on poverty and homelessness.

Once the students understood the basis of the column in the context of political satire, they found it to be as hilarious and pointedly critical of Buchanan and the far Right as I did.

Leonard got it right. Royko's column was mean-spirited, but throughout his career he had fired shots at the pomposity or posturing of one group or another trying to achieve power or status at the expense of someone else. Paul O'Connor, who worked for Royko in the 1970s, said the controversial column was simply a Royko trademark. "It was a rite of passage for immigrants to be ridiculed. Royko grew up with that and he wrote about it. It happened to every ethnic group. You knew you were on the way to being a full fledged American when people started making fun of you and calling you ethnic names. He did it to everyone, the Irish, the Italians, the Poles. When he did it to the Mexicans the times were different and they went berserk. To him it was being part of the club."

When the noise died down, Howard Tyner, by then editor of the *Tribune*, assigned Royko a permanent editor. Royko was offended. He thought it was a signal that after thirty-three years he could no longer be trusted, and that the *Tribune* was appeasing the Generation X staffers who considered him a dinosaur. After a brief period, the column was again sent through the city desk. But more and more, young editors whom Royko had never heard of began calling him at home at night to ask questions about the day's column.

Royko loved movies and memorized huge swatches of dialogue from dozens of John Wayne films and from other movies he watched over and over. But the line he quoted most was from *The Godfather*, the Mafia classic where one character tries to explain to his brother that the shooting of their father was not something to become emotional about. "It's business, it's not personal."

For all the idealism that cloaked his words, Royko was a practical man. The son of an immigrant, he grew up in the depression and found himself earning more money than any other newspaper columnist in America. He quoted the line from *The Godfather* many times to rationalize the idiosyncrasies of editors and corporate bosses who were lining his portfolio with stock options. But by the 1990s, the resentment he felt among his newspaper peers was not just business. It was personal.

"He was devastated by the way the people at the Tribune reacted," said his wife, Judy. "I went up to his office on Thursday, the day before the protest, to get him to go out for dinner and he was at his desk. He was depressed and disappointed. He couldn't understand how anyone could think after all the years of writing columns that he was a racist. He was very hurt."

On the Friday of the demonstration, when Royko returned from the Sunset Ridge driving range, he had a few drinks and called Howard Tyner, who had also had a few drinks after the long week of dissent.

"Royko called me and said he was quitting," Tyner recalled. "After all the crap of that week I wasn't about to say, 'Please don't.' I just said something like, 'Let's talk,' and we hung up.

"The next day was Saturday and I called his house and asked if it was all right if I came over. I went and we talked for several hours about everything except what we wanted to talk about. Finally, I said, 'You're not quitting?' Royko said, 'No, in fact, I'm going to write a column for Monday and go back to writing five columns a week just to show everybody I'm not fired.'

For the next two weeks, Royko wrote Monday columns, which he had not been doing for the past two years. Then he reverted to his four-a-week schedule. Four columns a week is a monumental load, but Royko always resisted it because he was so proud that he had written five columns a week for thirty-plus years while his peers often did only one or two.

Besides answering telephone calls from Mexican-Americans,

Tyner had been battling the dissent in the newsroom. One of his top editors, a Hispanic, had demanded that he be allowed to write a piece for the op-ed page denouncing Royko on behalf of the *Tribune's* Hispanic staffers.

The newsroom resentment was partly caused by Royko himself. Royko never felt comfortable at the *Tribune*. He was not really a newsman by the time he arrived in 1984. He was a legend. Few of the 500 *Tribune* newsroom employees dared approach him, and when someone ventured a hesitant "Hello," it was likely to be greeted by a grunt or a scowl. Most people thought it was arrogance. But it was a combination of traits that the his old colleagues at the *Daily News* had learned to accept. Royko, despite his celebrity, his success, and his moments of arrogance, was shy. Royko also hated mornings. Some mornings he was hungover. Every morning he was facing the demon of the column. As a result, Royko made few friends at the *Tribune*. His pals were the old *Daily News* or *Sun-Times* veterans, several of whom he had helped get jobs at the *Tribune*. He simply didn't respect most people at the *Tribune*. He had spent too much of his professional life "pissing all over those people" to suddenly turn around and accept them as good newsmen. And there were several colleagues he actively disliked and disdained.

It was also a generational passage. By the time Royko arrived at the *Tribune* he was fifty-two years old, not an age to enjoy the barroom camaraderie of cub reporters. And the barroom revelry that was so much a part of the journalism of his youth was gone. Reporters no longer hung around to bitch about assignments or editors, replay their investigation, trade ideas, or gloat over a scoop. Reporters in the 1980s ate salads at their desks and went to health clubs after work. Royko didn't like much about them. They no longer spent all their hours at their jobs. They didn't drink. He didn't think they worked very hard, and he was certain they had no idea how hard he worked.

For Royko, the column was everything. It came before his wife, before his children, before his friends, before his golf game. A good column could put him in a euphoric mood. A finished column could send him to his favorite bar in search of a reward. Mike Royko, writer, was not the same as Mike Royko, regular guy, the sweet person who loved shooting off fireworks on July Fourth, scaring trick or treaters on Halloween, and slyly explaining the difference between a Von Karajan and Solti recording of the same symphony. The column created Mike Royko, the tough, witty celebrity.

"The column took a terrible toll on him," said Hanke Gratteau, one of Royko's closest friends. "He paid a great price. Everyday it had to be good. Everyday it had to be important to everyone. Everything else had to come second. The family had to come second. His health had to come second. He worked when he was sick, when he was exhausted, when he was hung over."

Royko dedicated *Boss*, his 1971 best-selling book on the Daley machine, to his sons, David and Robby, "for all the Sundays missed." David wondered about that. "There were never any Sundays, not while he was writing *Boss*, not before it, and not after. Every Sunday of my life he was working on the column."

After the ruckus over the Mexican column, Royko began to work more and more from the third-floor office in his prairie-style home overlooking the ravines of Winnetka. Each day he would go to his computer and look over his financial statement, figuring out if he had enough money to retire. In 1990 he told *Chicago Magazine*, "In two years I expect to be out of the column. I expect to call it quits. I know my craft. I don't have to prove anything." He had said the same thing to interviewers two years earlier and would repeat it two years later. Some of his friends doubted he would ever retire.

In 1995 he told Tyner he was quitting at the end of the year. In 1996 he talked of quitting in the fall. But he never quit. He had come into the newspaper business when police reporters were on the take from mobsters, and editors paid more attention to advertisers than readers.

He had one newspaper, the *Daily News*, folded under him and another, the *Sun-Times*, sold to a man he could not respect or work for. He insulted and taunted the most vicious killers in Chicago and ridiculed the most powerful big-city political boss of the century. Royko had not only seen the city and its newspapers change for the better; in small and large ways he was responsible for both. If, in the fourth decade of writing a daily column he was not as appreciated by his peers, if newspaper readers were a dwindling crowd and if newspapers were becoming homogenized and bland, it did not affect Royko's passion for damning hypocrisy and stupidity in the same breath. He never waved the flag of the First Amendment, although he was its best example for forty years.

"He was the soul of journalism," said James D. Squires, editor of the *Tribune* from 1981 until 1989. "He inherited from H. L. Mencken the mantle of reminding all of us all the time of our responsibilities to seek out the truth and represent all the people. It was Mencken, and then Royko, who took the load on their shoulders and defined peer review, who made us all watchdogs of one another and in doing so made newspapers their very best. Royko was the one who inspired my generation and this generation to act and think on behalf of the interests of the little people. Hopefully that spirit will live on because Royko was the last one who could inspire it. There is no one out there today like him."

Royko would have scoffed. He refused to be labeled or praised for anything other than his hard work. He liked to call himself "the tallest midget in the circus." He was far more interested in preserving the rights of everyone to live as they pleased so long as they did not disturb his right to do the same. Beneath the millions of words that created so much laughter was a toughness that drove him. It was far deeper than the facade of rudeness that sparked his well-publicized barroom brawls or disappointed admirers.

That toughness was in his blood.

chapter 2

Blood has flowed freely in Galicia for centuries. This land at the western edge of the great central European steppes was a trading center and crossroads that dates back to the thirteenth century, and it has been a battleground for Mongols, Tatars, Turks, Cossacks, Slovaks, Poles, Lithuanians, Czechs, Austrians, Hungarians, and Russians.

From the fourteenth century until the dawn of World War II, Galicia was an Eastern European melting pot, with as many Poles as Ukrainians, and a good smattering of Russians, Bohemians, Slovaks, and Hungarians. They spoke each other's languages freely and interchangeably. They traveled from country to country even easier than these countries changed their names and allegiances through the tumult of two twentieth-century conflagrations and the onslaught of both fascist and communist horrors.

The Ukrainian city of Lvov was the traditional capital of Galacia. It sits just north of the foothills of the Carpathian Mountains, and is only a few hundred miles in each direction from Romania and Hungary to the south, Bohemia and Slovakia and Poland to the west, and Byelorussia to the north. Just sixty miles southwest of Lvov lies the small town of Dolina. In the late nineteeth century, Dolina was a lumber town and attracted workers from other countries to cut the beech and oak that grew at the lower levels of the Carpathians. It was also a place where many Ukrainians in the east sent their young daughters to stay with relatives in the hopes that a passing Polish or Slovak woodcutter might marry them.

When Georg Rojko arrived in Dolina in the early 1890s, he not only found a job cutting lumber, he was also hired as a security guard by the lumber company. Rojko was a fortunate man to have two jobs. In the early 1890s, a great drought lay over eastern Europe and many people starved, but Rojko prospered. He was able to read and write, and the local Dolina residents speculated on whether he came from Poland or Slovakia because the latter country had a reputation for fine education. Or perhaps he was Hungarian owing to the fact that the name Rojko was not common to Russia or Poland. Rojko was obviously a prize catch as a husband, and in a short time he was betrothed to Anna Wolkowski, the daughter of a farmer who probably came from Poland. They were married in 1896. While Rojko labored at the lumberyard, Anna worked in the fields with other farmwives. In 1897, she was digging up potatoes when she went into labor. "The baby was born right there in the field. I cut the cord, put him in my apron, and took him to the house and cleaned him and put him on a bed. Then I went back to pick potatoes," she would tell her grandchildren years later.

The child was a boy and was named Mikhail. The next year, a daughter, Caroline, was born, and in 1900, another daughter, Louise.

In the winter of 1902, Rojko was called out in the bitter cold to fight a fire at the lumber mill. He got sick, developed pneumonia, and died in a matter of days. Anna was left with no means of support. She decided to join a brother in America. She left her three young children with an older sister, Maria, who lived on a farm and had twelve children of her own. Anna promised her five-year-old son, Mikhail, she would send for him and his sisters as soon as she could save enough money. She had heard there were many jobs in the burgeoning city of Chicago.

On their first day at Maria's farm, the children were told to go into the barn and join the family and farm laborers at a large table for dinner. When young Mikhail and his sisters went to the table they were pushed aside by the older children. By the time the others had finished there was no food left for them. They went to sleep hungry.

The next morning the same thing happened. That evening, the five-year-old Mikhail was determined to eat. He waited as the food bowls were brought from the house and set on the large table. He then dived between the older children so he could get his head close to the bowls. Before the older children could push him back, he spat several times into the food. The older children beat him viciously, but they would not eat the food he had fouled. Mikhail took the bowls to a corner and shared them with his little sisters. He was pleased to have gotten himself and his little sisters dinner, but he also felt ashamed of what he had done. He knew his mother would be angry if she knew how he had behaved. He went to the farmhouse and told his aunt what he had done, showing her the bruises and splotches of his beating. The next morning, the aunt took a separate bowl of food to the barn and told the others they must not eat it because Mikhail had already spit in it. That was how the Rojko children dined. And that was when young Mikhail Rojko learned that he had to fight for whatever he was going to get in this world.

By 1906, Anna had saved enough money working twelve hours a day in a laundry to bring her son to Chicago. By this time, the family was spelling their name Royko, probably the courtesy of an Ellis Island immigration official. The nine-year-old Mikhail was quickly sent to work in the steel mills on the South Side, running errands, carrying water, and unloading wagons. He and his mother lived with his uncle, saving every penny they didn't need for food. In 1914 Mikhail's sisters came to America, but immigration authorities in New York sent Louise back to Europe because of an eye infection. Caroline moved in with her mother and brother and soon married. In 1918 America entered World War I and Mike Royko joined the army. He volunteered for a cavalry unit like many of his Polish friends. Royko never talked much about his brief army career. He did not go overseas, and he apparently served his brief tour at a U.S. training base. The war also prevented Louise from leaving Europe. She didn't join her family in Chicago until 1920.

Growing up poor and tough, Royko hung out with his Ukrainian and Polish friends. They all lived near each other and worked in the mills together. They played cards and went to picnics, weddings, and dances. These were the social highlights for most immigrants, and there was always homemade wine and beer. Royko loved gambling and dancing, and was always ready for a brawl. He never backed down from anyone. Living in a city where everyone from merchant princes to the newest arrival was looking for a fast way to make a buck, he began to measure men by how hard they worked and how much money they made. He was determined to work harder than anyone and get his share of the money. And his share of the fun.

Dance contests were one of the fads of post–World War I America, and Royko entered a lot of them. One of his partners was Pauline Zak. Pauline and Royko claimed several contests before Royko warned her they had to stop winning or no one else would enter and there would be no more prize money. At one dance, Pauline introduced Mike to her older sister, Helen, a vivacious, cheerful girl who had, in Royko's words, "cement feet."

Helen Zak was eighteen, the second child born to Anton and Amelia Zak. Her mother, Amelia Bielski, was a native of Poland who was living in Newark, New Jersey, when she met Anton Zak, a former circus performer from Russia. Anton was a small, sprightly man. Like so many Eastern Europeans who had worked the land as *moujiks,* he dreamed of owning his own farm. In the early 1890s, he traveled to downstate Illinois and purchased a small farm in De Pue, about 100 miles southwest of Chicago. When Anton told Amelia about his farm she had visions of the wealthy estates owned by noblemen in Poland, and she agreed to marry him. She quickly discovered that she was not mistress of an estate but rather a simple farmer's wife who had to help feed the pigs and clean out the chicken coop. She gave birth to a son, Henry, in 1901 and decided she had had enough of life on the farm. She convinced Anton they should move back to Poland where her family had some businesses. She said they could save money there and return and purchase a bigger farm. Anton agreed to sell his farm, and they booked passage to Europe. Amelia was pregnant again, and Helen was born aboard ship in 1901.

In Poland, near Krakow, Amelia opened a saloon, a restaurant, a bakery, and a small hotel and ran all of them prosperously. She also gave birth to another daughter, Sophie. But Anton yearned for his Illinois farm. She persuaded him to return to America and gave him the money to buy back his farm, promising she would follow him in a brief time. Anton kissed her good-bye and returned to Illinois. He never saw her again.

Amelia then fell in love with a Russian count and bore another daughter, Pauline. Later, she began an affair with a former Russian cavalry officer who had to leave Russian in a hurry after getting caught supplementing his income by selling army supplies in the villages. When Amelia decided to return to America with her new lover, he had to borrow her brother's passport because he couldn't return to Russia and apply for one in his name. They settled in Chicago, and he became "Mike Bielski." No one in the family knew him by any other name. Bielski became a housepainter Royko would later describe as difficult: "If you're used to having people salute you, painting some-body's house is a bit of a comedown...He hit the juice pretty good. He'd paint and he'd drink."

Amelia and her Russian cavalry officer never married, perhaps because he didn't have proper identification, or perhaps because, despite bearing him two daughters, Adele and Betty, Amelia was a feminist long before it was popular. She smoked and read constantly, and disdained domestic chores.

Anton Zak remained on his farm for the rest of his life. He married his housekeeper, and although he never saw Amelia again his children and grandchildren visited him often. During summer vacations Zak taught them Russian dances while he played the violin or concertina or accordion, a few of the many instruments he had learned to play in his circus days.

In 1920, Anton and Amelia's daughter, Helen Zak, married Anna and Georg's son, Mike Royko. On April 9, 1921, their first child, Eleanor, was born. On December 2, 1922, they had another daughter, Dorothy. Royko now had a growing family and was looking for a better way to earn money than laboring in factories. The family lived above a grocery store at 2020 N. Hoyne.

When the Roaring Twenties arrived, Royko saw an opportunity to make good money driving a taxi. The taxi wars were raging in Chicago. John Hertz had started Yellow Cab Company before he moved on to create the largest rental car company in the world. When a rival company, Checker Cab Company, started up, the two companies engaged in acts of terrorism against each other. Drivers were beaten, taxies were burned. Someone set fire to Hertz's horse barn in McHenry County, and the owner of Checker Cab had his home sprayed by machine-gun fire. But Royko was tough enough to take and give out beatings. He was also still drinking and dancing, and his eye for women hadn't closed after his marriage. If there was a time and place for a young man with a few dollars in his pocket to have fun it was Chicago in the 1920s. There were speakeasies on every corner, and Louis Armstrong was inventing jazz on the South Side. There were flappers and hookers, and there was no reason for a dancing man to go home after a hard night driving. Mike and Helen's marriage broke up; they divorced.

Mike's fancy-free lifestyle didn't last. As the Roaring Twenties ended and the Great Depression began, people stopped taking taxis. The party was over. Mike Royko came home. He and Helen remarried and moved into a basement flat on the northwest corner of Potomac and Wolcott Avenues on the near Northwest Side.

Several blocks south, at the corner of Augusta and Damen, stood the Pure Farm Dairy, which was owned by a Ukrainian immigrant. Royko went to ask for a job. The dairy owner said he had no jobs but that he would hire Royko if Royko could sign up 200 people to order milk.

Royko was neither the best husband in the world nor the best father in the world, but he was a great friend. He would do anything for his friends, and they knew it. So he went and talked to all of them, Ukrainians and Poles. He told them that because they were all on relief and that they got milk free, it didn't matter who delivered it. He signed up 200 customers and promptly entered the dairy business.

Royko was an inventive dairy deliveryman. If he was on the milk wagon and passed a store early in the morning when the bakery delivery was outside, he would help himself to several loaves of bread and hand them out to his customers. If anything extra happened to be on the wagon—some cheese, butter, or milk that was spoiling—he gave it to them. They turned it into sour cream, which they used for salad dressing and soups. He usually helped himself, but the dairy owner, realizing Royko was good for his business, said he could buy a certain amount of sour cream and cheese at cost. "The rest you can steal from me," he said.

Royko had never been to school and didn't know how to read.

When his daughter, Eleanor, was eleven, he asked her to teach him to read. She also helped with the bookkeeping at the dairy and baby-sat for her new brother.

Nineteen thirty-two was a year that marked the beginnings and the endings of some of America's most colorful and important careers. Some of them were significant and some were notorious. Some of them began with the U.S. presidency and some of them began in a two-flat on the Northwest Side.

On September 19, 1932, Mike Royko was born in St. Mary of Nazareth Hospital at Division and Leavitt Streets. It was the depth of the Great Depression. Fittingly, Royko and the Democratic machine of Chicago arrived at the same time. Their lives would be entwined in their journeys to American celebrity.

In November 1932, only six weeks after Royko's birth, the Chicago Democrats took hold of the city. Chicago gave Franklin Roosevelt huge majorities. In the Twenty-fourth Ward, not far from the Royko household, 98 percent of the vote went to Roosevelt. It was the biggest margin in any ward in America. The Democrats elected Ed Kelly mayor in such a landslide they clamped a hold on city hall for the remainder of the century. In their seventy years of power the Democrats became renowned for playing pivotal roles in the elections of Harry Truman and John F. Kennedy, and they became notorious for ghost voting and double-dipping, ballot-box theft and whatever form of chicanery delivered votes, cash, and sometimes both. The Chicago machine was so adept at winning or stealing elections that it became the envy of every American big-city boss, and it created the greatest boss of them all, Richard J. Daley. Daley ruled Chicago for twenty-one years. The only voice decrying his autocracy and insensitivity was raised by Mike Royko, who was only a baby crying in his crib when it all began.

In 1932, Chicago was less than 100 years old, but it had already grown from a simple trading post set up by a black man, Jean Baptiste Du Sable, to a massive melting pot of nearly 3.5 million people.

In 1932, the Chicago school system was broke, forcing its newly elected mayor to travel to Miami to beg President-elect Roosevelt for a federal bailout. Chicago got the money, but the mayor, Anton Cermak, was standing too close to Roosevelt when a gunman took a shot at the president. Cermak was killed.

In 1932, Samuel Insull, the creator of Commonwealth Edison and the magnate who controlled electric power in twenty-seven states, became the most despised man in Chicago. Insull's utility empire had been one of the casualties of the stock market crash. Many of the banks that had closed in Chicago had invested in Insull's holding companies, and depositors who lost their money made him the scapegoat.

In 1932, Al Capone went to jail. His legacy— a criminal organization that dominated the saloon and restaurant industries, the labor unions, and the Democratic party—would subsist far into the future. In fact, young Mike Royko would make his first mark as a newspaperman covering the remnants of Capone's mob in the 1960s.

The early years were happy ones for Mickey. In December 1935, his kid brother, Robert, was born. The family was stable, and some of Mickey's earliest memories were of his father leaving at 3 A.M. on his dairy route. He knew how hard the work was, carrying milk up to the third and fourth floors, working late into the afternoon. He learned at an early age about the virtue of hard work, and he never forgot it.

From the beginning, he was called "Mickey," and he was a precocious child. "I remember," Eleanor Cronin, Royko's sister, said, "when Dorothy was dating a guy who talked real Chicago—you know, dese, dems, and dose—and while he was talking to us Mickey came up and said, 'It's there, not dere,' and the guy said, 'What do we have here, a little professor?' For a while we called him the 'professor.'"

Robert was soon sleeping in the bunk bed beneath Mickey, and doing all the things little brothers do to big brothers. "I was always getting him into fights although he always talked his way out of things. He was always playing practical jokes but I don't ever remember him being harsh with me or putting a hand on me. My clearest image of him was lying on the couch with a book. I thought it was strange on a summer day when I was outside playing he'd be on the couch with a book. He was always reading, everything, the classics, whatever he could find. That's what I remember about Mickey as a kid, always reading," Bob Royko recalled.

When he was four, Mickey had to have surgery for a hernia. His sister, Dorothy, recalled, "I told mama it was because he cried so much as a baby. But I went to the hospital and read to him. I always read to him. Mama always read but she read fairy tales. I read Hemingway and Steinbeck and I remember when Mickey was older I read a book by James Baldwin, *Another Country*, and it was so powerful. We talked about it."

In those years, Mickey's baby-sitter was often the man he called Grandpa Bielski. The family trips to the farm in De Pue had ended, and Mickey never saw his real grandfather, Anton Zak. But Grandpa Bielski provided Royko with wonderful memories that he would later use in columnns.

> We never walked far. Only one block up Wolcott Avenue to the nearest tavern on Division Street…As he sat in the tavern drinking his Saturday steins of beer, he would always say: "Tell your mother we went to Wicker Park."
>
> I asked him why we didn't go to Wicker Park.
>
> "Because I don't like Wicker Park. Too many flies and bugs. This is better."
>
> So I would tell my mother that we had gone to Wicker Park, and that seemed to please her. And not going to Wicker Park pleased my grandfather. It also pleased me since I preferred drinking ginger ale

in the cool bar. That meant that all three of us were pleased...So I learned that freedom of choice and movement is what makes people happy.

He would also say: "Always take care of your brushes."

"If you don't take care of your brushes you won't get work and you'll be a bum."

The year 1938 was a big one for Chicago and for the Royko family. The Cubs won the National League pennant after catcher Gabby Hartnett struck a home run on a September afternoon so dreary the ball disappeared into the fog that had settled over Wrigley Field. It was forever after called the "Homer in the Gloaming" and became part of young Mickey Royko's folklore. It may well have been that glorious season that started Royko's life-long love affair with the Cubs.

It was also the year Mickey Royko began school at the Hans Christian Andersen elementary school, and the year that Mike Royko Sr. went into the saloon business.

"He wanted to get into the bar business because he thought there was money in it," Eleanor Cronin recalled. "He had learned if you really want to make money you should own your own business. It was something you didn't need an education for, and he wasn't afraid to try anything. He had no fears. And he was an outgoing man, friendly with everyone, especially with the women. Women loved him. All he needed was a license and the keys to the door, and both were easy to get in those days," she said.

Mike Royko opened the Blue Sky Lounge at 2122 N. Milwaukee, and the family moved in upstairs. Mickey Royko had gone from the "flat in the basement kid" to the "flat above the tavern kid"—phrases he used to describe himself throughout his career. He also began second grade at the Salmon P. Chase school, 2021 N. Point Street, almost across the street from the saloon.

"Once the business started," Eleanor Cronin said, "my mother did all the cooking. Every Friday she would make potato pancakes, because that was the day the men got paid and came in to cash their checks. She would put a few potato pancakes on the bar and tell the guys that's what she was making for Friday night dinner and suggest they should bring their wives over. She was pretty good at marketing."

Years later, young Mickey would write of his first impressions of the saloon:

> From the day the place opened, one work of art was tacked to the wall of the Blue Sky Lounge on Milwaukee Avenue—a huge portrait of President Franklin Delano Roosevelt. It had been donated to the tavern's owners by Stanley the Mooch, who was an assistant precinct captain.

And while his formal schooling in the second grade had shifted to the Salmon P. Chase School almost across the street, it was clear Mickey's earliest lessons in partisan politics came on a barstool.

> To the patrons of the Blue Sky—the name stemmed from the cheap, blue, crepe-paper fake ceiling held in place by star-tipped nails— politics was quite simple. First, Republicans were all rich, greedy no-good WASP bastards....Democrats were all hard-working, two-fisted family men.

Over the next fifty years Royko would confess to voting for a few Republicans, but only a few.

During their adolescence, Mickey and Bob would explore their neighborhood, especially on Saturday trips to the movies—to the Strand, the Royal, the Congress, Harding, Oak, and Rio—down Division Street, where Mickey knew all of the bars because of his grandfather's unique approach to baby-sitting.

Their father took them to Lake Michigan to net smelt and sometimes to Wrigley Field to see the Cubs. Mickey's first visit to Wrigley was on opening day in 1939 when his dad hit a daily double and took Mickey to the ball game along with Dutch Louie and Shakey Tony, loyal customers. The first Cub hitter slammed a pitch off the ivy-covered wall and wound up on third base with a triple. The six-year-old Mickey became a Cub fan for life.

By the summers of World War II, Mickey and Bob would go to the games on their own. They walked the five miles from Milwaukee and Armitage to save streetcar money for a Coke. They carried fried egg sandwiches in wax paper and waited outside the left-field gate until they were allowed in free in exchange for setting up the folding chairs that were used in the box seats.

On hot nights they slept in Humboldt Park with many of their neighbors, and on hot days they played baseball and sixteen-inch softball, a unique Chicago game which Royko loved as much as the Cubs.

Mike Royko Sr. had opened the Blue Sky in partnership with his cousin Gus, whose family also moved into the flat above the bar. Royko's sister Dorothy remembered the apartment was crowded and violent. "They had three little girls and then Uncle Gus was always beating up Aunt Catherine. He was such a wonderful man, I loved him, but when he started to drink, he got mean." Finally, the cousins ended the partnership, but Mike Sr. was also drinking a lot and had started an affair with a woman who would eventually become his second wife. This was a painful period for the Royko family, especially when Mike Sr. hit and kicked Helen in front of Mickey and little Bob.

"He broke her ribs," Dorothy said, "and when he came around a few days later I told him to go away, that Mama didn't want to see him again, that she was going to divorce him. He began to scream at me but he went away. I don't think Mickey ever felt the same about him after he saw him beat Mama."

When Mike and Helen divorced for the second and final time in 1940, it was difficult for Mickey. He adored both parents and was much closer to his father than were the other Royko children. It was a difficult few years. In September 1940, Amelia Bielski was killed by a hit-and-run driver as she left a movie theater on Milwaukee Avenue. In the spring of 1941, Anton Zak died at his farm, and a year later, Mike Bielski, who had disappeared for several years and hadn't lived with Amelia since the 1920s, was found dead on skid row. In later years, Royko would write that Bielski was stabbed to death in a bar fight.

Early high-school classes bored Mickey. He had skipped two grades, which made him younger and smaller than his classmates. He began getting in trouble in the neighborhood. He was angry and frustrated by his parents' separation and divorce. His father was now operating a bar on the South Side, and his mother had opened a cleaning shop. With the post–World War II boom underway, the family suddenly had some money. Chicago was a wide-open town. Every bartender was a bookmaker. The cops were ready to be bought and stay bought, and the precinct captain was the local "Godfather" as long as everyone voted Democratic. In Chicago, they did.

Mickey's parents could afford boarding school so they sent him to Morgan Park Military Academy on the South Side.

"Every Saturday," Bob Royko recalled, "my mother would load up a bag with candy and fruit, and I'd ride out there with her. I'd take the Red Rocket, the old street car. I'd get on at Western and North Avenue and go all the way to 111th Street, then get off and go east two blocks. He would be happy to see me. He was so sad. He hated the whole regimen of the place and a lot of the kids there had problems."

During that lonely fall of 1946, Royko found a hero named Johnny Lujack. At that time, the greatest college football team in the country, Army, was filled with all-Americans like Doc Blanchard and Glen Davis, who were exempted from World War II service until their graduation. In 1944 and 1945 Army had crushed Notre Dame. Now with the war over, Notre Dame had many veterans returning. Key among them was Lujack. Both teams were unbeaten when they met in New York City. Late in the game Blanchard broke in the clear and appeared headed for the winning touchdown when Lujack cut him down thirty yards from the goal line to preserve the 0–0 tie.

In the contrarian fashion that would distinguish his writing, Royko was alone among all the young Morgan Park cadets in rooting against Army.

"It was the only time at that place that I was happy, when Lujack made the tackle," Royko said. "I hated it [the academy], so I finally persuaded my parents to take me out of it. I flunked several courses. It was my way of getting out of that hell hole."

But the teenager's educational odyssey was just beginning. He was next enrolled at Tuley High School, which he attended for a month. "I just stopped going. There was nothing to do. It was regimented and boring."

Royko went to live with his father, who was running the Cullerton Inn on the South Side. It was during this period that many of the saloon fables that would fill his columns were formed. He would write:

> Additional duties: Accepting bets on horses. Preventing customers from falling asleep with head in toilet. Admitting regular patrons through the side door at 8:00 A.M. on Sunday so they could get over the shakes and go to church, answering phones and telling wives that husbands had not been there all evening; appraising wristwatches for payment of drinks in lieu of cash; dispensing hard-boiled eggs, pickled pigs feet, beef jerky and other gourmet delights; breaking up fights by unleashing a Doberman named Death; and, finally, giving monthly cash-stuffed envelope to police bagman for assorted favors such as overlooking a 14-year-old bartender.

Royko met Nick, the bookie from Cermak Road who wore silk shirts, a genuine panama, and a big pinkie ring. "My father couldn't afford to carry any big losses so most of the bets were laid off with Nick. It was my job to carry the bets in a brown paper bag to the cocktail lounge run as a front by Nick."

But Royko also met the truant officers who showed up to escort him to his third try at high school, this time at Harrison High where he remained "about an hour, maybe two hours. I thought I'd take a look at it."

Royko was now jumping back and forth between his father's place on the South Side and his mother's apartment near California and Armitage, where she lived with Bob. While he avoided school, Royko never skipped work. He worked as a pin boy at the Congress Bowling Alley on Milwaukee, long before bowling alleys had automatic pinsetters, and he worked at the Marshall Square Theater on Cermak Road, running the popcorn machine and keeping the other usher awake.

Since nothing epitomizes the blue-collar ethnic as much as bowling, and the fact that Royko would seize any opportunity to enhance his kid-above-a-tavern image, he often wrote of being a pinboy.

"The pay rate, as I recall, was about 7 or 8 cents a game. So if you set "double" (adjoining alleys) for two leagues, you worked 60 games in about four or five hours and earned $4.80 plus maybe 50 cents or a dollar in tips. Pinsetting was also a fine physical fitness program. To earn the $4.80 you bent over about 2,000 times. There weren't many fat pinboys. I don't know what the labor laws were, but many of us started setting pins when we were about 12. You could tell a pinboy by the joints on his first two fingers. Hoisting the pins between the fingers made them big."

In later years, Royko made veiled references to running with a group of teens that rolled drunks in alleys during those years, but the sly grin, which accompanied so many of his stories, betrayed a bit of invention.

There was nothing imaginative about the truant officers. The next time they found young Royko he was hauled off to Montefiore, the day reform school where he remained until he was sixteen and could legally drop out again. The Montefiore experience was not wasted. It provided Royko with sobering moments and hilarious material.

"They had an entirely different approach to education. The teachers were chosen for their size," he told *Tribune* reporter Jerry Crimmins in a 1988 interview.

The fifteen-year-old Royko found he was "too smart and too small" and that without allies he could not always eat his lunch. "A guy would say, 'Gimme a bite,' and he would take the whole sandwich."

The student body at Montefiore was divided into black gangs, Irish gangs, Spanish gangs, and Italian gangs. There were no Ukrainian-Polish gangs. Royko informed the leader of the Italian gang, Angie Boscarino, that his mother was Italian and became Big Angie's adviser. "Big Angie could beat the shit out of anybody and was the mental equal of the average ape."

Royko once advised Big Angie against killing the eel that lived in a fish tank in the science classroom. Against Royko's advice, Big Angie killed the eel with a knife. As a result, the entire class "got the shit beat out of us" by the teachers.

Years later Angie became a killer for the Chicago crime syndicate, according to a column Royko wrote on the day after Thanksgiving 1965. It was two days after the police found Big Angie in a gutter with an ice pick in his chest.

When Royko dropped out of Montefiore he got a job as an usher at the Chicago Theater on State Street and worked at Marshall Field & Company's nearby store, picking up packages customers had purchased to be delivered and taking them to the shipping room. He was living again with his mother and Bob and helping pay the rent and buy food. Bob had just been enrolled at the new Gordon Tech High School on the Northwest Side. When he needed medical attention for his eyes, Royko paid for it.

In 1949, Royko tried high school again, and this time it worked. He enrolled at the Central YMCA High School, where all the students paid tuition. Many of them were war veterans. There were no extracurricular activities, and you could smoke anywhere outside the classroom. Royko finished four years of high school in two years. He was elected senior class president and got his degree in 1951.

Perhaps emboldened by his recent successes, Royko tried more school, enrolling at Wright Junior College. He found it full of what he scorned as the "bobby sox crowd," and too much like the public high schools he found stifling. He was enrolled in sociology and years later recalled, "I wanted to be a social worker until I read a couple of books and thought, 'Jesus Christ,' if I have to read this dribble, forget it."

It was during these years that he spent time at Riverview Park, the amusement mecca at Addison and the north branch of the Chicago River, and the nearby Mid-City Golf Course where the driving range charged a quarter. At Riverview he paused to watch a game called the Dips. For a dime, a customer got three chances to pitch a ball at a bull's-eye that, when struck, would drop a Negro from his seat into a pool of water. He wondered about that. At Mid-City he began to hit golf balls.

He wondered, too, about this place called Korea, where most young men his age were going. He did not wonder long. He would later write, "When the Korean War began I wouldn't have minded if they had singled me out to stay behind...I didn't know where Korea was or what Koreans looked like, and I didn't want to learn...I also recall that I didn't have any special feelings about the Chinese either. Except fear."

Those feelings were oddly similar to the famous Muhammad Ali quote, "I got no war with them Viet Congs," an unpopular stand that Royko defended. Yet Royko's disinterest in getting involved in a shooting war for his country was on one hand contradictory to the awe in which he always held World War II veterans. In the early 1950s, most young men had the same reservations about the draft, death, and Korea as their peers of the 1960s would have about Vietnam. The great war hero Dwight Eisenhower had come to Chicago's International Amphitheatre to accept the Republican party's presidential nomination and had pledged to end the war. Even President Truman insisted on calling it a police action. Korea was not viewed as a valiant effort to save the world.

"A couple of my friends went in the Army. One went to Europe, drank a lot of beer and met a lot of lovely ladies. Another got as far as Japan." Royko told an interviewer. "I figured the best place to rest would be in the Air Force. I figured the chances were that in the Air Force you'd go to Bermuda or something."

In the spring of 1952, Royko went to Lackland Air Force Base in Texas for basic training and was then sent to radio technician school in Mississippi where he spent eight months learning dots and dashes and getting a firsthand look at the stereotypical southern segregationist he would excoriate in print a decade later.

After radio-operator school he was ready for permanent assignment. He received orders to Korea, where he arrived at a fighter base south of Seoul in January 1953.

"I was issued a carbine, and I said, 'What the hell is this for? I'm a technician.'" Royko's combat experienced was limited. "One night they hit a point in the fence close to where I was. We shot at them. They shot at us. One of them was killed and a half dozen wounded." Royko never claimed he shot anyone.

Six months after he arrived, the Korean armistice was signed on July 27, 1953. Royko's next base was in Blaine, Washington, north of Seattle. After a year's duty there he learned his mother had cancer, and he applied for a compassionate transfer to the small air force unit based at O'Hare Field.

When he reported to the O'Hare base for assignment, he heard the duty officer on the telephone telling someone that the editor of the base newspaper was about to be discharged and that they would need a replacement.

"So when the man hung up the phone, I said casually, 'Being around a newspaper office sure brings back memories.' I always did think fast. He said, 'You mean you're a newspaperman?' I said, 'Sure, I was a reporter for the *Chicago Daily News.*' It struck me that any goof could write a newspaper story."

"He grabbed me and literally dragged me into the personnel office, gasping. 'This man's a professional newspaper man!' Immediately I was assigned to the public information office as editor of the paper which meant I had one man under me."

When he left the personnel office, Royko had second thoughts. He spent the weekend at the public library, looking through books on reporting and the makeup and layout of a newspaper. He made sketches of what the books described as ideal layout for tabloid newspapers.

Shortly after Royko took over as editor another man joined the staff. He was Stanley Koven, and he had worked at United Press before joining the air force. Koven watched Royko at work for a few days and said, "Don't con me. You never worked for a newspaper, did you?" Royko confessed his lie, and both men laughed.

Royko's first decision as editor was to assign himself a column called, "Mike's View." His first column attacked the American Legion for supporting the communist witch hunts of the late senator from Wisconsin, Joseph McCarthy. Appearing in a military newspaper, the column must have raised a few epaulets, but Royko was not reprimanded. Then he wrote his second column. He rebuked the officers' wives for coming on the base with their hair in curlers and wearing casual clothes while their husbands had to go around spick-and-span. He concluded the appearance of these women was bad for morale. Three wives burst into the public information office demanding to see "this Mike Royko." In those days servicemen did not wear name tags, and the grinning Royko blithely said, "That guy just left on a 30-

day leave."

But Royko's first and last stint as an editor ironically ended when he unwittingly printed his first exposé.

"The base commander was a nut on sports, and his pride and joy was the base softball team and the pride and joy of the team was the pitcher. But the pitcher was due for his discharge, so to keep him the commander extended his enlistment. I ran a big front-page story about it, thinking it was great news for the commander and the team. What I didn't know was that you can't extend a man's enlistment except during time of war or emergency. Well, the thing really blew up in my face. It went through all layers of command, clear on past the generals to our congressman. One man got sent to the Aleutians.

"As for me, nobody would talk to me. They simply decided they didn't need a base paper anymore. But they did need someone to clean up the officers' quarters. So I spent my last three months in the Air Force there as chief janitor and hotel clerk."

After the air force, Royko wasn't sure what he wanted to do. "Stan Koven kept after me to try for a job on a newspaper," Royko said. "But I was told you can't make money at that. I wanted to go back to college and become a lawyer and make a lot of money."

Koven, who joined the Associated Press after his discharge and later became a writer for television, told Royko he should try to get with City News Bureau, the legendary training ground for reporters in Chicago. Or, he suggested, Royko should try one of the many neighborhood chains that published weekly editions.

Royko decided he would give it a try. He sent an application to City News in March 1956, the month he was discharged. He didn't receive a reply. He found a job with the *Lincoln-Belmont Booster*, a twice-weekly publication of the Lerner chain. He was hired for forty dollars a week to report on local police, politics, and sports in the Northwest Side neighborhood. He also worked part-time in a machine shop to make ends meet and took a few courses at Northwestern University at night. He had heard that most newspapers were only hiring reporters with college degrees.

After six months, he applied again to City News. The application required a brief biographical sketch. Royko wrote:

"I think I can write. I've got a pair of strong legs, lots of energy and I don't plan on being rich. And I can type. What I lack in formal education, I've made up, to a certain point, in personal experience and a heck of a lot of reading.... Unfortunately, a recently acquired wife squelched my plans to get a college degree so I'm faced with doing it the hard way."

chapter 3

Mike Royko was nine years old when he fell in love with Carol Duckman. She was six. He was in the third grade and she was in the first. He lived on Armitage and she lived just down the street on Francis Place. He wrote of her house, nostalgically: "It had been the prettiest house in the neighborhood—red shingles, white trim on the windows, a high white picket fence, and a small backyard filled with flowers. The tall, pipe-puffing man of the house always had his toolbox out. I used to lean on the fence talking to the blonde girl who lived there."

They grew up together and were playmates. By their early teens she was taller than he. He usually got stuck with her short friend Donna LeBrock. Had he not grown up, Royko would often say, he would have married Donna LeBrock.

The "tall, pipe-puffing man" was Fred Duckman, the son of German-Alsatian parents, who was an electrician but just as adept at carpentry, masonry, or tinkering with a carburetor. He had married a

Scandinavian girl, Mildred Jensen, and they had a son, Robert, in 1929. Their only daughter, Carol, was born in 1935.

"She was beautiful," said Bob Royko. "Everybody was in love with her. Mike was in love with her. I was in love with her."

Carol Duckman was blonde, tall, and happy. The Duckman family was happy. Mickey Royko was never happy. He spent most of his adolescence being in love with Carol Duckman, but he never told her or asked her out. He used to brag about her beauty to his friends. They would ask her for a date, and Mickey would sit home alone and brood. She always thought they were just friends. But good friends.

After he escaped from Morgan Park Military Academy, Royko would sometimes bicycle from the Cullerton Inn on the South Side all the way to the Northwest Side. Riding with Carol in the fields not far from her home was one of the happiest times of his teen years. But he could never tell her how he felt about her. He was certain that this beautiful girl had no romantic interest in a skinny guy with a big nose and horn-rimmed glasses and not much of a future. He was also painfully shy.

"My grandparents loved to tell me," David Royko said, "about the time Carol had a party for a bunch of kids and my dad was invited. All the kids were in the house, but my grandparents could see my dad outside, walking back and forth in front of the house for about an hour. Then he left. He never came in."

While he was at Central YMCA and working at Marshall Field's, Royko dated other girls. But in that lovesick embrace of adolescence, he thought none of them worthy of the feelings he privately vowed to keep only for Carol.

When he joined the air force, Royko wrote to the Duckman family from Texas and Mississippi. Before sailing for Korea he went home on leave for Christmas 1952 and stopped by the Duckman home to see Carol. Her parents told him that seventeen-year-old Carol had suddenly married Larry Wozny, whose family owned a neighborhood tavern.

To welcome the New Year of 1953 Mickey Royko got a broken heart and a year in Korea. His family saw him off at the station, where he caught the train to the West Coast to board the ship. He remembered that his mother and Bob, now seventeen, and his sisters put up a brave front. "They were trying hard not to cry so I wouldn't feel so bad." What amazed Royko was the reaction of his father, the tough, brawling, drinking, womanizing immigrant who seemed to fear nothing. "He couldn't hold back the tears. I always thought of him as a pretty rough type of guy. He was the last one I expected to break down."

Royko returned from Korea on a troop ship that landed in Seattle on December 23, 1953. After a night of partying he used his last dollars for a train ticket and spent Christmas Eve and Christmas Day riding the Great Northern to Chicago. It was a lonely leave. He went nowhere near the Duckman house. He bought a 1950 Ford and planned to drive to Blaine, Washington, but the car had too many problems to make a long trip so he sold it and returned by train.

On February 1, in a lighthearted letter to the Duckman family, he wrote: "I left Korea intact due to the fact that on guard duty, when sighting a Korean, I yelled 'help' instead of 'halt.' If they try to send me back there will be three men at the dock, me and the two cops who will have to escort me." He did not inquire about Carol but she wrote him back. She told him that she was separated and living back at home.

The twenty-one-year-old Royko, a veteran of Korean fleshpots and military saloons, finally fought past his shyness and his fear of rejection. In a love letter to Carol he wrote: "Writing this letter is going to be the toughest thing I've ever done. A statement like that warrants an explanation. I'm in love with you. The result has been mental hell. For a couple of years I've been wondering when I'd stop thinking about you every day. I've come to the conclusion that I won't. I have been in love with you for so long, I don't remember when it started but when I decided to do something about it, it was too late.

"Love, Mick"

That letter was the first of more than 100 he would write over the next several months in a literary courtship as passionate and as prolific as the 8,000 columns he would later create. Though Carol did not reciprocate his feelings, he was exhilarated:

"I can live with one sided love and be happy to have experienced it," he wrote. He offered sympathy for her broken marriage. "You have been dealt a bad hand but you are young, attractive and intelligent and life will go on."

Royko also wrote that until he learned of Carol's separation he was thinking of volunteering for another tour overseas, and when Carol wrote back a thirty-page letter, Royko replied she was not "indefatigable," and shouldn't stay up late writing him. While the proper use of "indefatigable" was probably Royko showing off, it illustrated a sophisticated vocabulary that he had built up through incessant reading.

He also proposed. "When we were kids I used to cut my own throat by bringing you boy friends, then I just stayed away. Now, when I'm across the continent I find my voice. I'm proposing to you and I expect you to say no but please don't. Just say, not right now, maybe later. But if it's irrevocably no, send a rope and a 10-foot building."

He was excited about the prospect of working on a new magazine on the base, "like one of those college humor things" that another airman had proposed to him. The idea must have withered, because he never mentioned it again. Yet Royko must have discussed writing or impressed his colleagues with his memos. At one point he wrote Carol, "one of the guys thinks I'll be another Hemingway."

He also got encouragement about writing from his family. Bob Royko said, "We used to get letters, a lot of letters from Mickey while he was in the service, which was a very sad time for us. He wrote volumes while he was in Korea, beautiful funny stuff, and the tragedy is we didn't save those letters. Here was a nineteen-twenty-year-old kid on the other side of the world in a war, writing letters that were making his family laugh. It should have been the other way around. We should have been doing him some good, but the letters were just magnificent, and it was at that time that we all realized how funny he was, sitting one-on-one locally, but the way he put this stuff into words, we were all telling him the same thing. One, come home safe, and two, go to journalism school so you can become a writer."

Carol did save all his letters, even those filled with the trivialities and frustrations of life in the service. And he reflected his interest in current literature by suggesting parodies of two well-known World War II novels. "I think I will write about the lunacy of the Air Force, maybe 'From Here to the Golf Course' or 'The Blaine Mutiny.'"

For a week he wrote of his plans to call her long-distance, which was an intricate and expensive project in the early 1950s. He gave her instructions as to where and when she should wait for his call. Then he wrote in dismay, "I don't know why I froze on the telephone. I just couldn't say to you the things I wanted to say."

In May he wrote there were only 100 days until his leave in September, and he dreamed about going home and taking Carol dancing at the Edgewater Beach Hotel overlooking Lake Michigan. Carol wrote that she was filing for a divorce, spurring Royko into a frantic litany of devotion. He wrote of reading a horoscope that said Virgos were due for a new romance with an old childhood sweetheart, and he wrote that "lately I've been reading the works of Schopenhauer and he states that nature, in trying to create a balanced effect of color, has caused a greater attraction between blondes and brunettes than between people of similar coloring. You must admit stars and philosophers can't be wrong."

He told her about the night he learned she had gotten married and included a prescription for humor that he relied on many times in the future:

> When I left your home I sat on the bus and added up everything in my mind. Things can look so bad that after a while they get funny. I actually sat and laughed. Any type of humor is based directly or indirectly on someone elses [sic] troubles and I guess I was laughing at my own. Things can look just so bad then after a while they become ridiculous. That's how I felt. I was going to Korea. You were married and the future was hopeless. It's kind of foolish to be twenty and think that your whole life is ruined but at the time it looked that way.

On May 19, he wrote: "Sure it's possible for people to fall in love in three months. I fell in love with you in three minutes. Would I stop loving you if I could? I doubt it. But if I had to another year like 1953 I'd darn sure look for a way to forget you and that's the truth. I could never stop loving you. No matter what happens I'll always feel the way I do now. I close my eyes and see you standing before me, then I think about deserting."

The anguish of waiting until September for a leave was driving Royko to distraction. He wrote of going AWOL because "it would only cost him a stripe," but then he wrote he really didn't want to get into trouble. This conflict of romance and responsibility, idealism and pragmatism always haunted him, making his future work so remarkable and his life so chaotic.

Meanwhile, Royko made the base drill team, which often performed in Seattle or Vancouver, less than 100 miles distant south and north, respectively. He wrote about how good he looked in the drill uniform, and he bragged about his golf victories and winnings. Carol wrote that she had gone to Riverview with his brother, Bobby, and Bobby wrote that Carol turned down every guy who tried to flirt with her. "I don't know what she sees in you," Bobby wrote, and that made Royko happy.

He didn't write for a week. His evening job as bartender at the golf course clubhouse resulted in a lot of hangovers, which may have accounted for the interruption of his courtship. Carol wrote that she was angry about not receiving any letters and doubted he really loved her. Royko went into a panic, writing that she should never doubt his love and that he would go AWOL immediately if she asked. And he again played the rejected suitor role:

> Sure, I know that you don't feel the way I do and I've never expected you to and maybe you never will. That's one of the facets of the future that frightens me. But if ever you say you do, I'll probably be so exultant I'll probably bust. That's my dream for the future. If some other sap dreams about making a million or being president I've got more to hope for than any of them. Despite it's being one sided I'll always feel that way—I love you.

He began running with a group of guys who had also returned from Korea. They held KV (Korea vets) parties on a regular basis, and Royko wrote: "Hit every snake pit in town and then tested base security by crawling under the fences and captured the mess hall at 2 A.M. We frightened the cook into making us scrambled eggs and I overslept for the promotion board."

On June 19, Royko was promoted to Airman First Class, and his commanding officer told him he could enlist at Western Washington University in the fall semester. "I can carry 10 hours and get a year's credit before my discharge. This is my lucky day. I'd probably make a hole in one if I could play today."

On July 2, he reported that "there was another KV party, much drinking, some arrests (not me). I am off scotch. It will be just milkshakes until September. Last time something like this happened I didn't drink for three months. Your [*sic*] probably wondering what it is with this guy who writes letters to you. Look, I'm no lush. Usually I'm a pretty sociable, moderate drinker of good scotch but I just got carried away this time.

"You can use A/1C anytime now," he gloated over his promotion to Airman First Class.

Carol was again angry at the interruption of his letters and probably concerned about what seemed like a lot of drinking. Royko dipped into his insecurity. "It would have been better if I had sailed off and forgotten you. I'm not good enough for you. I know that...My family always seemed to wonder why I had so little to do with the fairer sex but I never told them I continually made comparisons and the girl of the moment was always woefully lacking.

"With you I could not only set the world on fire—I'd melt it. You'd be the incentive. I could do anything."

What he got to do was four days of KP (kitchen patrol) and the possibility of a transfer to a missile site in Oregon, which would have canceled his leave in September. Besides the grueling duties of washing 600 plates, salad bowls, and desert dishes, and scrubbing all the pots and pans in the mess, Carol had not written for three days, and he threatened again to go AWOL.

But the proposed transfer was canceled, the KP duty ended, and Carol wrote that she told all her prospective suitors that she would be unavailable the entire time Royko was home on leave.

He wrote back, "I can't describe how happy I feel. When Columbus saw land, when the Wrights flew, they couldn't have felt any better. I read your letters and I want to jump in the air and click my heels."

In September 1954 the radio was playing "Autumn Leaves," "Because of You," and "Secret Love," and a few upbeat songs like "Rock Around the Clock" and "Sh-Boom." Mike Royko and Carol Duckman liked "Dream" best of all. They played it in the 1954 Ford that Mike borrowed from his father for their dates to the fancy night club, Villa Venice, on Milwaukee Avenue. They drove it to Lincoln Park to sit and look at the lake, and they drove it to the Edgewater Beach Hotel for their promised dance. Sometimes they listened to the Cub games, and Royko got excited about two young infielders who were the Cubs' first black players: Gene Baker and a skinny shortstop named Ernie Banks. The movie houses, where Royko spent so much of his youth, were showing *The High and the Mighty*, with John Wayne; *The Country Girl*, with Grace Kelly; and *Sabrina*, with Audrey Hepburn.

They listened to music while sitting on the back porch of his sister Eleanor's house. And Royko told Carol how much he loved her and how he wanted to be a lawyer and have lots of money and a big house in the suburbs and two tall blonde children.

He took her to a restaurant at Dempster and McCormick, which was virtually in the country, although in a few years it would be a cluttered intersection surrounded by shopping malls and apartment buildings. The restaurant was the Mark, which eventually became the Mark II, then the Mark III, and the Mark IV.

He asked her to marry him. She said yes.

He asked when. She said they would have to wait until her divorce was final. He said the sooner the better because his air force pay would almost double if he were married. She could continue to live at home and work at her receptionist job in a doctor's office, and he could live on base and they could save more than $2,000 before his discharge in 1956. They felt it would be enough for a down payment on a house.

On September 22, Royko boarded a train for Seattle, and when he arrived at the base he immediately wrote Carol. All his letters had been addressed to Carol Wozny. This was addressed to Carol Duckman. All his letters had started, "Dear Carol." This one began, "Hi, Sweetheart, my sweet babe." With adolescent jubilation he scribbled romantic acronyms on the outside of the envelopes. "MYMTICS" meant "Missing You More Than I Can Stand." And he continued to woo her. "Some men find the main theme of their lives in some art, science or in the service or in a job. You're my main reason for being alive. Anything I'd ever accomplish, any achievements I'd presume, they'll be because of you and for you."

They had agreed that Carol would fly to Seattle to be married as soon as she received her divorce, which they hoped would be in late October. The pragmatic Royko mingled with the lovesick Royko for the next six weeks. His letters were filled with logistics for the wedding and lengthy discussions of money and the possibility of a transfer to a base closer to Chicago.

Royko also was trying to figure out how they could raise $300 for the wedding plus her airfare. He decided to borrow money from Eleanor's husband and to cash in the savings bonds he had taken out of his pay. He also made plans to supplement his savings by gambling. He won $2 on the New York Giants' sweep of Cleveland in the World Series, and $44 on a season-long baseball pool. Years later he told his son, David, a richly embellished story about his gambling prowess:

I didn't have any money. I mean, I didn't have enough money to finance a wedding, the honeymoon, the airline tickets for her [Carol] to fly from Chicago out there and back, the hotels and everything like that. I mentioned my problem to a friend of mine who was a very good card player, as well as a pimp—he had three girls working for him in Vancouver—he was a golf hustler, a pimp, and a card shark. All around good guy, I think he ended up working for the CIA. So we decided that since we were going to win the money, we would just accelerate the process, and we worked out a really stupid but very simple system of signals, so that I would know if he had a really good hand, if he had an unbeatable hand, and he would know the same about me. So if you get 6 or 7 players in a game and two of them know... We were able to run the pots up, sandbag, we were able to whipsaw guys, and so very quickly I had enough money for all the wedding expenses. I guess it was cheating, I say I guess, I know it was cheating, but I can look at it as a way of putting these guys out of their misery a lot faster, in one evening, rather than suffering night after night watching their pay envelopes shrink. That was the only time I ever did anything like that, but it was for a good cause. I never told [Carol] until later. She might have thought it was a character flaw.

According to his closest friend at Blaine, much of what Royko wrote and said about those days was as embellished as his gambling tales. On the day Royko arrived at Blaine, he showed up near closing time at the enlisted men's club. The bartender, a second-generation Greek American from Akron named Don Karaiskos, was getting ready to close.

"I was washing glasses and straightening everything up and this guy walks in with his duffel bag and I figured he had just arrived on base," Don remembered. "The base only had about 120 enlisted men and 20 officers so everybody knew everybody. He walks up and asks for a beer. He said, how much is beer. I said 20 cents. He said, how much is scotch. I said 50 cents. He said, 'I'll have scotch.'

"I asked if he had been assigned a place to stay and told him my roommate had just transferred and I had an empty spot. We all were assigned to two-man rooms. He accepted, and after I closed we went to my room and we were roommates until September when I was discharged," said Karaiskos, who had enlisted in 1950, and who later became a United Airlines pilot.

"Actually, I knew about Royko before I met him that night. I had been assigned the chief clerk's job because I knew how to type, and I always got the records of the guys who were transferring in. When I got his I saw that he had gotten a perfect 100 on the air force qualification test, which was like an IQ test. I had never seen a perfect score and I thought either this is a mistake or the guy's a genius. It turned out he was a genius. I remember months after he came, one of the officers was standing by my desk and he said to another officer, 'The smartest guy on this base is that guy there, Royko.' Everybody knew how smart he was."

Royko did start a base newspaper. "It was a just a four-page mimeograph thing," Karaiskos recalled, "but he would write about the base baseball team and the softball team. You could see he was very funny then. I remember he wrote about our softball team and said I was the only left-handed Greek shortstop in existence.

"We played a lot of golf. It cost only fifteen dollars for a service-man to get a season's pass to the Birch Bay golf course. We checked out clubs from base special services. We never had a base golf team, and I don't remember Mike playing in any tournaments but he could really hit the ball."

Karaiskos eventually got a job as a waiter at a restaurant in Blaine, and Royko got a job at the Birch Bay golf club. "We chipped in and bought a 1937 Graham Paige, and we would take turns driving and picking each other up. I remember one night we had been out drink-ing and we were driving back to the base along Puget Sound and he had a snoot full. He was driving with his head down, with his chin on his chest and his eyes were glazed and he had it floored. I said, 'Slow Down,' which meant he only pushed it down harder. I finally reached over and pulled the key out of the ignition and threw it out the win-dow. The car finally stopped and we went to look for the key but couldn't find it. I had another one in my pocket but I wasn't going to tell him. We hitchhiked back to the base and went and got the car the next day."

Royko didn't drink any more than anyone else in the crowd, but Karaiskos noted, "He really couldn't hold it too well. Two or three beers and he was gone. And he never got anywhere on time. He was always late. If I made up his bunk once I made it fifty times for him so he wouldn't get in trouble. And the way he dressed? He never had his fatigues buttoned, his blouse was always out, and he had these big brogans that were never tied. You'd see him walking with fifty-four inches of shoelace trailing and wonder why he never tripped himself."

Although he was professing his undying love in his daily letters to Carol, he was a twenty-two-year-old male and, like his peers, on the prowl.

Don added: "He kept dating this one girl, Daphne Snow, and every night he'd come back swearing, 'I can't score, what am I doing wrong?'"

Still, for nearly a year he had pictures of Carol pasted up in his room and in his office. One day a newly arrived second lieutenant saw one of Royko's favorites, tall, blonde Carol wearing a tightly fitting imitation leopard-skin dress. The young officer stared at the picture, then stared at Mike. Royko broke the silence. "I can't understand it, either." He wrote Carol, "The guy was so embarrassed he finally said, 'Your [sic] nice looking too' and then he got redder than before when he realized how silly that was." Although Royko enjoyed the incident it reflected an awe he felt all his life that someone as beautiful as Carol would choose him.

Instead of joining KV soirees, the soon-to-be bridegroom spent much of his time reading. He got a collection of short stories by W. Somerset Maugham from the base mobile library. "He says that some people fall in love many times but the fortunate, or unfortunate, fall in love only once. That's us." He also tried to rationalize the separation they would endure for his remaining sixteen months in the air force. "I've been reading Emerson, an essay called 'Compensations,' and he says 'for every good there's an evil, for every gain a loss.'

"Can you understand the feelings I experienced when, after years and years of loving you, hoping and yet hoping without hope, seeing dreams shattered, dreams that I'd had since I was a small boy, become sources of pain rather than pleasure, when after all this, you said you loved me? Honey, it was like being brought to life."

He became despondent over the dim prospects of a transfer. The air force only granted compassionate transfers if a stricken family member could show evidence of being cured or dead within a year, and neither diagnosis applied to his mother, who had cancer. Additionally, such a transfer could only be granted if there were adequate replacements to do the airman's job, which wasn't the case.

Meanwhile, he continued with the wedding preparations. He spent a weekend in Seattle and booked the University Lutheran Church on the campus of the University of Washington for Saturday, November 6. He asked a friend, Ralph Peterson, an engineer who maintained the base's radar equipment, to be his best man. Bill Varns, the captain whom Royko worked for, had invited Carol to spend the night before the wedding at his parents' house in Seattle. Royko booked a cottage on a lake outside of Seattle for a five-day honeymoon.

Apparently, Carol's relatives were shocked that she was planning to marry again so soon after her first marriage splintered. Royko reassured her. "So your [sic] relatives are shocked, but we've known each other 12 years."

On October 27, Carol Wozny was granted a divorce from Larry Wozny. On Friday, November 5, Carol left her shocked relatives and boarded a Northwest Airlines flight for Seattle. When she arrived she took a limousine to the Olympic Hotel, which Royko had booked. He had also given her instructions: "Say your [sic] Mrs. Michael Royko. It's Seattle's best hotel."

At 11 A.M. Saturday, November 6, the wedding took place in the University Lutheran Church. Ralph Petersen was the best man. Margo Wade, a friend of Royko's boss, Captain Bill Varns, stood up for Carol. Royko wore a blue suit.

Mr. and Mrs. Royko spent a five-day honeymoon at the lodge at Lake Wilderness twenty miles outside Seattle. They used a 1950 Ford that Royko rented from another airman for twenty dollars. She then returned home to Chicago.

On November 12, Royko wrote his first letter to Mrs. Carol Royko.

He noted that it was exactly five months and eight days until he could get another leave in April.

Two weeks later, he received a letter from Carol informing him that she had written U.S. Representative Timothy Sheehan, a Republican from the Northwest Side, for help in obtaining a compassionate transfer. Sheehan, one of the most honorable and long-suffering Republicans in a city dominated by the Democrats, included the Duckman family of 5408 N. Central Avenue among his constituents. He immediately requested that the air force transfer Royko to the base at O'Hare Field on the northwestern edge of the city, a small airstrip that was being considered as a site for a second airport to augment Midway, the city's major airfield at the time.

On January 3, 1955, Carol received a telegram from Congressman Sheehan:

I AM HAPPY TO INFORM YOU THAT I HAVE BEEN NOTIFIED VIA TELE-PHONE TODAY OF DECISION OF DEPARTMENT OF AIR FORCE REGARD-ING AIRMAN MICHAEL ROYKO'S APPLICATION FOR TRANSFER FROM HIS PRESENT STATION TO ONE NEAR HIS HOME DUE TO CRITICAL CON-DITION OF HIS MOTHER. DEPARTMENT OF AIR FORCE WASHINGTON HEADQUARTERS TODAY APPROVED THIS TRANSFER.

AM INFORMED THAT AIRMAN ROYKO WILL BE STATIONED AT O'HARE FIELD.

On January 14, 1955, Carol received another telegram.

ARRIVING SUNDAY AT 2. GREAT NORTHERN RR.
LOVE, MICK

chapter 4

Mike Royko began his career in journalism as the "Front Page" era came to a close. From the turn of the century until World War II, Chicago journalism was characterized by bawdiness, conniving, drinking, whoring, and almost savage competition. Journalism in Chicago was different than in other places because of the personal animosity the newspaper publishers and editors felt between one another. By the end of the nineteenth century the eastern heavyweights in New York City, Boston, and Philadelphia had accepted their war would be decided by the bottom line of their accountants' journals. Residents of these cities were also more refined than Midwest city dwellers, and did not need daily scoldings by editors to remind them of their civic duties.

Chicago had nine newspapers at the end of the nineteenth century. The two best and biggest in terms of circulation and and advertising revenues were Joseph Medill's *Tribune* and Victor Lawson's *Daily News*. Medill was one of two men responsible for nominating

Abraham Lincoln in 1860 and had turned his *Tribune* into the voice of the Republican party in the Midwest. It would flourish for another half century under the autocratic, eccentric, and sometimes brilliant leadership of Medill's grandson, Colonel Robert R. McCormick, and would go on to become one of the most successful media conglomerates of the second half of the century.

Lawson was also a shrewd businessman. He had developed an uncanny instinct for responding to the sprawling, brawling city of immigrants. Lawson and such spirited editors as Henry Justin Smith led Chicago journalism from cut-and-dried dispatches and acutely partisan editorial positions to high entertainment. The *Daily News's* literary talent was unmatched. It began the first newspaper column, "Sharps and Flats," by Eugene Field, who filled his space with bits of pungent editorial comment, humor, and verse. He excelled at children's poetry, scripting his famous "Little Boy Blue," the child's ode about tin soldiers gathering dust after the death of their owner.

At the *Chicago Post*, Finley Peter Dunne created the first enduring fictional newspaper commentator, Mr. Dooley, an Irish bartender who dosed out humorous homilies on the state of everything to a customer called "Mr. Hinnissy." When Theodore Roosevelt wrote a self-serving book on his experiences in the Spanish-American War, Mr. Dooley said wryly, "If I was him I'd call th' book *Alone in Cuba*."

Royko's favorite Dooley line was, "Thrust ivrybody, but cut th' ca'ards."

The newspaper column would take other forms in the years to come, but it was not what created the myth and reality of Chicago journalism. That was created by William Randolph Hearst. The millionaire Californian had introduced "yellow journalism" to New York when he started the *New York Journal* to rival Joseph Pulitzer's *World*, and now he wanted to be the king of newspaper publishers.

Hearst had supported William Jennings Bryan for the Democratic presidential nomination in 1896 after Bryan delivered the most famous

phrase in American political convention history at the Chicago Audi-
torium: "You shall not crucify mankind upon a cross of gold."

Hearst was for Bryan again in 1900. He wanted a midwestern
newspaper to launch the campaign, so he ordered one to be created
in Chicago. On July 2, 1900, the *Chicago American*, an afternoon rival
of the *Daily News*, came off the press, and two days later it flooded the
Democratic convention at Kansas City where Bryan was again nominated. In 1902, Hearst established the morning *Examiner* to go head to
head with the *Tribune*. As he had done in New York and as later
depicted in the classic movie *Citizen Kane*, which was based on the life
of Hearst, he tried to buy every talented newspaperman in town.
Salaries skyrocketed in Chicago as staffs were raided and raided back.
Beyond that, the circulation wars. Dion O'Banion, who briefly rivaled
Al Capone for the bootleg empire of the 1920s until two of Capone's
men killed him in his flower shop, worked for Hearst. So did the
tough Annenberg brothers, Moses and Max, until McCormick hired
them away to the *Tribune*. Newsboys were beaten and killed, and rival
papers were burned in their wagons.

Peace eventually prevailed on the street corners but never in the
newsrooms. In those days, reporters and editors were a shiftless
bunch. Newspapers offered no benefits. People changed jobs while
having a drink. It wasn't unusual for the *American* or *Examiner* offices
to be filled with ex-employees of the *Tribune* or *Daily News* and vice
versa.

There was little that was highminded about Chicago newspapers
from 1900 to 1930. They feasted on scandal, mayhem, and sex. Still,
there were intervals of brilliant journalism. James Keeley, the *Tribune*
managing editor, spearheaded an investigation in 1910 that led to the
ousting of a U.S. senator who had bought his office by bribing legislators to vote for him. The *Tribune* stories ultimately led to the direct
election of U.S. senators in 1916. In 1919, McCormick's reporters
got an advance copy of President Wilson's unsigned peace-treaty

agreement with its proposal for a League of Nations. The *Tribune's* world beat spurred America to reject participation in the League.

For the most part it was mayhem and magic. In 1910, seventeen-year-old Ben Hecht turned up in Chicago and got a job through an uncle on the scandal-prone *Chicago Journal*. He became a "picture chaser" whose job was to cajole or steal photos or portraits of any noteworthy deceased person before any other paper could get them. On one occasion the only portrait he could find of the deceased was hanging above the coffin. Hecht climbed on top the coffin, removed the portrait, and fled.

He became the leading practitioner of "never let the facts interfere with a good story." He moved to the *Chicago Daily News* and wrote the first story column, "1001 Afternoons in Chicago." Hecht fascinated readers with the tales he uncovered, many of them pure fiction. His most outlandish effort concerned a fissure he found (or dug) in Lincoln Park that led to a column about an earthquake that had supposedly struck Chicago during the night.

Later he joined with Charles MacArthur to write *The Front Page*, with its hilarious caricature of Hildy Johnson, the unscrupulous reporter whose "Hello Sweetheart, Get Me Rewrite" became a poster in every newsroom in America during the 1960s and 1970s. But Hildy Johnson was not a caricature at all. If anything, those who remembered Hildy said the Hecht-MacArthur caricature was too tame. Hildy Johnson was neither the most famous, most colorful, or least ethical reporter in Chicago's newspaper heyday. Nor was he the best. But, like his peers, he was always out for a buck.

One of the biggest stories in Chicago in 1919 was the tale of Carl Wanderer, a war hero who could not settle down to his life as a butcher, husband, and expectant father. He paid a derelict five dollars to shoot his wife. Police determined the gun used in the killing belonged to Wanderer, who was caught, convicted, and sentenced to be hanged.

Hearst's *Examiner* paid Wanderer $200 for his life story. He spent his last hours at the Cook County Jail playing cards with Hildy Johnson, who collected the last pot and the last of Wanderer's $200 just as the hangman summoned.

When he reached the gallows, Wanderer was asked if he had any last words. "Don't play rummy with Hildy Johnson. I think he cheats," he said.

Newsmen often participated in investigations rather than reporting on them. When a fugitive wanted for murder in Illinois was captured in Wisconsin, the *Examiner* reporter at Chicago Police Headquarters notified his desk that the cops were taking their time selecting a team to go collect the suspect. The *Examiner* sent its own team, which arrived at the small Wisconsin jail, flashed a few phony badges, picked up the murder suspect, returned him to a Chicago hotel, and had his interview all over page one the next day. Only then did they turn him over to police.

When the first "Crime of the Century" took place in 1924, two brilliant University of Chicago students, Nathan Leopold and Richard Loeb, were the key suspects. Police took them along the route the young victim, fourteen-year-old Bobby Franks, had walked the night he disappeared only to be found the next day bludgeoned to death, his naked body dumped near a railroad culvert. In the car with the suspects and police were two *Daily News* reporters, who did most of the interviewing.

A sensational murder did not have to take place in Chicago for Chicago newsmen to scoop the world. Two of the "telephone" legends of American journalism worked for the Hearst papers. In 1934, Harry Reutlinger, city editor of the *American*, telephoned America's greatest hero, Charles Lindbergh, at his Hopewell, New Jersey, estate and became the first reporter in the country to verify that the aviator's infant son had been kidnapped and that a ransom note demanded $50,000 for his return. No one knew how Reutlinger got

Lindbergh on the telephone. He may have said he was Franklin Roosevelt. A few years later, Reutlinger got a big beat when he made a ship-to-shore call to the burning luxury vessel *Morro Castle* off the New Jersey coast. Identifying himself as the owner, he convinced a young steward to supply him with all the details of the fire and had an exclusive in print before the New York papers, which were only a few miles from the scene.

Reutlinger posed as a policeman, a sheriff, a coroner, or anyone else who could help him get the story. His successes were remarkable, because on another floor of the Hearst Building at 326 West Madison Street was an even more practiced telephone magician, Harry Romanov, the *Examiner* city editor.

Collier's Magazine named Romanov as the world's greatest telephone reporter. Once he posed as the police commissioner and got through to a hospital where several dead and injured had been taken. The man who answered the phone provided all the details Romanov asked for and then said, "If you will get a paper and pencil, I'll give you names, ages, addresses, and extent of injuries of all concerned." Romanov was so astonished at the degree of cooperation he was receiving that he blurted, "Who is this anyway?"

"Police Commissioner Fitzmorris, Romy. I knew you'd be calling."

Reporters of the 1920s were not the media stars of the 1990s. They lived a shadow life between respectability and the underworld. They were probably the best customers of the speakeasies. They were often close friends of the mobsters who provided so many front-page dramas. They were not paid very well, and much of their drinking and "upstairs" social life was on the house. Nevertheless, Chicago newspapers in general and the *Tribune* in particular lashed out in indignation when *Tribune* reporter Jake Lingle was gunned down in 1929 in a mob-style execution. It was later revealed that Lingle had been on Al Capone's payroll. That was hardly the first or last time a Chicago reporter was on the take.

As late as the 1950s and 1960s, it was fashionable for politicians to tuck a ten dollar bill into a reporter's jacket pocket. When television arrived, a cameraman who focused on an obscure county commissioner or member of the sanitary district board might get a twenty dollar bill "for lunch." Even as the new breed of reporters were waving critical fingers at the monolithic administration of Mayor Richard J. Daley, most of the city hall press room lined up expectantly at Christmas for their presents, usually a twenty-five-dollar gift certificate from Marshall Field's or an expensive tie that could be exchanged for cash.

This was the tradition and lingering reality of Chicago journalism when twenty-four-year-old Mike Royko joined the City News Bureau. Royko was hired by the general manager, Isaac Gershman, who filled that post from 1931 until 1964, ably assisted by Larry Mulay, who spent fifty-five years at CNB, and Arnold Dornfield, who was there forty-four years. City News Bureau was established in 1890 by the publishers of Chicago newspapers. It trained more than 3,000 would-be reporters. The ones who suffered and persevered filled newsrooms all over the world. Others found more suitable endeavors, such as sculptor Claes Oldenburg and actor Melvyn Douglas. City News was founded to coordinate coverage of various neighborhood courts, police stations, and firehouses. Eventually its beats included hotels, train stations, and airports. CNB reporters sent brief dispatches to newspapers through pneumatic tubes that wound their way under the city streets. It wasn't until 1961 that they began using printers like the major wire services. CNB reporters were the first to sound alarms for such landmark stories as the Iroquois Theater fire of 1906, the sinking of the *Eastland* in 1915, and the St. Valentine's Day Massacre of 1929. They never got bylines and were not encouraged to hone writing skills. They rarely wrote anything of literary value. Their reports were terse and factual. It was Dornfield's customary advice to young reporters that became a legend in American newsrooms: "If your mother says she loves you, check it out."

In his early years, Mulay once was sent to interview a woman who police said had called to report a missing child. Mulay became suspicious when the woman wouldn't open the door to talk to him. He relayed his suspicions to a City News editor who ordered, "Get in there if you have to burn the place down." Shortly thereafter, firemen were called to put out a fire burning on the porch of the woman's house. When firemen broke down the door, Mulay followed them in. He was there when police discovered the body of the child and that of the woman, who had committed suicide. Those were the kinds of editors who ran City News.

City News was the toughest boot camp in journalism. Reporters were badgered to find out everything. One recalled being screamed at for not knowing the color of a dead infant's eyes. Reporters were sent back to crime scenes over and over until they got every detail even though the editor had far more than he needed. City News reporters memorized every police call sign so they could rapidly identify a burglary or homicide. They deciphered the city Fire Department's bell codes.

A *Chicago Tribune* story in 1940 described the typical CNB rookie. "The West Side police reporter is expected to cover 12 police stations, three branches of the municipal court, the county hospital, the morgue, the psychiatric hospital and court, contagious disease hospital, Dunning asylum, the juvenile home and juvenile court, and the House of Correction. In his spare time, he runs to fires and gets the verdicts at perhaps a dozen coroner's inquests."

One CNB veteran recalled covering the death of a child who choked on a tree ornament on Christmas Eve. "I thought I had it all wrapped up but the desk made me call the family back: What color was the ornament your baby choked on?"

Royko began like everyone else on cops and courts, chasing minor arrests and verdicts, learning to get every name and address right, checking street directories and telephone books to verify the information, and enduring the harassing questions of the desk. He mastered the art of having the answer to every possible question. He also expanded his knowledge of how the police units functioned, which desk sergeant would supply the most information, which lieutenant loved to be quoted, which captain would leak the name of a suspect.

He learned that every judge had a political sponsor called "Chinaman" or "Rabbi," who had gotten him on the bench, that every lawyer who wanted his case called made a "drop" of a few dollars in a drawer at the bailiff's desk, that payoffs were as routine as windy days.

Shortly after Christmas, he got his first big story.

On the night of December 28, 1956, the Grimes sisters, Barbara, fifteen, and Patricia, twelve, left their home in the near Southwest Side Back of the Yards neighborhood—Mayor Daley's neighborhood—to see the rock 'n' roll sensation Elvis Presley in his first movie, *Love Me Tender,* playing at an Archer Avenue theater. They never came home.

Chicago parents began living another nightmare, one that had appeared for the first time only a year earlier when the city was shocked by the murders of Robert Peterson, fourteen, and the Schuessler brothers John, thirteen, and Anton, eleven, who, like the Grimes sisters, had gone to a movie and never returned. Their naked and molested bodies were found October 18, 1955, in a forest preserve on the Northwest Side. Chicago was horrified. Although the city had weathered the violence and a myriad of grisly crimes during Prohibition, not since the Leopold-Loeb aberration had young children been targets of what police and newspapers were describing as a "fiendish murderer." When the Grimes sisters disappeared, the nightmare began again.

Policemen went door-to-door looking for clues in the case in what was then one of the most massive police efforts in Chicago history. At one point, police said 43,740 people had been questioned and 3,270 suspects interrogated.

After several weeks of a city-wide search, their naked, frozen bodies were found January 22, along an isolated road near Willow Springs, fifteen miles southwest of their neighborhood.

During and after the search, the newspapers were in a frenzy trying to dig up fresh information for their daily front-page spreads. A skid row dishwasher from Tennessee, Edward "Benny" Bedwell, twenty-one, became a prime suspect. Authorities said he had signed a confession. Some newspapers reported he had led police to the bodies. Others knocked down the story. Editors all over the city watched the CNB wire for any new information. In the middle of the City News coverage was Royko. Armed with a shiny star he bought from Woolworth's Five & Ten Cent Store, he interviewed neighbors of the victims and, posing as a coroner's investigator, went up and down Skid Row on West Madison Street talking to people who said they knew Bedwell.

Bedwell reportedly told police that the Grimes sisters were runaways looking for sexual activity and that he and another man got them drunk one night in mid-January and ditched them after knocking them unconscious in a scuffle. Cook County Sheriff Joseph Lohman and State's Attorney Ben Adamowski were quoted widely as believing Bedwell. But Cook County Coroner Walter McCarron said the autopsies showed that the girls had been killed the night they disappeared and Bedwell's story was fake.

The newspapers took sides. While the *Tribune* quoted Lohman and Adamowski and virtually convicted Bedwell, the *American* was supporting the McCarron theory and charged that the investigation was bungled and that Bedwell was being railroaded. The *Sun-Times* and *Daily News* changed sides often. Chicagoans may have been mortified at the slayings, but they were also baffled by the differing newspaper accounts.

Bedwell eventually was released and drifted off to anonymity. The Grimes case has never been solved. And Mike Royko got a raise. He was now making fifty-five dollars a month.

The turnover at City News was rapid as reporters eagerly sought newspaper jobs as soon as they felt they had paid their dues in the CNB trenches. Royko was assigned as a desk editor. Now he could chew out the rookies for getting the wrong middle initial or calling a "street" an "avenue."

Just as Royko entered Chicago journalism, the Hearst era came to an abrupt end. When Hearst died at eighty-eight in 1951, his publishing empire began to disintegrate. By 1956 his company wanted to sell the *American*. John S. Knight, then-owner of the *Chicago Daily News*, wanted to buy the *American* to establish an afternoon monopoly. Marshall Field III wanted it to link with his morning *Sun-Times*. The *Tribune*, mindful that the *Daily News* had once been Chicago's circulation leader, outbid the competitors and on October 21, 1956, paid $9 million for the *American*. The old Hearst hands moved into a seven-story addition on the north side of Tribune Tower. Chicago still had four newspapers, but only three owners.

In the spring of 1957, with six months' experience at City News, Royko applied to all four Chicago newspapers. The *Tribune,* the *Sun-Times,* and the *American* told him they weren't hiring people without college degrees. It may or may not have been the same at the *Daily News*. In some of the many interviews Royko gave over the years he said that he was offered a job by city editor Maurice "Ritz" Fischer, but that he turned it down. "I was overwhelmed, looking around the room at all these great reporters. I told Fischer, 'I don't think there's any point in continuing this interview. I don't think I can do it. I just don't have enough experience. I'm going to fall flat on my face.'"

In other interviews, Royko said the *Daily News*, like the others, was not interested in hiring people without degrees.

It's difficult to see how an untrained, inexperienced Royko could lie himself into a job as an air force newspaper editor, and, a few years later, after passing muster in the best training school around, believe he wasn't ready for the *Daily News*. Yet mixed with the supreme ego that nurtured his powerful voice, Royko always harbored some fear, some touch of unworthiness and insecurity.

Whether or not he had invented his own modesty, Royko was still manning the phones at CNB and still looking for a better job in the summer of 1957. He applied for a position at a television station in Fort Wayne, Indiana, auditioning as a combination news director, reporter, and anchorman. However, Royko "failed to project." This rejection would haunt him in future years when he alternately embraced and avoided television, sometimes scoffing at its value and other times trying to become a talk show celebrity.

At City News, Royko began hanging out in saloons with his pals. He also held parties for CNB people at his home. At night, he joined other staffers in looking out the Randolph Street windows across to the upper floors of the Bismarck Hotel, where anyone who left up the shades provided cheap entertainment for the newsmen. On weekends, after a snowfall, most of the skeleton staff at CNB went downstairs and played football in the parking lot at the northwest corner of Randolph and Wells. One man was left to watch the telephones and summon the others if news occurred. One of the players often was Bill Garrett, who had joined CNB as a copyboy in 1957 in what he would always refer to as a "diversity hire."

"I was seventeen, had just gotten out of high school, and somehow my mother knew Larry Mulay and got me an interview. I didn't think I had a prayer of getting a job, but when I went there he asked if I could start the next day. I said sure. Then I found that Mulay was mad because Gershman was always hiring Jews. Gershman was on vacation when I showed up for the interview, so Mulay hired me on the spot and got me started before Gershman returned."

One of the "Jews" Gershman had hired was Sy Hersh, whom Royko liked to recall in later years after Hersh had broken one of the biggest stories of the 1960s, the massacre of Vietnamese women and children by U.S. Army troops at the village of My Lai. "I used to beat the shit out of Sy Hersh at golf," was a frequent Royko utterance.

When Garrett was promoted finally to reporter, he often worked for Royko. "He was a good editor, a good teacher," Garrett recalled. "He was always sarcastic and he was tough but after Dornfeld nobody was that tough. Everyone knew even then that Royko was going to be a success, a prominent newsman. No one knew how good he would be but everyone knew he would be good."

One of the men Royko became closest to at CNB was Bob Billings, a brilliant brooder who could match Royko in intellect and disdain for all things and people he considered stupid. The two became constant golfing companions, although Billings never shared Royko's enthusiasm for the latest technological improvements in golf clubs. Royko kept scorecards from rounds he and Billings played at obscure courses, many of which had disappeared by the 1980s. All the cards Royko kept showed him winning.

Although he couldn't crack the four dailies, Royko was valued as an editor and a fast rewrite man at CNB. He received eleven raises during his two-and-one-half years there, bringing his weekly pay to $105 at the beginning of 1959. Although Carol still worked as a receptionist in a doctor's office, the couple was still living on the second floor of the Duckman's house on Central Avenue. During the 1950s, Carol's mother, Mildred, was diagnosed with multiple sclerosis and was often confined to a walker or a wheelchair. Carol's brother, Bob, opened a mortuary in the house, and Mike began selling tombstones for him to make some money on the side.

He was not a super salesman. His sister, Eleanor, said, "I think Mike could have been a good salesman if he believed he was really selling something people could use. I don't think he put much heart in selling tombstones along Milwaukee Avenue."

Royko also cashed in, as did many other underpaid newsmen, on one of Chicago's major civic and cultural events, the International Livestock Exposition. For most of the century, Chicago hosted the country's biggest livestock show on the day after Thanksgiving, an event tied to its "hog-butcher of the world" heritage. Until corporate farming and more fanciful pastimes such as professional football and basketball came along, the livestock show was a big story.

Each year, the *Tribune* would banner on the front page the opening of the show. And each year for weeks in advance, newsmen would climb to some dusty office building and turn out four-inch releases for the hometown newspapers of the hundreds of 4-H Club members from all over Middle America who would be awarded prizes in Chicago. The pay was three dollars an hour, and the job was the most tedious imaginable.

Royko desperately wanted to go to the *Daily News,* which was clearly the writer's newspaper in Chicago. He believed there was a conspiracy to keep him out. The city editor, Ritz Fischer, was adamant about hiring college graduates, and Fischer's predecessor, Clem Lane, had liked to hire only Irishmen. One of them was a fellow named Jay McMullen, who later covered city hall for the *Daily News* and the *Chicago Sun-Times* and struck up a romance with one of Mayor Daley's department heads, a young widow named Jane Byrne, whom he married on St. Patrick's Day, 1979, only a few weeks after her shocking victory in the mayoral election.

"But McMullen was really an Orangeman," Lois Wille said, "and when Lane found out he really wanted to fire him because McMullen hadn't told him.

"Mike was sure that Ritz Fischer, who was Jewish but converted to Catholicsm, was prejudiced against him because he was Polish. Mike felt he couldn't get hired during Lane's Irish period and then there was, as Mike saw it, Ritz's German period. I came in under his German theory and so did Gee Gee [Georgie Anne Geyer]. He used to tease us about Slav-hating Germanic roots. And Bob Schultz was there and Harry Swegle. He included Nick Shuman in the German group until he realized Nick was Russian. Then there was the managing editor, Everett Norlander, who was Swedish but Mike said he was just another Aryan. After the German theory was the conspiracy of the Northwestern clique. Mike got to hate some Northwestern grads because he felt they were getting jobs he should have gotten."

Royko went to his boss, Larry Mulay, for advice. Mulay made a personal appeal to Basil "Stuffy" Walters, then filling the revolving-door position of editor at the *Daily News*.

Whatever Mulay told Walters, and whatever Walters said to Fischer it worked—because in May 1959 Royko was hired by the *Chicago Daily News*. He crossed town to the Daily News Building at Madison on the west side of the south branch of the Chicago River, a historic structure designed by Holabird and Root in the 1920s.

Royko's first job was night rewrite. He worked on a desk that included several future newspaper stars, including Ray Sons, who became sports editor of both the *Daily News* and *Sun-Times*, and Richard Christiansen, whose criticism and essays on the theater, dance, music, and culture would enlighten and delight Chicago readers of the *Daily News* and *Tribune* for four decades.

After several months of working nights he was assigned to police reporting, where he was out of the office covering stories of mob bombings and the usual Chicago mayhem.

In September 1959, the Chicago White Sox won their first American League pennant in forty years. Their last champions had entered sports history as the "Black Sox" who threw the 1919 series. Ever since, Chicago baseball fans had wondered if the Almighty was punishing them in perpetuity. So there was reason for excitement, and the most excited person of all was Mayor Daley's fire chief and crony, William J. Quinn, who, without warning, blew all the city's fire sirens, sending more than a few people scurrying to basements, certain that Russian warplanes had appeared over Lake Michigan.

Royko was assigned by the City Desk to call the state civil defense director for comment. "I said, 'Look, the man [Quinn] is a complete fool. He shouldn't be an elevator operator in city hall, he's such a jackass.' And he was. He was a political hack. I don't mind a political hack as long as he's a smart hack. But this guy was a dumbbell. And I thought, what I'd really like to write is what a jackass that man is...," Royko recalled in a 1973 interview.

But Royko couldn't just write anything he wanted to. He was a reporter, not a columnist. There were, of course, columnists at the time. The hottest newspaperwoman or -man in town was the *Sun-Times'* advice columnist, Eppie Lederer, who had inherited the old Ann Landers column and was picking up record syndication across the country. Irv Kupcinet had begun his half-century career as the *Sun-Times's* gossip columnist, and Herb Lyon had the "Tower Ticker" in the *Tribune*. Those columns were patterned after Walter Winchell's dot-and-dash style of who was eating where and who was divorcing whom. They were the most popular columns in most newspapers long before celebrity news on rock stars, athletes, news anchors, and movie stars moved to the front pages.

The sports columnists were the big guns; John Carmichael in the *Daily News*, Warren Brown and Leo Fischer, a brother of *Daily News* City Editor Ritz Fischer, in the *American*, and David Condon writing the "Wake of the News" column in the *Tribune*. The political writers were also heavyweights: John Drieske at the *Sun-Times*, Charles Cleveland at the *Daily News*, and the powerful George Tagge, who many believed ran the Illinois Republican party under the auspices of his *Tribune* position. They had just covered the reelection of Richard J. Daley in a landslide over the only person that Republican party chairman Timothy Sheehan could find to run—Timothy Sheehan. It is without question that the normally Democratic-voting Royko repaid

Sheehan for arranging his air force transfer to O'Hare by voting against Daley, a habit that became rather routine.

The *Daily News's* star was columnist Jack Mabley, who appeared on page three with bits of political gossip, back-of-the-fence wisdom, and occasional crusades. Everyone covered the Rush Street cabarets, although television was beginning to take its toll on such night spots as Chez Paree, the Boulevard Room, and the dance pavilion at the Edgewater Beach.

Newspaper circulation was steady if not increasing. Television news was still in its infancy, with people like John Cameron Swayze offering little more than a reading of wire-service reports with a still picture in the background. For news, people needed newspapers.

In newsrooms, Royko was hardly the only reporter with a persecution complex.

Lois Wille grew up in suburban Arlington Heights, got a master's degree from Northwestern in 1954, and married another aspiring journalist, Wayne Wille. She was hired in 1956 by Peg Zwecker, the fashion editor. "I'm not sure why she hired me," Wille said. "The managing editor, Everett Norlander, didn't want me hired because I knew nothing and had no experience in fashions. I was mostly a gofer.

"I spent nine months in fashions until there was an opening in the news room for 'the' woman reporter. My predecessor had just quit to join the marines. This was the era of 'our gal' stories. My first big story was flying with the Blue Angels. The city desk insisted I get my picture taken in my flight suit while I was powdering my nose.

"The first day I was on news I was sitting next to a big Irish cop reporter. He opened his desk drawer and bent over. I thought he was looking for something and the next thing I know I feel his hand rubbing my ankle."

Wille was responsible for getting a second woman, Georgie Anne Geyer, on news.

"Gee Gee had joined the paper in the features department. I was trying to get her to news side. She was an assistant to the society editor but she had been a Fulbright scholar and eventually wanted to do foreign reporting.

"I went to Ritz Fischer about Gee Gee and he looked at me and said, 'Well, you've been here almost a year now and the thing I like about you is that you have never cried. Do you think Gee Gee would cry?'

"This was at a time when one of the men staffers, a guy who had been a prisoner of war in Stalag 17, would go off drunk for three days and we all had to rehearse what to tell his wife when she called, as she always did. We had men who threw tantrums and smashed typewriters on the floor. But Ritz worried that women would cry.

"That was about the time that Mike came to the paper, but he disappeared on to nights and in those days women weren't allowed to work nights so I didn't really see him or get to know him then. But I remember when he was hired, someone who knew him at City News described him as this Polish kid who really knows the city and has this gorgeous, sexy blonde wife."

When Royko joined the *Daily News* staff, Carol was gorgeous and blonde and seven months pregnant.

On June 8, 1959, she delivered a baby boy, whom the parents planned to name David Michael Royko. "The nurse assumed they made a mistake and so I was officially named Michael David Royko. My father told me years later that he didn't want me named Michael, and he took a certain delight that I was always called David, which he said my grandfather thought was a 'sheeny name,'" David Royko recalled.

But the year was not all joyous. On December 5, Royko's mother died at age fifty-seven, after suffering from cancer for four years. After her divorce, Helen Royko had opened a little cleaning store on Armitage near California and married Joe Pardell, who had changed his name from Pardolfsky.

In one of the first columns he ever wrote, Royko eulogized his mother's cleaning shop. "In a good year, she'd make $2,000 after paying her rent....There was a rack of clothes which were for sale. The sign said 'unclaimed' but in truth she bought them used at the Salvation Army outlet store. Her customers knew this but the subterfuge gave a little class to buying used clothes. The store hours were 9 A.M. to 6 P.M. but that didn't mean anything. If somebody needed their clothes at 8 o'clock on a Saturday night they just rapped on the door and she opened the place.

"When the lady's grown-up children visited her, they'd often find her still sewing late at night and ask what the heck she was working for at that hour....When the children asked why the heck she didn't sell the place for the little it was worth and come live with one of them, her lips would tighten....If one of them said that it was a rather small business and not worth all the work, she would answer that it had put food in their mouths, shelter over their heads and clothes on their backs, hadn't it?

"She kept her store open long after she had cancer—long after it became painful."

Throughout his career Royko filled his columns with autobiographical detail but always to make a point, usually about the value of hard work and individual responsibility that he learned from both his mother and father.

He once contrasted his early days at the *Daily News* with the expectations of the modern million-dollar athletes: "My first day, I was seated at a typewriter in the most remote corner of the newsroom and spent my time writing obituaries. A week or so later, I was put on the night shift. And that's where I stayed for the next 2 1/2 years. I didn't complain because it was understood that new people would work nights. We thought of it as paying our dues."

After "paying his dues," Royko was assigned in 1963 to cover the county building. The county beat was not as prestigious as the criminal courts, the Federal Building, or lofty City Hall. Other states cover governors and senators, but in Chicago reporters lusted after the mayor. Chicago reporters drooled over every word uttered by the mayor. They expected the wide array of aldermen who strolled the corridors of City Hall to give them enough front-page copy to make them stars in the city room. From Richard J. Daley to Michael Bilandic, to Jane Byrne, Harold Washington, and Richard M. Daley, the governmental quota of news for any Chicago daily or broadcast station usually began and ended at City Hall. The presidents of the Cook County Board of Commissioners usually were regarded as mere handpicked cronies of the mayor. The commissioners were not regarded at all.

But Royko was all over the county beat. He saw it as a "building beat," an opportunity to step into the spotlight. He covered the treasurer, the clerk of the courts, the commissioners, and the sheriff. No one before or since has covered Cook County the way Royko did.

At that time, and since, the Democrats owned every city office, but Republicans, helped by the growing suburban vote, were often able to capture a county office. Shortly after Royko took over the county beat, Richard B. Ogilvie, a Republican, was elected sheriff, and Royko became his fast friend. Ogilvie had become a public figure in the late 1950s when he prosecuted Anthony Accardo, a former Capone slugger who had become head of the Chicago mob. Royko viewed Ogilvie as a bright, honest alternative to the motley array of pols who filled most county jobs.

In his later years, Royko derided "access" journalism. He was highly critical of reporters who had close relationships with news sources, particularly politicians, and who tended to close their eyes to possible wrongdoings. In fairness to other reporters, the personal conduct of public figures did not matter much in the 1960s. Reporters did not feel it was their role to discuss the private habits of politicians. Even if reporters knew that public officials were sleeping around or making last call every night, they didn't report it. If they had, their editors would have chased them out of the newsroom.

While Royko was learning his way around the county building, John F. Kennedy was entertaining various starlets and mob playmates in the White House. Defenders of politicians whose dalliances were made public in the 1980s and 1990s used media bias as the excuse, and pointed to Kennedy as the example. But during the Kennedy presidency and all the years before it, accusing anyone, including public officials, of something as defaming as an affair with a girlfriend of Chicago mob boss Sam Giancana was an invitation to a libel suit that could be extremely embarrassing and financially disastrous unless the newspaper could prove the allegation was true.

That would all change in 1964, with the Supreme Court ruling in the *New York Times* vs. Sullivan case, when the high court ruled that a public official could only win a libel case against a newspaper if the newspaper had knowingly and recklessly printed false information. The ruling gave newspapers room to make errors and even allowed for incompetency. If a newspaper believed what it said about a public official was true, it became almost impossible for that official to win a libel suit.

The *New York Times* vs. Sullivan decision opened the door for the news media to say just about anything it wanted about a public official. It was soon expanded to include anyone who could be considered a public figure. In newsrooms, "public figure" was quickly translated to mean anyone caught in a bedroom with someone not his wife.

As much as Royko later detested snooping into bedroom windows, he was appalled as a young newsman by the coziness of newspapermen and the people they covered. The City Hall crowd reported on Richard J. Daley's political empire in the most flattering tones. And they showed a real affinity for mobsters. In the late 1950s, a *Chicago Tribune* police reporter covering night court decided he could make a few extra dollars by going into the bail-bond business, and he persuaded one of the Chicago mob's zany characters, Sam

DeStefano, to bankroll him. But the partnership didn't go well, and DeStefano wound up taking a few shots at the reporter and notifying the *Tribune* of the odd business deal.

But Royko wasn't quite ready to define new standards for Chicago journalism.

He was content to finally have a beat, get regular bylines, and experiment from time to time with drawing word caricatures of some of the people he covered.

He also continued his education with the city as his classroom. He began to stop regularly at the saloon operated by one of Chicago's most colorful politicians of the period, Matthias "Paddy" Bauler, a North Side alderman whose reaction to Richard J. Daley's first mayoral election in 1955 was the classic: "Chicago ain't ready for reform."

Bauler was one of the pack of powerful and independent aldermen who ruled the Democratic party power structure until Daley reduced them to a rubber stamp city council. He didn't like Daley's brand of monolithic power and became even less enamored when Daley did, in fact, make several reforms to end the traditional B-girl and bookmaking sidelines of Chicago saloon keepers.

Paddy Bauler did have a past that colored his image. In 1933 he won a gunfight with an off-duty policeman outside his saloon. He was put on trial for wounding the cop and testified that the policeman had started the fight when Bauler refused to admit him to the tavern after closing hours. "Ahern [the policeman] made noises at me," Bauler testified. "And he called me a big Dutch pig who ought to be back on the garbage wagon. Then he hit me." Paddy was acquitted.

Bauler enjoyed the company of newsmen and liberals who were beginning to grouse at the Kremlin-like regime of city hall. "Paddy loved anyone who took on the establishment and right away he liked Mike," said Phil Krone, a precocious Chicago political figure who ran for alderman in 1963 when he was barely twenty-one years old. "I ran in the Forty-fifth Ward and Paddy gave me thousands of dollars—

they didn't have campaign disclosure then—to run against the regular organization's guy. I got 2,000 votes but I lost."

That was when Krone met Royko, "an irascible sort of fellow who covered the county beat for the *Daily News*." Over the next thirty years he became Royko's eyes and ears inside Chicago politics. They would see each other at Bauler's, where other habitues included Robert Merriam and Leon Despres from the South Side neighborhood of Hyde Park, which included the University of Chicago and was the spawning ground of the liberal movement started by U.S. Senator Paul Douglas when he was the ward's alderman. Merriam had run unsuccessfully against Daley in 1955, and Despres was the current alderman who would be a public, albeit ineffective, thorn in Daley's side for more than a decade. There were newsmen like John Dienhart and Len O'Connor of WMAQ radio, who had been a pioneer in Chicago broadcasting by carrying a bulky wire recorder to interviews and news conferences in the 1950s.

Royko's political education and his innate skepticism of Chicago government, honed by the payoff days at his father's saloon, were polished in Bauler's bar by these older, more passionate liberals who provided prejudicial insights about City Hall. Royko quickly learned to absorb the insights and dismiss the prejudice. He became a friend and admirer of some of the liberals who fought fiercely against the Daley regime, such as Abner Mikva, who would go on to Congress and later sit on the U.S. Court of Appeals in Washington, D.C. But as a group he usually disdained what he referred to as "limousine liberals," who took time out from their comfortable lives to rail against machine politics without doing much to overturn them. "Goo-goos," he called them.

During the early 1960s Royko expanded his love for all kinds of music. He began playing guitar in the midst of the folk music fad that swept America in the early 1960s. He made visits to an off–Rush Street club called the Gate of Horn, where America's top folksingers appeared. He visited Win Stracke's Old Town cafe and would later take guitar lessons from Stracke. Chicago was still a center of traditional nightlife, and jazz musicians like Erroll Garner and Oscar Peterson played rooms such as Mr. Kelly's and the London House, where even low-paid reporters like Royko could stand at the bar and nurse a fifty-cent beer through the show. At home he listened to classical music, once boasting that he could hum all nine of Beethoven's symphonies. Some of his appreciation for the classics came from his old City News boss, Arnold Dornfeld, who would invite friends to his Wheaton home for long sessions of recorded Beethoven and Bach.

Royko was still scrounging for spare money. He edited the monthly newsletter for Medinah Country Club, which allowed him extended golfing privileges at the prestigious No. 3 course. Although Royko was already casting flinty glances at colleagues who were taking money from news sources, neither he nor most of the sports writers saw any ethical conflict in such "freebies." This was still an era when most of the top baseball-beat writers supplemented their income by sharing official scoring duties at Cub and Sox games for fifty dollars a game. It was not until the 1970s that such practices as being paid by the people you covered were halted.

On his way home from Medinah, Royko would often stop at his sister Eleanor's home in Elmwood Park on the western edge of the city. "Dorothy and I used to go to the Salvation Army and pitch in and buy Mike a dozen pairs of shorts and socks," his sister recalled. "We'd wash them and fold them just like our mother did, and I'd give them to him to take home. He and Carol didn't have much. He'd ask how much he owed us and I'd say something like two dollars and he'd say it had to be more so I'd take four dollars but it cost more than that. We were happy to do it for him. We never said how much it really cost."

Most of the Royko family entertainment in those days was a family songfest with Carol, her brother, Bob, and Bob Royko sitting on the porch while Mike or his brother strummed a guitar, with all four singing songs ranging from old standards to folk hits. The refreshment was usually a jug of wine.

A close friendship developed between the Roykos and Carol's brother, Bob, who had married Eleanor's daughter, Barbara. In effect, Royko's niece was also his sister-in-law. "After Bob and I got married, we became a foursome every Saturday," said Barbara. Either we would go to their apartment or they would come to ours every Saturday in 1959 and 1960. We would play cards, canasta or pinochle. If they lost he would start shouting at Carol. You could see steam coming out of his ears. He was a terrible sore loser.

"Carol and I were both pregnant at the time, and we had a bet on who would have the bigger baby. I won and Mike had to buy dinner. We were pregnant together again the next time and I won again," Barbara said.

Mike and Carol's second son, called Robbie, arrived June 3, 1963. "I think Mike loved him from the beginning because he was a such a homely baby, with a wide forehead and that same big nose," Eleanor Cronin said.

In 1959, John Knight sold the *Daily News* to Marshall Field IV, chairman of Field Enterprises, owner of the *Chicago Sun-Times*. The sale made Field and the *Tribune* competitors in the morning and afternoon. The following year, the *Daily News* moved from its building to the new Sun-Times Building on the north side of the river at Wabash, only a few blocks from the *Tribune*.

The *Daily News* crowd liked the new neighborhood. There were newspaper watering holes like the Wrigley Bar on the ground floor of the handsome terra-cotta-faced Wrigley Building opposite the *Tribune*; there was Riccardo's at the foot of Rush Street, just out the door of the *Sun-Times*, the Corona a few blocks north, the Radio Grill on the lower level under Michigan Avenue, and the Boul Mich on the upper level. Most of these spots were filled with reporters, pressmen, drivers and copyboys, editors, columnists, rewrite men, and printers.

"There was a lot of drinking," said Ray Coffey, who was hired at the *Daily News* in the autumn of 1960 and would cover the civil rights struggle in the South, Vietnam, Europe, Asia, and Washington, D.C., in his career at three Chicago papers.

"We worked hard but there was plenty of leisure time on newspapers in those days," Coffey recalled. "I remember the great John Carmichael would come in about 10 o'clock and spend the next couple of hours with the guy who handicapped races figuring out his bets for the day. In those days you could bet with any of the circulation drivers who were all bookies on the side. There were also a lot of World War II veterans, and they didn't take a lot of things too seriously. They had a different attitude about life, and they didn't go home at 5 o'clock. There were many liquid lunches at Riccardo's. A lot of people lived in the south suburbs and took the Illinois Central to work. There was a bar at the Randolph Street station, the Bomb Shelter, and some of these guys wouldn't think of coming to work without having a few belts first."

If some of them had looked around the Illinois Central station while having their eye-openers, they might have discovered a story that the nation's news media totally missed in the 1950s. Not one Chicago reporter, nor for that matter any reporters in Detroit, Cleveland, or New York City, was noticing the number of African Americans who were getting off trains at places like the IC station. African Americans came to Chicago in the early 1950s at the rate of a thousand a day. Chicago's black population, which had been less than a quarter-million in 1940, had jumped by 500,000 when Mike Royko arrived at O'Hare Field in 1955 to try his hand at putting out a base newspaper. By 1960 there were nearly a million blacks in Chicago. They lived in highly segregated areas on the South and West Sides. As Martin Luther King Jr. would later say, "Chicago is as segregated as any Southern city."

Chicago was segregated mentally as well. Blacks who did not grab one of the factory jobs in the South Side steel mills were left menial service jobs as busboys, waiters, bus drivers, and taxi drivers. Blacks were never seen in Loop restaurants or Rush Street cabarets or meeting in the big, downtown hotels. Blacks were rarely seen after dark in Chicago's nightlife areas unless they were picking up litter or parking cars.

No editors or reporters in Chicago wrote about what this phenomenon meant or would mean to Chicago. They did not write about the attendant civil rights movement that was stirring in the South and would eventually erupt in urban unrest. The Chicago newspapers seemed oblivious to most of this until October 1962, when the *Daily News* appointed a new executive editor, Larry Fanning, who intentionally set out to change the way the *Daily News* covered Chicago. He would create the most influential newspaper columnist of the next thirty years.

chapter 5

A native of Minneapolis, Larry Fanning had gone west in 1932 to attend college at the University of San Francisco. He got a job as copyboy at the *San Francisco Chronicle* and rose to copy editor, telegraph editor, assistant news editor, assistant managing editor, and, in 1945, at age thirty-one, managing editor. He remained in that position until he was hired in 1955 as editor of the *Chicago Sun-Times*'s syndicate division. One of his first assignments was to find a replacement for the syndicate's advice columnist. It was his decision to hire an untried housewife named Esther Lederer, who became "Ann Landers," the most widely syndicated newspaper columnist of the next forty years. Fanning not only "discovered" Lederer, he coached, mentored, and edited her until the column took on the familiar snappy retorts and sage advice that made it a hit.

Fanning took over the *Daily News* from Tom Collins, who, after the retirement of Stuffy Walters, had held the job for only a year. Lois Wille recalled that one of Collins's least contributions was "hiring twin women photographers and sending them out to cover stories together and then poor Bob Schultz had to follow them and write stories about how people reacted. It was a very foolish period at the *Daily News*."

But the *Daily News* was still selling more than 400,000 newspapers every day and winning the afternoon battle against the *American*. The *Daily News*'s great strength was its foreign staff, which had been started during the 1890s by the legendary Raymond Stannard Baker. On any given day, *Daily News* readers could read dispatches from Pulitzer Prize–winners George Weller from Jerusalem, Keyes Beech from Asia, or Paul Scott Mowrer from Europe. William Stoneman wrote from London, Milt Freudenheim from Europe and Africa, and Gerry Robichaud covered Latin America. The *Daily News* had more foreign correspondents than any other newspaper except the *New York Times*, and more than the other three Chicago papers combined.

Fanning wanted the paper's national and local coverage to match its dominance in foreign news. One of the first things he did was to cover civil rights. Ray Coffey, who had been hired from United Press in 1960 after covering Illinois politics in Springfield and Chicago sports, had already been sent to cover the events at the University of Mississippi, where federal troops had been dispatched to enforce desegregation. "It was sort of funny," Coffey said about the *Daily News*'s situation. "I was down in Mississippi for a couple of weeks and this guy Collins is sending me a telegram a day saying 'nice job,' and when I get back I go to thank him and he's gone. Fanning is the new editor and he certainly became one of the favorite editors I ever worked for."

Coffey was not back for long. He spent most of 1963 and 1964 covering church bombings in Birmingham, segregation at the University of Alabama, the murder of civil rights leader Medgar Evers, and the slayings of three young men trying to register black voters in Mississippi.

He was not the only *Daily News* reporter in Dixie. Fanning boldly hired Nick Von Hoffman and sent him to Mississippi, where Von Hoffman delivered stories that angered readers on both sides of the civil rights issue. Ed Rooney, an insistent, sometimes irritating, always aggressive reporter, who looked like the Chicago cop that his father, grandfather, and brother were, also traveled the civil rights trail. Prize-winning photographer Henry Herr Gill took the pictures for their stories.

Lois Wille went to work on a series that examined how blacks lived in Chicago. "It was the first time in my career that I had a chance to do that kind of reporting," Wille said. But it was not the first time Wille had proved she was a reporter and writer of rare distinction.

"In 1962," Wille said, "there was a movement to provide birth control services for welfare recipients. In those days a public aid recipient couldn't get any information or contraceptives from Cook County Hospital because abortions were illegal and public-funded institutions couldn't provide any of that information. We began to do a series of stories on the subject and how difficult it was for poor women. Bob Rose, the first assistant city editor, was in charge of the series because Ritz Fischer was ill. When Fischer, a Jew who converted to Catholicism, came back he saw the series and just left it on his desk for several weeks. I tried to nudge him but it didn't work until I told him the *American* was planning a big series on the subject. He told me he was afraid I didn't understand the Catholic church positions accurately so he had given it to a friend to read. I was too naive to blow my stack. Later I found out he had given it to Monsignor Jack Egan who I kidded for my years about being my editor."

Egan, one of the most progressive churchmen in Chicago history, gave Wille some valid suggestions, which she incorporated into the series.

"Before it was published I was called in by Marshall Field IV who I barely knew. He was concerned that Catholics would boycott the paper. But the series ran with all kinds of follow-up stories on various actions being taken by public agencies. There was no backlash at all. The editor, Tom Collins, never got involved. I don't know what editors did in those days.

"The next spring, my husband Wayne and I were in Egypt. It was a trip we had planned for a long time. We were just boarding a sailboat on the Nile when one of the hotel maids—a man in a long flowing Arab garb—came running toward the dock shouting: 'Cables, Cables.' But the boat left the dock. It ruined our day. In those days, the only time you got a cable was if someone was sick or dead. When we finally returned, Wayne went and picked up the cables and came back smiling so I knew no one was dead. The first one I read was from the Planned Parenthood people in Chicago and said simply, 'Congratulations.' The next was from Ritz Fischer saying the *Daily News* had won the Pulitzer Prize public service award for my series. I thought it was great but I really didn't think it was all that great."

Fanning and his staff reveled in the prize, gloating mostly at the elephantine *Tribune* which, with all its resources, had only a handful of Pulitzers compared to the thirteen the *Daily News* had accumulated. But Fanning was also busy trying to find a replacement for the popular columnist Jack Mabley, who had been hired away by the *Daily News*'s afternoon rival, the *Chicago American*. Fanning finally offered an op-ed page column to John Justin Smith, the charming nephew of legendary editor Henry Justin Smith, but not before a public argument broke out when the *Daily News* refused to let Mabley take his favorite chair across Michigan Avenue to his new office.

In the 1960s and even into the 1970s, the raiding of star columnists by rival newspapers was commonplace. In the 1960s, almost every city of any size in America had at least two newspapers battling for dominance. Swiping columnists was a tactic in the circulation wars as were the host of silly games offering cash prizes to lure new subscribers. By the 1990s only about thirty American cities had more than one newspaper, and competing newspapers were usually put out under joint operating agreements with both papers produced in the same printing plant, sometimes by joint staffs. Local columnists could never again enjoy both the egomaniac and monetary rewards of being wooed by the competition, because the competition no longer existed.

Royko was attracting attention. He brought life to a beat that usually constituted a few turgid paragraphs on the inside pages of the Chicago papers. Several of the building's beat reporters were given weekly columns to supply tidbits about City Hall, the Federal Building, and the County Building. Royko used his Saturday space differently. He saw the natural comedy in the oafs who populated county government. He wrote stories about how they frittered away money, not because he wanted to expose these payrollers, but because he wanted to ridicule them. He began to quote exactly how they talked and what they said. His cheeky stories contrasted with other news reports that prized brevity and tedium.

But the copy desk didn't get it. They began editing out everything but the tedium, and Chicago came perilously close to losing the voice that became its minstrel. Royko complained to assistant city editor Bob Rose:

> After three attempts at writing a column, I think it is an unfortunate venture. I don't think I can write what the desk wants, as is evidenced by the heavy editing required. I can't produce the type of column Jay (McMullen) does because I don't find the same things amusing, i.e., Sid Holzman running around with a sheet saying "look, a ghost voter"; Paddy Bauler munching German food and damning blue stockings, etc. My only interest in these people is showing them up for what they are and I can't combine the two approaches. Since the

thing has only run three weeks, my relatives, co-workers and Seymour Simon are about the only people who have noticed it, so it would not be noted if it just faded out.

Royko.

Royko was persuaded to keep doing the column, and the desk editors became more lenient in letting him do it his way. He wrote irreverently and turned inconsequential items that were overlooked by other reporters into delightful little stories.

— "County board employees can have an extra week of paid vacation if they are delegates to the American Legion convention in Las Vegas. The decision illustrates one of the important fringe benefits of public employment—plenty of days off plus vacation time."

—"Dr. Edward F. Khuen, Cook County's chief rabies inspector, has made his mid-year survey into who is biting whom. This, of course, doesn't include back-biting, biting-the-hand-that-feeds-you, and nipping-at-the-heels, all of which occur in the County Building. His records show that last year, three persons were bitten by swine. He's happy to report this year there have been none."

His word pictures made obscure politicians real. "Commissioner Charles F. Chaplin has settled a ticklish question raised by observers at board meetings: Who has the bushier eyebrows—Chaplin or Board President Seymour Simon? Both have such lush growths that it was feared they might become joined at the brow if they brushed in a corridor. To settle the issue, Chaplin plunged a cigarette into his left eyebrow. It stuck there."

People were beginning to notice Royko. One of them was the county board president that he covered, Seymour Simon. Simon offered Royko a job as his public relations spokesman at a salary much higher than the $9,000 a year he was making at the *Daily News*. But Royko had already decided he wanted to be a columnist and turned Simon down. He was more interested when the *Chicago Amer-*

ican tried to hire him. He went to Fanning and told him about the other opportunities he had and said he would stay at the *Daily News* if he could have a shot at a general column. It was not an arrogant or unreasonable proposal. Most people at the *Daily News* knew that John Justin Smith was not working out as a replacement for Mabley, and quite a few of his collleagues were impressed with the Polish kid with the big nose who was churning out funny stuff.

In an interview for the *Chicago Tribune*'s 150th anniversary in 1997, Royko recalled, "I found that Chicago politics as it was being covered didn't really reflect reality. The beat reporters would clean up the language. They would make really idiotic people sound reasonably normal. And I wanted to write about them the way they really were.... There was no shortage of investigative reporting. Trying to nail the bad guys. But there was something I thought was lacking; the color of it. The humor. The comic scene. That's the way I felt about pols. They were comedic material."

"I started it with the understanding it would be a Chicago column. That was it. They wanted a Chicago column, a Chicago voice, and I'd covered politics and city and county government. I'd lived here all my life. I was probably as equipped as anybody my age at the paper to do a column at that time."

Fanning agreed. He gave Royko a chance to do three columns a week on the op-ed page. On Friday, September 6, the first one appeared.

"It was a sad surprise to hop into a cab and find that the driver was a man we once knew as a neighborhood tavern keeper," the column began.

The short, direct sentence was the hallmark of Royko's writing. He wrote clear, declarative phrases "to get into the story as fast as possible." He never wasted the reader's time with metaphors or irrelevant description. He always used simple language. From his first column to his last he never tried to overwhelm the reader with imagery or dazzle him with literary flourishes. The strength of his writing was in Royko's highly original material, but Royko was equally dogged about finding just the right word, the right phrase to set the column's tone.

The first column appeared when residents of the rust belt cities were grappling with new government language intended to fix their dirty and dilapidated inner cities. The ex-tavern keeper of Royko's first column explained his problem in a way everyone understood: "They came in and renewed us. You know, the urban."

Throughout 1963, Royko wrote either two or three columns a week, feeling his way around subjects as general as the World Series and as personal as the death of his father-in-law's pet parakeet. The latter showcased Royko's distinctive personal tone.

Fred Duckman, at his wife's insistence, had taken her pet parakeet to a bird hospital because the bird just sat, hunched in his cage. The nurse at the bird hospital told him the bird had poor circulation. "I told her," Duckman said and Royko wrote, "his feet are like tooth-picks; what can circulate in them…She told me to give her five bucks and said it would cost more if we wanted to keep him in his own cage. I said anyplace would do." As the reader suspected, the parakeet died, but then came the kind of ending that made Royko famous:

"I asked her about my five bucks," Duckman said. "I said I wanted to buy my wife another bird with it. Do I get it back?

"She said, 'No, we had to give him oxygen.'

When Royko did personal journalism it was personal to everyone.

Almost immediately he searched out the themes he would repeat for the next thirty-four years. One of them was the ethnic column. On December 3, 1963, he wrote: "They say that when the first Pole— or maybe he was a Bohemian—landed in the United States, he ran down the gangplank and headed for the nearest forest with a bushel basket in his hand. Someone shouted: 'Stop, there are hostile Indians in there.'

'I don't care,' answered the Pole or Bohemian. 'It is mushroom season.'"

Royko had an instinctive feel for the right mix: humor, indignation, triviality, baseball, politicians, mobsters, and music. In his first weeks he wrote about Senate hearings where Mafia hit man Joe Valachi first uttered the phrase, "Cosa Nostra." He wrote about Frankie the Elevator Man in the County Building, who got "viced" because his "Chinaman" was a Democrat and the Republicans won the 1962 county elections. He took the first of many salty swings at Ed Quigley, the city's sewer boss and chief patronage thug. But he did not step out of his own backyard.

When John F. Kennedy was assassinated, Royko did not write about it. He told coworkers he wanted to leave the space for the coverage of the story and that he was not qualified to write about national news, an odd reflection of the insecurity that always lurked behind his scowling face. It was a refrain he would revive many times to explain his rejection of job offers around the country.

After three months of Royko, John Justin Smith disappeared from the *Daily News*.

"I was a better columnist than John right from the beginning," Royko said in a 1993 interview with WBEZ on the thirtieth anniversary of his column. "And, uh, it was a tough position for him to be in so he took a job as a TV commentator. No one ever told him to leave....It was just the subtle way they have—sticking my column underneath John's. He was unhappy doing the column."

Royko was an immediate hit with readers. His tough, fearless approach to the political structure that ruled Chicago was innovative. His columns on Bohemian mushroom pickers, pet parakeets, and baseball were read with grinning approval. At the start of 1964, Royko began writing five columns a week, and his pay was raised from $173 a week to $192.

It didn't take him long to get around to golf. "These are terrible months for thousands of men who suffer from a yearly ailment called the twitching overlap grip," he wrote. "Their eyes get glassy; their bodies suddenly twist in a slight crouch; their arms shoot forward, and their hands come together. These tormented, miserable weak-willed wretches are golf addicts."

In February he wrote about a tamale shop, declaring it to be the best in Chicago. Over the years readers would chase madly about the city hunting for whichever chili parlor, pizza place, ribs oven, or steak house that Royko had declared, with impunity, "the best." Royko's palate might not have been impeccable, but his humorous restaurant reviews were irresistible.

Of all the saloons in all the world, the only one Royko loved as much as Rick loved Casablanca was the Billy Goat on Hubbard Street at the lower level of Michigan Avenue, a shortstop's throw from the Tribune Tower and a few blocks north of the *Sun-Times* and *Daily News* building. It was Royko's den, the place he went first after a hard day at work, and often the place he stayed last after a night of hard drinking. The original Billy Goat had opened in 1934 on West Madison Street across from the Chicago Stadium but got "renewed" for a parking lot and moved to Hubbard Street in 1964. There it became home to newspapermen, and in time, its walls were littered with photographs and reprints of stories by Chicago journalists. Royko held the preeminent spaces.

He wrote about the joint many times. The first time was November 27, 1964, when Royko bawled out the owner, William Sianis, whose trimmed white beard had given him the nickname of "Billy Goat." Royko chided Sianis for being robbed of $1,500 because he refused to keep a tavern dog.

"Without tavern dogs, tavern keepers would have to sleep in their taverns at night instead of going home and beating their wives. Without tavern dogs, there would be few taverns in this city and everybody would have to move to the suburbs."

Royko also wrote about his interest in drinking. "It was my pleasure, recently, to spend an evening examining one of the world's biggest collections of rare Scotch whiskies. This story is difficult to write, however, because my notes are blurred."

Royko always liked Scotch. He grew to like vodka, fine wine, various beers, some brandies, and most things that came out of bottles. His most empathetic columns often dealt with hangovers, and his most hilarious columns often reflected the evils that befell people who were overserved. On a few occasions the columns were not so funny, because Royko was forced to write about himself.

In the very beginning, before he ever dreamed he would write 8,000 columns, Royko vowed to himself that he would never fall back on the old reliable trick of filling column space with letters from readers. He reneged on March 5, 1964:

"Publishing letters from readers is a habit I had vowed to avoid. It is a dangerous habit because some letters are written so well, editors start wondering why they are paying the columnist. They also make it appear that a columnist is too lazy to write. I am publishing two letters today because I am too lazy to write."

Of course nothing could have been further from the truth. Royko sweated over every column he wrote. When he wasn't telling humorous anecdotes about Chicago life, he was talking serious political issues. Even as he was starting out, Royko always seemed to be on the edge. He was among the very first journalists to become passionate about gun control in America. On August 10, he wrote the first of dozens of antigun columns, a Swiftian effort inspired by a mail order catalog for weapons: "If you're a bit short of money, you can pick up a nice French anti-tank gun for $125. The 25mm armor-piercing shells cost $2.50 each. One of the problems with cannons is that they are nice in the living room, but they are bulky." He closed the column: "Incidentally, there is a movement among law enforcement people and some legislators to tighten up mail order gun laws. The gun interests oppose any such laws. They assert that they direct their merchandise at sportsmen, serious gun collectors and people who want weapons for home protection.

"After all-most people need anti-tank guns."

He could listen to people talk and hear Chicago in a way that no other writer had done. Perhaps James T. Farrell and Nelson Algren did it in fiction, but Royko did it for real, every day. He wrote words that could never make their way into *Webster's,* but his readers never had to reach for the dictionary.

And he filled his columns with wonderful parodies of speech, as he did in talking about Milwaukee Braves fans:

"Wudda Braves do?

"Won. Fordwun.

"Who pitched?

"Spahn.

"Neehomers?

"Aaron. Wunon.

"Putsum two out.

"Toonahaf."

Royko was funny, but he was also making some readers nervous. In one acidic column, he wrote that Cook County public golf courses used a variety of gimmicks to keep Negroes from playing: "No evidence has been found that grass, trees, sand, air, sunshine, ponds or players have suffered as a result of integrated golf." He wrote that courses on the North Shore seemed less restrictive than elsewhere. "'I guess they don't mind us playing a game up there just so long as we don't try to buy a house,' laughed a Negro."

He also did a piece on a game at the beloved Riverview amusement park of his youth. It was called the "Dips." For a dime, a player got three balls to throw at a little paddle that would dump three Negroes in a cage into a pool of water. When Royko noticed the dunking machine was gone he contacted the man who operated the game and asked what happened. The operator claimed the park was afraid of being boycotted even though it was an "integrated" amusement stand.

"We had white men working at the counter and taking in the money and we had colored boys in the traps," he said. He said the colored boys made $2 an hour, more than they could make anywhere else. Royko asked why he didn't put a white boy in the cages.

"I would have done that, maybe, if I had two dressing rooms. But I only have one dressing room. You can't have whites and Negroes using the same dressing room." The man said he was sorry he had to give up the game because it was started by his father-in-law in 1909.

Royko concluded: "And this, as others have discovered, is 1964."

It was the subtle Royko, using a device he mastered to let people make fools of themselves by quoting them, letting their own words disclose their stupidity or cruelty or racism, and to indirectly but knowingly make all his readers who felt the same way squirm at their kitchen tables.

In 1964, Chicago was undergoing a metamorphosis, both physically and spiritually. It was the eighth year of the reign of Richard J. Daley, the mayor who was getting national magazine coverage for the great building boom which would give Michigan Avenue the sobriquet of "Magnificent Mile," and sprout such new structures as the Prudential Building and the rusty-looking Civic Center Plaza (which would be renamed the Daley Building and replace the dusky County Building where Royko had his first beat). It was also the year that blacks began asserting themselves throughout the north and in Chicago. They began to picket Daley's house. The Committee on Racial Equality (CORE) left dead rats on the mayor's desk. While Chicagoans expressed public remorse over the slayings of civil rights activists in Alabama and Mississippi, they privately hoped that blacks weren't planning to move into their Southwest and Northwest side enclaves.

Nineteen sixty-four was also a presidential election year and Lyndon Johnson was going to be matched against conservative Senator Barry Goldwater of Arizona unless a more moderate Republican, either Governor William Scranton of Pennsylvania or the highly popular Governor Nelson Rockefeller of New York, could take the nomination at the GOP National Convention in San Francisco in August. Fanning thought Royko should cover the convention and write his columns from San Francisco. Royko begged off.

"He always felt insecure about covering national events," Lois Wille said. "He thought he wouldn't be able to do it as well as the reporters who were used to covering conventions and if he couldn't be the best he didn't want to do it."

In the 1993 WBEZ interview, Royko said, "I didn't want to go to the Republican convention. First of all, I didn't like Barry Goldwater, and I thought I'll just alienate some of our readers. Why should I go? I write about local stuff. My readers want me to write about stuff that I know about and stuff they know about."

Instead, Royko focused on the familiar. In September, his son, David, began school:

"Get up."

"What for? Mothers take their kids to school on their first day. I'm a father."

"I know, but you should give him some advice. He's nervous and doesn't want to go."

"Tell him he's got to go because he can't drop out until he's 16. Federal training programs don't cover five-year-old dropouts....Tell

him if he doesn't go I'll belt him. Tell him the teachers don't belt him
so he'll be safer there....Then tell him that if he finishes kinder-
garten I'll buy him a used car."

"But he's too young for a car."

"I know, but we can back out of the deal later. It'll be a good lesson for him—always get things in writing."

For a rookie columnist Royko was remarkable in the diversity of
his topics. Many columnists turn into Johnny-One-Note, rehashing
the same subjects and unleashing tedious attacks on the same people
or peeves. Royko wrote about City Hall and the Beatles, his family
and civil rights, baseball and golf, topless bathing suits, CTA (Chicago
Transit Authority) buses, and the joys of fishing in the Humboldt
Park pond.

But his specialty was political hypocrisy. On the last Friday of October 1964, Daley hosted the Democratic party's traditional torchlight
parade to close the presidential campaign. Four years earlier, Royko
had been one of dozens of reporters lining the three-mile route from
Grant Park to the Chicago Stadium to see John Kennedy riding in an
open car with Daley. Kennedy was awed by the turnout—nearly a million patronage workers were there cheering—and said to the mayor, "I
hope they all vote." The taciturn Daley only said, "They will."

In 1960, Royko had only filed notes to the city desk to be incorporated in the main story on the Kennedy parade, but, in 1964, he set
out in search of column material on the torchlight salute to Lyndon
Johnson.

The *Daily News* was the only Chicago newspaper that did not pub-
lish seven days. It published Monday through Friday and then ran a
weekend edition which came off the presses early Saturday and
remained on the newsstands through Sunday.

Royko covered the Democratic rally intending to use it for his
Monday column. But when Fanning read the start of Royko's column

on Sunday he decided it belonged on page one. Royko expanded his piece. His take on the parade was unusual. While the other papers indulged in their standard boosterism, Royko took a different tack. Rather than recounting how many people cheered the president and how many times Daley promised a Democratic victory, Royko counted the number of skid row bums who were hauled off the streets and taken to police stations so the mayor wouldn't be embarrassed by the Madison Street winos. Daley also had every abandoned car and truck towed from the neighborhood, but as Royko noted, "There was still a pretty good turnout of bums to see their President."

Royko wrote about the crowd in front of the House of Rothschild liquor store, which advertised "McGee's Corn Whiskey—20 cents a shot." He watched a scraggly bunch of men fight for what they thought were dimes being tossed from a float by a pretty girl. They were LBJ pins, and the bums left them in the street.

He wrote, "They looked ready for the Great Society as they stood on the curb and in the gutter watching the bands, the floats, the pretty girls and waiting for the President. Anything would help. Even two bits. Few of them had any front teeth. If a man had front teeth, he probably lacked half an ear. If he had both ears, then some fingers were missing....They didn't cheer much, but they did a lot of coughing."

When the presidential limousine finally passed, Johnson was looking at the other side of the street, and the crowd in front of the House of Rothschild didn't get to see much.

"The Great Society had come and gone. Now the rest of the boys could drift back to their street. In four years they'll see another parade," Royko ended his piece.

Lois Wille recalled the excitement of the *Daily News* newsroom when Fanning ran the piece on the front page. "It was a watershed event. Looking back, it may not seem like much, but it was such a different way than we had covered those kinds of things. And taking a columnist to the front page was unheard of at the *Daily News*."

During his first years at the *Daily News*, while he was assigned to general reporting and the overnight beat, Royko had struck up a friendship with a lawyer named John King. One night the two of them decided to have fun with all the pompous politicians and self-important lawyers who ran around city hall. They created the LaSalle Street Rod & Gun Club. One of the first members was Wayne Wille, who had worked at the *Sun-Times* and was married to Lois Wille.

"Mike had stationery printed with a letterhead and in between assignments he would send out letters to all sorts of important people. He would write, 'Dear So-and-So, I am sorry to inform you that you have been rejected for membership in the LaSalle Street Rod & Gun Club.'" Wille, who included his membership in his listing in *Who's Who*, said that sometimes Royko would urge people to apply at a later date or inform them they could never apply again.

"Some of the biggest guys in town must have been shocked to learn they had been rejected from something, especially since they had no idea what it was."

Near the end of his column's first year, Royko wrote a spoof of a news story on December 8, 1964, about the then-innovative idea of a poor woman being appointed as a national adviser to a federal antipoverty program. The fictional setting was a special emergency meeting of the fictional LaSalle Street Rod & Gun Club, with the participants including Big Red, the president; Fat Mary, the secretary; Sidney Bilge, an unemployed liberal; and Tony from the sewer department.

And in that column one other club member appeared for the first time, and he disagreed that being poor qualified the woman to serve as national adviser.

"I don't see any logic in that thinking. I have been drunk as long as I can remember and I ain't never been asked to serve on the board of directors of Seagram."

That was the debut of Slats Grobnik.

Slats Grobnik became Royko's alter ego. He served up nostalgic memories of boyhood during the depression, Polish weddings, broken hearts, Halloween pranks, July Fourth mischief, and a thousand lines of laughter and wisdom ranging from the foolishness of marriage and sobriety to the proper punishment for Richard Nixon.

Associated Press writer Jules Loh described Grobnik as "the archetypal Chicago neighborhood denizen, an urban Huck Finn who lives in a three-flat, hangs out in a tavern, cracks his knuckles, spits through his teeth, pitches pennies and disdains suburban grass as 'Stuff worms live in.'"

Born in a flat above a tavern, Slats was often crude and boorish. He was skinny with a big beak, and he was both a bully and a peacemaker. He was cranky, insensitive, and chauvinistic. He was witty, wise, sentimental, and sweet. He sounded exactly like the guy who created him.

"His parents took him with them on all their social outings, most of which were to the Happy Times Tavern on Armitage, which was owned by Slats' uncle, Beer Belly Frank Grobnik.... They would sit him on the bar or the pinball machine and he was the most contented child you ever saw, chewing on a hard-boiled egg, washing it down with a tiny tumbler of beer.... Customers enjoyed having Slats around and would bounce him on their knees and give him snacks, such as herring, pickled pigs' feet, pepperoni and beer foam. By the time he was three the only way Mrs. Grobnik could persuade him to drink milk was put it in a stein and say, 'Have a snort, little Slats.'"

Many of the columns that caused Chicagoans on buses and El trains to chortle aloud were filled with the sagacity and antics of Slats. There was the one about Slats learning to play the accordion: "He preferred the violin because it was small and he was lazy. But his mother said, 'You can't play the violin. People will think we're Jewish. Besides, if you learn the accordion you can earn money playing on weekends at the taverns. And that will you give a chance to get to know your father.'"

Slats was also a skeptical youth. "He was just a mere child when he told his mother he doubted that an Easter bunny went around leaving candy-filled baskets for children. 'No rabbit would come in this neighborhood. He'd be hit by a beer truck.'"

And Slats learned the hard way that there was no Santa:

He was awakened during the night by the sound of somebody moving about in the kitchen. Slats crept from his bed, hoping at last to catch a glimpse of Santa. But there, by the kitchen stove, stood his father in his long underwear, his arms loaded with gifts. Slats bounded through the kitchen into his parents' bedroom, howling: "Ma, get up quick—pa's filching every damn present Santa left for us!" So that's when his parents decided —when Slats picked up the phone and started yelling for the cops—that he ought to know the truth.

Slats never really got in trouble:

"Most of the youths in my neighborhood had it drummed into their heads that just one arrest would result in a life-long black mark that would make it impossible to get a job....Mr. Grobnik was always telling his boy Slats: 'Just one pinch, kid, and you'll never get no job.'

"Slats always answered: 'That won't interfere with my plans.' But he never got in trouble anyway because it would have kept him awake."

Slats was too shy to have much to do with girls, but once in while he had the urge: "He tried passionate love letters. But he didn't write well, so several of the girls turned the letters over to the authorities and Slats was questioned in a couple of unsolved sex crimes."

Another time he tried a love potion, which he obtained from his Aunt Wanda Grobnik, who was also famous for her mystical readings and the things she could see in coffee grounds. She gave Slats some black powder and told Slats that if he put it in a girl's food or drink, the girl would fling herself at him. On his next date, Slats furtively sprinkled it on the girl's bowl of chili. She was sick for three days and the Greek who owned the diner was closed down by the health department.

Royko got into a routine of using Slats to create holiday themes: hangovers for the Christmas season, nostalgia for the Halloweens and July Fourths of his childhood, and unrequited love for Valentine's Day.

Valentine's Day didn't mean nothing to Slats because he didn't think much of girls. As he put it, "When you hit 'em they cry. They must be queer."…Naturally he was the only boy who didn't receive any valentines. Once, a teacher mistakenly felt sorry for him and gave him a card. He promptly reported her to the principal, saying, "Next, she'll be molesting me."

Slats's mother used to tease him about his indifference to girls. "'Why do you think your father comes home at night?' Slats thought about that and said, 'Because the tavern closes?'"

By the end of 1964, Royko had fully established himself as the new, fresh voice of Chicago journalism. He was the accepted star of the *Daily News,* and he roamed the city room constantly during the day, building the routine he would follow for years: arrive with coffee at 10 A.M., scowl, say nothing, and retreat to his cubicle, a small area with a cluttered desk and a three-foot surrounding wall that separated him from the rows of reporters' desks, but afforded no privacy. He would pore over the other newspapers and scan large rolls of wire service copy to see if anything worth poking fun at had occurred outside Chicago. Then he would get an idea and prowl the room.

"Sometimes," Lois Wille said, "he'd come over to our area and begin acting out his column thoughts. There was me and Betty Flynn, who later covered the United Nations for the *Daily News.* Mike really liked her; her father was a Chicago cop. He liked people who came from working-class backgrounds. He'd act out these funny anecdotes and if any of us laughed they would become his column for that day.

"He had male friends, but he always seemed to feel more at ease with women. I think women became his closest friends, at least the people he would confide in over the years."

In his new status, Royko began hanging out with some of the stars of the Field papers: a pair of Pulitzer Prize cartoonists, Bill Mauldin and John Fischetti, and Herman Kogan, who may not have been the finest newspaperman in Chicago history but who ranked close to the top. These men represented the ideal Royko admired more than anything. They came from humble backgrounds, went through difficult times as kids, and became tough enough to survive the depression and whip the Nazis and the "Japs." They were talented and dedicated to their craft, and adult enough not to take their lives or their jobs too seriously.

"When I broke into newspapers," Royko said in a 1997 Tower Pro ductions interview, "most of my mentors—the senior reporters and editors—they didn't take themselves quite as seriously. They liked being newspapermen. You never heard the word 'career.' Everyone had worked in one kind of grungy job or another. At the *Daily News*, one of the rewrite men had been a prisoner of war. Another guy limped because he had taken a Japanese bullet in his heel. Our music critic, a very quiet, sensitive man—his wife once showed me a picture of him sitting in his fighter plane. He was a World War II ace. These guys had a different attitude about life. They were more forgiving. If we nailed a politician, nobody wanted him to go to jail. We nailed him. We got a story. Throw him back. It's like a fish. We'll catch him again later."

Herman Kogan was the son of a Russian immigrant who ran a newsstand during the depression. Kogan attended Wright Junior College and got a job at City News Bureau, a career path Royko followed two decades later. That's where the similarities ended. Kogan spent his nights at the University of Chicago and graduated Phi Beta Kappa in three years and was hired at the *Chicago Tribune,* where he sat on the rewrite desk with a couple of other notable Chicago newsmen, Lloyd Wendt, a future publisher of Chicago's *American*, and Robert Cromie, a gifted and versatile war correspondent, critic, and television show host. When World War II began, Kogan enlisted in the marine corps, which must have shocked a few salty drill instructors wondering why a sheltered Jewish intellectual had volunteered to become a jarhead.

Kogan was eventually hired by the *Sun-Times* and *Daily News*, created the *Daily News*'s popular "Panorama" section for "Arts and Entertainment," authored several books about Chicago, and was the father of a son, Rick, who grew up to be a newspaperman, protégé, late-night companion, and golfing partner of Royko.

"Mike always said Herman should have been made editor of the *Daily News*," Kogan recalls. "He said he was the best and smartest newspaperman he ever knew. When I met Mike I was about twenty-six, although he had been to my house when I was a young kid. He looked at me and growled, 'You're Herman's kid, uh? You're a good-looking guy. Must've had a different dad. Anything you need, come over. I love your dad.'"

Kogan, Fischetti, and others usually lunched daily at Riccardo's, an Italian restaurant tucked at the bottom of Rush Street, a block north of the river, and shrouded by the shadows of the Wrigley and Sun-Times buildings. It was a newspaper and advertising hangout during the 1960s and 1970s, and its owner, Ric Riccardo, had decorated the large bar area with murals whose shades grew smokier over the years. Royko became a regular at lunch and at the bar, although Ric's was never as comfortable for him as the Goat. Royko rarely drank at lunch with the column beckoning; he did, however, begin keeping a six-pack of beer in one of his file cabinets, a habit that never changed even though there would be weeks or even months of dry periods in his life.

It was Kogan who first brought Royko to the attention of Chicago's resident liberal, television talk-show pioneer and future Pulitzer Prize–winning author Studs Terkel. Terkel, a native New Yorker, hosted what was probably television's first talk show in 1950, featuring his folk-singing, guitar-strumming pal, Win Strake. The show was canceled when Terkel signed a petition protesting Senator Joseph McCarthy's communist witch hunts. Station bosses asked Terkel to say he was duped into signing the petition, but Terkel refused and was booted off the air. He landed at WFMT, a classical music station where he introduced Chicago listeners to the sounds of Woody Guthrie, Big Bill Broonzy, and gospel goddess Mahalia Jackson.

By the 1960s, with its rioting and marching and assassinations, the white-haired Terkel may have seemed a tame radical. But never a quiet one. He was always mouthing off about the rule of Richard J. Daley, or the conduct of police and anyone else he viewed as oppressive.

"Herman Kogan called me," Terkel recalled, "and said 'Something's happening at the *Daily News*. There's a new editor, Larry Fanning, and he's given columns to some people, Nick Von Hoffman and a guy named Mike Royko. I remember reading one of the first columns I saw from Royko, and I said to myself there's a lot of nice guys writing columns in this town but this guy is different. He tells it in a different way. Then Nelson Algren called and said, 'Have you seen this guy Royko? He tells it with all the warts and the carbuncles, and he hits hard but he's pretty funny.' That was something coming from Algren. He hated newspaper people.

"I never saw a columnist who had the impact he did. I don't drive. I take buses. I used to take the 146 bus, which was full of the Lake Shore Drive people or sometime I'd take the 151 bus which had the Nelson Algren people on board and either way you'd see five, six, seven people all reading page 3 of the *Daily News* and later page three of the *Sun-Times.*

"He was a caricaturist, like the great political cartoonists he exaggerated to make a point. He did that with his writing but there was also his great style. He wrote and it seemed he was talking to you. His columns had an oral flavor. And he had a great change of pace. One day it was Slats and then a great expose and then one of what I called Miss Lonelyhearts, giving out advice to readers who would call him."

Soon Terkel became part of Royko's evening entourage, meeting at Riccardo's, moving to the Goat and then up to Rush Street or Division Street for a nightcap. Nelson Algren joined them along the way.

Algren and Royko formed an immediate mutual admiration society. Royko told Algren he thought his best work was the short story collection, *Neon Jungle,* which Royko read when another airman recommended it to him during his stint in Korea; Algren told Royko he was the best columnist in town. They drank to that.

By the 1960s, Algren was an embittered man. He felt, justifiably, that Chicago did not appreciate him. It had been a decade since his *Man with the Golden Arm* had won the first National Book Award for fiction in 1950, but Chicagoans were still relatively uninterested in writers and artists. One of the reasons Algren enjoyed hanging out with Royko was that he got some of the overflow attention being paid to the town's hot, new columnist. Algren had just come out of his steamy affair with the French writer, Simone de Beauvoir, but most Chicagoans didn't know and cared less about the guy who had warned, "Never play cards with a guy named Doc."

Chicago was not a place for the finer arts. The Chicago Symphony had to rely on the donations of the few wealthy patrons and could never scrape up enough money to do a tour. The Lyric Opera went belly up. For some odd reason the only heroes Chicago applauded were the hapless players of the Cubs and the Sox, and the only people the city seemed to admire were the politicians who could get them jobs, fix a traffic ticket, or get the streets paved. It was still a time when most Chicagoans were rather proud of Al Capone.

But Royko suited them fine. In one short year, he went from being the nearly obscure reporter covering the County Building and stopping for a few beers at Paddy Bauler's saloon to the funny, fearless, and hard-nosed sage of a city that suffered a great shortage of them. He was on his way to the top.

chapter 6

After years of paying sympathetic lip service to the plight of blacks in the South, northern whites were getting a closer look at the harsh reality of race relations in the South. A decade after Rosa Parks refused to move to the back of the bus in Montgomery, Alabama had flared into a blaze of church burnings, lynchings, shootings, beatings, stabbings and bombings. Television cameras had invaded the small hamlets of the South, recording the bigotry and prejudice that barricaded blacks from the white world of education and the white prerogative of the voting booth. Northern newspapers, which had conveniently ignored the subject for most of the twentieth century, found a new voice. In 1960, the *New York Times* sent Harrison Salisbury to Alabama, where he wrote one of the first indictments of white Southern resistance to Martin Luther King's dream:

No New Yorker can readily measure the climate of Birmingham today. Whites and blacks still walk the same streets. But the streets, the water supply and the sewer system are about the only public facilities they share. Ball parks and taxicabs are segregated. So are libraries. A book featuring black rabbits and white rabbits was banned. A drive is on to forbid "Negro music" on "white" radio stations. Every channel of communication, every medium of mutual interest, every reasoned approach, every inch of middle ground has been fragmented by the emotional dynamite of racism, reinforced by the whip, the razor, the gun, the bomb, the torch, the club, the knife, the mob, the police and many branches of the state's apparatus.

But this kind of reporting was not common in American newspapers. Only as King's crusade built and culminated with his 1963 march on Washington, D.C., did newspapers begin to send reporters to the South.

On March 14, 1965, Mike Royko became one of them. Royko would travel rarely in his career. Except for the national political conventions where he could keep his eye on the local political contingents, and except for columns written from vacation destinations or the odd visit to Washington or New York, he made only one trip to specifically cover a story and a place. The place was Selma, Alabama.

Royko arrived at the end of a week in which a white minister who had wanted to join the protest march to the state capital, Montgomery, had been killed. Four men had been charged in the slaying. Royko went to the civil rights enclave and watched and listened. On Monday, March 15, he wrote about what he saw.

He wrote about a state trooper with a leery grin who said he wanted to turn fire hoses on civil rights marchers. He described handbills that had been dropped from a low-flying airplane, promoting a defense fund for the four men who had killed the minister. There was no opinion in Royko's writing. No decrying of racist rednecked troopers. Not a wisp of his trademark humor.

The next day Royko wrote another front-page column about the truth the way the folks in Selma saw it:

"It is true that the death of the white minister was tragic. But it is also true that he barged in here, where he wasn't invited."

"It is also true that he was killed by hotheads—not by the majority of the decent, law-abiding citizens. And the hotheads were agitated by agitators."

More Selma truths:

"Negroes and whites live together in great contentment."

"Negroes are happy with what they have and they don't want anything more."

And, as usual, Royko found just the right people to talk.

"We get along fine with the colored folks here. You talk about integration? It's nothing for me to be walking downtown and have some colored boy I know come up to me and say, 'Mister Jack, can you let me have a couple of dollars for a couple of days?' "

"I'll tell you, I like a lot of niggers....These niggers is like children. They need someone to take care of themselves."

"Now you take Sheriff (Jim) Clark. I know him real well. He's a good man, but he doesn't look that way on television. He's the one who's kept this from getting worse. When this whole thing started he was smart enough to round up some of the toughest hotheads in town and make them his deputies. He figured he could keep his eye on them that way."

Having lived through a generation of literature, film, and television about racism in the South, the dialogue may sound trite and predictable now, but the satire was stark and shocking to readers of the *Daily News*, not all of whom necessarily disagreed with the people Royko interviewed.

Two days later, Royko was back in another form:

"Two hours until the next demonstration."

"How pleasant. Not a single galloping horse in sight. No fleeing ministers. Nobody singing, 'We Shall Overcome.' Even the TV cameramen are resting somewhere.

"A warm Gulf breeze flaps the Confederate flag that flies high above the graceful white Capitol with its wide, gently rising marble stairs. The morning sun glints on the Confederate flag license plates on the front of the state police car bumpers. Even the troopers are relaxed, their billy clubs at rest."

"A police dog sleeps in the back seat of a squad car, dreaming of good things to eat—if he can catch them."

Royko spent only six days in Alabama. He had seen enough. He also felt distinctly unwelcome in Selma:

On my second day in Selma, I happened to get a haircut in a downtown shop. After asking me "where I was fum," the barber and the patrons just stared. I had parked my car outside the shop and I left it there while covering the day's civil rights activities. As I returned to the car that night, I noticed four men in a car a few spaces behind mine. A sheriff's helmet was stored near their rear window.

When I pulled out, they pulled out. When I turned right, they turned right. When I turned left, they turned left. They did everything I did except sweat and shake....Fortunately my rented car was the fastest thing for rent. Their car was old. I went past the motel, went over a small hill, cut my lights, made a rather spectacular u-turn and was going the other way before they could react.

I parked, reached the hotel bar in two great leaps, ordered enough drinks for everyone and drank them all myself.

Years later when Sheriff Jim Clark was invited by a branch of the John Birch Society to speak in Chicago, Royko suggested he discuss

"How to spot the soft spot on a skull while riding a horse," and "the 25 points on a person's body most sensitive to a cattle prod."

Royko raged against racism throughout 1965. He praised Martin Luther King's efforts to register black voters, he castigated Chicago pols for frittering away antipoverty funds in the black hole of machine patronage, and he hammered schools boss Ben Willis, who steadfastly resisted all efforts to bus black children to integrated schools.

In July, Martin Luther King joined local civil rights leader Al Raby in a march of 10,000 protesters to Chicago's City Hall. Royko wrote: "The city reacted the way people sometimes do when a bill collector, a salesman or an unwelcome guest shows up on the front porch. They turn out the lights, pretend not be home....Nobody ran out on the porch with a shotgun. Nobody sicced the dog on anybody. Nobody called the cops."

But in one of the wonderful twists that was his hallmark, Royko turned from lamenting the attitude of disinterested whites to the apathy of black voters:

> A few months ago an aldermanic election was held in a ward that is almost entirely Negro. The candidates were an unknown white City Hall lawyer and a Negro civil rights worker. The white City Hall lawyer drew almost as many votes as Dr. King drew marchers. That's the reason why the civil rights movement in Chicago is going badly— and why Mayor Daley and the city's top Democratic Negro leaders were part of the huge audience that didn't show up for Dr. King.

Columns like that and the ones from Selma there were being applauded in liberal Hyde Park and in newspaper gathering holes such as Riccardo's and the Billy Goat, but they weren't being applauded by many of Royko's ethnic neighbors on the Northwest Side or by the readers of the *Daily News* in its traditional North Shore suburban base.

Although America generally sympathized with the plight of the Selma marchers and martyrs, the other side of the civil rights struggle was turning ugly. Militant black leaders such as Stokely Carmichael were frightening whites with declarations of "Black Power." A light-skinned, frail black man named Elijah Muhammad was preaching to a group of Chicagoans called Black Muslims and decrying "white devils." In the summer of 1965 the Los Angeles section of Watts went up in flames and bullets. Lyndon Johnson was calling for a Great Society, but to most Chicagoans that still meant a white society.

"A man named Ken Johnson who was head of *Daily News* circulation used to storm up to Fanning's office about twice a week to complain about Mike's columns," Lois Wille said. "In the newsroom we were all so proud of what he was writing and what the paper was doing, but in the circulation department they were angry. The *Daily News* was losing circulation, all afternoon papers were."

Racism was not the only problem. By the middle of the 1960s, the expressways, which the newspapers had welcomed with effusive praise, were part of a great economic and societal change that would eventually kill most afternoon newspapers. The expressways had freed suburban commuters from public transportation. Subscribers no longer read their papers on the train. Now, they drove to work in automobiles, crawling from dawn to sunset on highways. The prosperity of the 1950s and the encroachment of black neighbors sent newspaper readers scurrying to older suburbs like Wheaton, Barrington, Arlington Heights, and to the new so-called bedroom communities such as Hoffman Estates, Elk Grove Village, and Bolingbrook. These suburbanites worked and played in Chicago and journeyed home only to sleep. Populations in older villages like Naperville doubled and tripled over the next two decades.

At the same time, the heavy manufacturing base was disappearing across the rust belt cities of Pittsburgh, Cleveland, Detroit, and

Chicago. The steelworkers who used to work 7 A.M. to 3 P.M. and read their afternoon newspaper before dinner were disappearing, as were blue-collar workers in other related industries. They were forced to move into service-industry jobs that ended at 5 P.M. and barely got home in time for the six o'clock network news.

And the network news was evolving. NBC had created the Chet Huntley and David Brinkley show to rival CBS's popular Walter Cronkite. Feeds from network affiliates in Atlanta, Chicago, and Los Angeles were providing film of events that had occurred only a few hours earlier. The networks were even showing some rudimentary combat film from Vietnam, where American troops were supporting the South Vietnamese army in its struggle against the communist North.

All these ingredients made it difficult for the *Daily News* and the *American* to retain their afternoon readers, and circulation began an inexorable dive. The situation in New York City was even worse. Exacerbated by a series of strikes in the early 1960s, several newspapers folded. The prestigious *Herald-Tribune* stayed alive only briefly by joining with two other newspapers before it finally died in 1965, tossing some of America's finest newspapermen, including Homer Bigart and sports columnist Red Smith, on the street.

The first casualty of the Chicago newspaper war in 1965 was Larry Fanning. On September 18, 1965, Marshall Field IV died suddenly. Ownership of the *Sun-Times* and *Daily News* was left principally to his sons, Marshall Field V, then twenty-four, and Frederick, thirteen. The younger Marshall Field had been working at the *New York Herald-Tribune* but returned to Chicago to become assistant to the general manager, which meant a short stewardship until he became publisher of both newspapers.

The death of Marshall Field IV signaled a shake-up, and Fanning, whose brief editorship was marked by journalistic brilliance, had done little for the dipping circulation numbers. He was replaced by

Roy Fisher, who had been an assistant city editor under Ritz Fischer and then became editorial director of World Book, also owned by Field Enterprises. Fisher was viewed by the staff as the type of editor who preferred attending civic events and formal dinners to taking a pencil to a piece of copy.

"Roy Fisher was a good, kindhearted person," said Ray Coffey, "but as an editor, well that was a different thing. I returned from Saigon in 1967 and Fisher told me to come to his house and make a speech to various people he said were important. I said, 'Sure, I'd be glad to do it,' and then he says, 'It's black-tie.' I said I didn't have a tuxedo and he said, 'Just go down to Marshall Field's and buy one. Put it on expenses.' Dinners were important to him.

"I remember between foreign assignments I was putting out the red streak which came out about 9 o'clock and didn't have much news in it but I was trying hard to get some fresh stories in it. Every morning Fisher would come in, pick up the paper, hold it at arm's length and say, 'Looks nice.' I thought, 'looks nice'? What about reading it?"

Whatever the rest of the staff thought of the change, Royko was traumatized. "Fanning was the guy who gave him the column, who knew how to handle him, how to coddle him," Lois Wille said. "Larry would sit and puff on his pipe when Mike got upset and act very fatherly toward him, protecting and assuring."

"A lot of us felt, and Fanning encouraged us to feel this way, that he was fired because of his courageousness," she said. "About the time that Fanning was spreading the word that he was leaving, Mike led a staff revolt, well, he tried to make it a revolt. He began passing around a petition to mail to Marshall Field protesting Fanning's firing. That consumed his time for about a week. A couple of people wouldn't sign it. I guess they thought management would be angry at them. Mike never forgot them."

All the bluster achieved little except to make Royko feel that he had once again fought the establishment. Fanning departed. Fisher arrived.

"One of the first things Roy did when he took over was to call Mike into his office and tell him how much he loved him. He said he knew about the petition and that he, too, regretted the circumstances," Wille said.

Royko responded by threatening to quit. This may have been the first of what probably were a few dozen tempestuous resignations, many of them made in indignation or embarrassment; most of them done while he was sober, a few not. It would also become a characteristic of Royko that while he privately ridiculed the many corporate bosses who would fawn over him in public, he never forgot who signed his paycheck. Fisher was just the first of many to fawn. He assured Royko that he would remain the most prized newsman at the *Daily News* and told Royko that he could do anything he wanted, cover anything, go anywhere.

To prove it, Fisher said, "Why don't you go and report from Vietnam?"

Royko told everyone in the city room about the offer, which was quite a compliment because the *Daily News* already had the great Keyes Beech and Coffey sharing tours in Vietnam.

Of course, Royko turned it down. Maybe he was reluctant to try his hand at anything that didn't involve Chicago. Maybe he couldn't get over his fear of flying. And maybe, after his Korean experience, Royko had had his fill of Asia.

"After all, David and Rob were small," Lois Wille said, "and Mike certainly didn't want to get killed. He agonized over going and made sure everyone in the office knew he was agonizing."

Nineteen sixty-six was a big year for reporting Vietnam. In February, the Senate approved $4.8 billion to fight the Vietnam War, and in March the Defense Department announced that 235,000 American troops were in South Vietnam. In April, *Time* magazine's cover asked,

"Is God Dead?" Peter Arnett of the Associated Press won the Pulitzer Prize for his Vietnam coverage.

Royko stayed home. On May 31, 1966, Roy Fisher moved Royko's column from the op-ed page to page 3. Royko objected, thinking his column wouldn't be able to compete with all the news. He didn't want to be on page 3. "I fought against it," Royko said in a WBEZ interview. "I thought a newspaper should have news on page 3. I thought if the column went there I couldn't write columns that were funny, that I would have to be serious if I were on a news page. Besides, the *Daily News* took a survey that showed more readers read page 14 than the front page."

Throughout his career Royko would toss out such survey results or circulation audits that demonstrated how many people bought whichever paper he was appearing in solely because of him. There was no question that he was the most valuable newspaper commodity of his era, but he also felt some compulsion to hype his universally accepted value with statistics.

He immediately set out to prove that he could be funny on page 3. His first column was a typically light question-and-answer session with himself:

"Now that your column is in a new location, will there be any change in your basic philosophy?

"No, it will remain the same—eating regularly."

The Royko column wasn't the only thing that moved in Chicago that summer. Martin Luther King Jr. moved into the city and began to lead a series of open housing marches in some of the toughest ethnic neighborhoods on the Southwest and Northwest sides. This was the heart of Chicago's "bungalow belt," and home to second- and third-generation Lithuanians, Poles, Ukrainians, Russians, Serbs, and Croats. They were scared that the value of their small homes would sink to nothing if blacks continued to move out of ghettoes of the South and West Sides and into their neighborhoods.

King was granted a summit meeting with Mayor Daley. The most memorable moment came when King asked the mayor if he realized how many people in Chicago went to bed hungry. Daley turned to an aide and said, "Get their names."

While King tried negotiation and persuasion, Daley responded with stony silence.

The summer dragged on. It was so hot in mid-July that kids starting turning on fire hydrants and splashing in the water to cool off. Lake Michigan was closed because of a pollution warning. On Tuesday, July 12, at the corner of Throop and Roosevelt, some black children turned on a hydrant. A fireman turned it off. The kids turned the hydrant on again. The police arrived and turned it off and arrested one youth. Someone threw a rock. Someone picked up a bottle. In a three-flat building, someone fired a gun. Police reinforcements were called. Suddenly the street rioting that had plagued Los Angeles a year earlier broke out on Chicago's West Side.

Lines of helmeted policemen, trailed by television crews, filled the streets. Porches and windows were crammed with jeering blacks. A phalanx of police suddenly charged toward a house, an alleyway, or an intersection, dragging one or two black youths toward a patrol wagon. At night, the Molotov cocktails started sailing and the looters came out.

On Wednesday Mayor Daley called out the National Guard, and by Friday the worst of the rioting was over. Two African Americans had been killed, 60 were injured, and 300 were arrested. The first death occurred Wednesday night, when a pregnant fourteen-year-old black woman was killed by stray gunfire.

Almost at the same time, a pockmarked merchant sailor named Richard Speck forced his way into a South Side townhouse that served as a dormitory and stabbed, strangled, and sexually assaulted eight student nurses over a period of five hours.

For two days, Chicagoans didn't know which terror was worse: a fiend on the loose who might invade their bedrooms, or angry hordes of black youths tossing firebombs and crashing store windows. On Saturday, July 16, Speck was arrested at Cook County Hospital after a suicide attempt. He was subsequently sentenced to life in prison and died there of a heart attack in 1991.

The civil rights movement did not dissipate that quickly. The open housing marches resumed at the end of July. They were often led by King's local apostles, Al Raby, James Bevel, and a young, tall, articulate recent arrival to Chicago, the Reverend Jesse Jackson. Jackson immediately became the most quotable.

"All these people got money. All these people can afford houses in good communities. All you white folk are always wondering why black people in the slums are always driving big, black Cadillacs. That's because GMAC is happy to take our money. Now, we will see if these real estate agents want to take it."

They marched in Gage Park and Marquette Park on the South Side, and the Cragin neighborhood on the Northwest Side. Royko wrote a biting column about the white youths who lined the sidewalks: "When a punk got off what he thought was a fine witticism such as 'Hey, you nigger-loving preacher, why ain't you in church?' he rushed about and told his friends what he had said and how the minister had flinched and how good he felt and what he would do later when he got his hands on a dirty so-and-so of a ..."

He continued throughout August and September blasting white racism, and when King promised to lead a march into suburban Cicero, a town whose history included sheltering Al Capone and burning out a black family, Royko suggested King walk alone because the white crowds were filled with nothing but vulgar cowards.

In the fall, David Royko came home from school and found his path blocked by people picketing his house. Carol came down from the porch and took his hand. "What are they doing out there?" the young boy asked. Carol explained that the picketers didn't like what his dad was writing about integration.

The year 1966 was a prelude to the racial anger and antiwar protests that would torment America for the remainder of the decade. It was also a milestone year for Royko. *Time* magazine profiled him:

> In a city where newspaper columnists are almost always civic boosters, Mike Royko, 33, is a constant critic. A foe of all forms of cant and pomp, he carries on a love-hate affair with his home town. He writes tenderly of its ethnic neighborhoods, its traditions and folkways; he fires at will at its politicians and their pretensions.

Nineteen sixty-six was also the year Royko realized writing five columns a week, raising a family, and raising hell most nights after work was killing him. For a while in July and August, he had actually been writing six columns. In September he went to Fisher and asked for help. He said he couldn't continue without some assistance. Fisher pointed out that Royko already had a secretary to help with the telephones and mail. Royko argued that he needed a newsman who would be able to spot a potential story and ask questions and go out and dig up facts. Fisher, like all editors before and since, was always wary about adding staff. It meant an explanation to the bookkeepers. He said no.

On October 3, Royko explained to readers that he was cutting back to three columns a week.

> One day recently I went to a bar I used to frequent. The bartender didn't know me. I hadn't been around for so long that he had forgotten me....I went home and tried to tell my wife what had happened. "I'm working too hard. My friends have forgotten me or drifted away." She didn't hear me because a couple of kids were huddled in the corner crying.
>
> "Why are those kids crying?" I demanded.

"They don't know you. They are afraid of you."

"Who are they? Are you baby-sitting?" I asked.

"They are ours."

There comes a time when a man must examine what the whole thing is about. The next day I went see the editor ...

"I'd like to write three columns instead of five."

"Impossible. Nobody does three days' work for five days' pay."

"Sandy Koufax does. He pitches every fourth day. And his arm hurts so they pack it in ice."

"Fine. You write a column every day and we will put your head in ice."

"I've tried that and it doesn't help."

"All right. How about this. You write three columns a week and sweep up around here the other two days?"

That is the story of why I will be appearing here on Mondays, Wednesdays and Fridays from now on.

But Fisher began to hear from readers who missed Royko's column, and in a few months he told Royko to hire an assistant and resume five columns by the first of the year. Royko went shopping and his first stop was his old training ground, City News Bureau. He called Larry Mulay and asked him to recommend someone who knew the city, had covered the courts, could field telephone calls, spot a possible story, and ask all the right questions. He was not looking for a clerk; he wanted someone to be his "legman," newsroom slang for a reporter who does all the running around while someone else writes the story.

chapter 7

The first legman Mulay recommended to Royko was Terry Shaffer, who had been at City News about eleven months, was assigned to cover City Hall, and was contemplating taking a job at the American.

"Royko called me in for an interview and we spent about an hour or two talking. Then we adjourned to the Goat where we continued talking and drinking for another few hours," Shaffer recalled.

Shaffer grew up in East Chicago, Indiana, a scruffy town on Chicago's southern edge, adjacent to the steel mills where Royko's father had labored as a child. He worked days and attended Indiana University's extension school in East Chicago at night. After two years, he switched to the main campus at Bloomington to get a degree in journalism. He drank a lot, never cracked a book, got married, dropped out of school, and enlisted in the army in 1963. Five months after he was married he got orders to Vietnam.

During his interview with Royko, Shaffer described his army

career. Prior to being sent to Vietnam, Shaffer was sent to one of the many Nike missile sites that dotted the country in the late 1950s and early 1960s. He was stationed in Arlington Heights, where he helped put out a tiny newspaper on the base.

When Shaffer arrived in Saigon in May 1964, there were only 4,000 American troops in the country. He was first given the job of caretaker at a general's house and then reassigned to work at the weekly newspaper the army published in Saigon. Royko relished the resemblance to his own military experience with newspapers.

"It was a great job," Shaffer said. "The paper was only about four pages and the total press corps in Saigon was about five people—the AP, the UPI, Agence France Press, the *New York Times*, and a few others. I had a jeep and a driver who took me from bar to bar and I wore civilian clothes. There wasn't a lot of shooting going on at the time but there were a few dust-ups.

"I remember once being told to cover a firefight that was taking place about fifteen miles west of Saigon. When I got to this little village, Peter Arnett of the AP was there. A South Vietnamese army major told us we couldn't move out to the scene of the fight and Arnett told him, 'This isn't your war. It's our war.' Then Arnett went and hired an ox cart driver to take us to the fight. The guy said no and Arnett gave him some money and we went about a mile outside the village and suddenly there's bullets dusting the ground. I asked Arnett, 'Is this how you earn a living?' We took some pictures of the wounded—I never did see any dead—and we got out of there."

When his army tour ended in October 1965, Shaffer applied to City News. Like Royko, his lack of a degree worried him. "I knew they were only hiring people with degrees but Larry Mulay said, 'You were a combat correspondent in Vietnam. I think you can handle this.'"

Shaffer went through the usual mill—police court, Federal Building, cop shop, and city hall. Eventually, Jack Mabley, managing editor as well as columnist for the *American*, offered him a job working on the paper's consumer column, "Action Line," which fielded complaints about bureaucratic and business foul-ups.

"Royko told me, 'You don't want to work at that fucking place. Come to work for me.'" Shaffer surely impressed Royko, who identified with his working-class background, his failure to finish college, and his capacity for beer.

Terry Shaffer became the first of sixteen legmen to work for Royko over the next thirty years. Ten were women and six were men. Some of them became Royko's close friends. Some of them became outstanding journalists. Some of them became unabashed worshipers, and most of them thought their apprenticeship under Royko was the most valuable tool in their journalism arsenal.

"The main difference between me and all the rest," Shaffer recalled, "is that they worked for him. I ran with him."

If Shaffer learned about journalism from Royko, Royko learned how to deal with his subsequent legmen. For the next twenty years he did not allow any of them, after Shaffer, to become social friends while they worked for him. It was too difficult.

Royko could be a martinet, although few of his legmen could recall him actually shouting. He demanded accuracy but gave only vague instructions. Sometimes he was disinterested in their reports, and he was usually oblivious to their ambitions. He did not encourage them to write.

"I would sometimes try and give him a column instead of notes," Shaffer said. "He always said, 'I can't use this. It's you, not me.' Shaffer was older than most of the people who would follow him, and his social presence gave him a freedom with Royko that the rest of the legmen never enjoyed until after their apprenticeship.

"But the one thing you learned early about Mike, you could never show him up, not in anything," Shaffer said. "I learned to read him pretty good after a while. There were periods when he wouldn't talk to me and I thought I was really screwing up and he was going to fire me. Then I learned there were those times when you had to give him a lot of space. His moods were tough."

Shaffer also spent more time acting as a "legman" than any of his successors. "I was out of the office about 70 percent of the time," he said. "One day he sent me out to get information about gypsies. So I found a gypsy girl in a bar and she introduced me to a bunch of her friends and I hung out with them about five days. When I got back Mike said, 'I told you to go and find some gypsies, not join them.' I don't think he got a column out of that."

Royko was often skeptical of his assistants. Once he insisted Shaffer return again and again to investigate a tavern owner that Royko thought was dirty. Shaffer became a regular and struck up many conversations with the owner. He finally told Royko he thought the guy was clean. "He said I screwed up. That his cop friends had told him the guy was dirty and that he was going to have to get the story himself. Finally, a couple of weeks later he said to me, 'You were right, the guy's okay.' He didn't admit he was wrong very often."

Shaffer did the longest single stint with Royko, working for three years until 1969, when he moved to the *Daily News* general assignment staff.

"We were out every night. He never went home, one bar after another, but a lot of other people were going out, too," Shaffer said.

From 1966 through the end of decade, the period Shaffer worked for Royko, Chicago was a newsman's dream. Spurred by Royko's example, reporters were actively looking for flaws in the once invulnerable Chicago triad of politics, business, and labor. Black leaders, such as U.S. Representative William Dawson, Daley's longtime plantation boss, were being ridiculed by younger black activists, and newspapers were printing what they said. On the national scene, every day seemed to turn up new evidence of the federal government's duplicity over the war in Vietnam. And every night more body bags appeared on the network news.

Everything and everyone seemed to be a potential story. Thirsty for a piece of the front page or the 10 o'clock local news, reporters got their fill in bars and taverns. They lingered in Riccardo's, the Goat, the Corona, and O'Rourke's. They invaded the politicians' watering holes such as the Garden Lounge at the Bismarck Hotel, the Club on 39, Binyon's near the Federal Courthouse in the south Loop, and the tonier Rush Street saloons. And the politicians or their lackeys stopped at the newspaper bars. It was an era of access journalism, and reporters rarely wondered how a $25,000-a-year traffic court judge could afford to pick up every tab as long as they might pick up a possible story or a tip.

By this time, Royko didn't need to hang around saloons to get tips. He hung around to drink. For the most part, he basked in the admiration of other newsmen.

One person who did not admire Royko was Ross Cascio. Ross Cascio ran a towing company that preyed on visitors to the North Side nightlife area. Royko did not object to towing cars that were parked in places reserved for people who paid for them, but he did object to Cascio's collection methods and his associates. As Royko said, "To intimidate those who objected, Cascio hung bats, blackjacks, chains and other pacifiers on his office wall. If a person tried to escape with his own car, Cascio's men would dance on his chest."

Cascio was the butt of Royko columns for a decade. What made Cascio unique was the fact that he was the only person in thirty-four years and nearly 8,000 columns who ever filed a libel suit against Royko. The suit was dismissed as frivolous, probably by a judge who had had his car towed.

The first Cascio columns began in 1967:

When I first wrote about what a mean, sly, no-good, greedy insensitive hulk [Cascio] was, I also predicted that these qualities would make him a great success in Chicago, which they have....I can't say that Cascio has always acted grateful for all that I did for him. He sued me a couple of times for several million dollars, saying I had damaged his reputation, as well as the reputation of one of his loved ones—a huge yellow-fanged, howling dog he used to keep at his side to intimidate car owners who refused to pay the ransom. He would tell them that if they didn't pay he would let the dog eat them. The dog would salivate, gnash his teeth and howl like a wolf. Most people paid. I won both court cases although the dog might have had a chance had he not been a known associate of Cascio.

After three years of Royko columns about Cascio the city finally decided to take action, but Royko remained skeptical:

The City Council is, at last, engaging Ross Cascio, the tow-truck pirate, in combat. It's hard to decide whom to cheer for. Based on past performances, both should be on the same side.

Royko was also skeptical when during a 1971 aldermanic election, a reform candidate, Dick Simpson, pledged to drive Cascio out of business.

Cascio won't object to all the attention. He's often said that every time he is attacked for his strong arm methods, somebody who owns a parking lot or a high-rise hires him....Cascio went on TV to deny that he is a rascal. As other aldermanic candidates berate him, he will come on TV even more to deny everything. Some viewers will assume that anyone who is fat and sly looking and keeps denying he is a crook, has to be running for alderman himself. If that happens, this being Chicago, Cascio will be elected in a landslide.

On one occasion, Cascio braced Royko in his office, armed with a public relations man who was a former cop. "How is your dog?"

Royko asked Cascio. "My dog is dead. I think the stuff you wrote about him made him nervous and shortened his life."

Another first in 1967 was Royko's annual paean to baseball's opening day, his Cubs quiz, which he repeated, with updates and changes, for more than twenty years.

He asked arcane but hilarious questions: "Name at least one Cub pitcher of the 1950s who wore a golden earring?" The answer was, "The immortal Fernando Pedro Rodriquez." He also asked which Cub pitcher of the 1950s was a thirty-eight-year-old rookie. The answer, of course, was the same "immortal Fernando Pedro Rodriquez." It was an example of Royko's ability to make readers laugh by wielding one precise word. In this instance, picking "immortal" to describe an obscure and rather ridiculous overage rookie set the tone for the entire column.

Some of Royko's questions were straight, such as naming Ernie Banks and Hank Sauer as the only Cubs to win home-run championships. Over the years, almost all the questions changed, but Royko always had one reference to a former newspaperman turned local television anchor, Walter Jacobson.

Walter Jacobson had been a rival of Royko's in the early 1960s when he covered the county beat for the *American*. Royko simply didn't like rivals. In his Cubs quizzes, he would always ask "What television figure was a former batboy?" and the answer would be, "Walter Jacobson who says it wasn't much fun because some players threw their underwear at him," or, "Walter Jacobson who says Hank Sauer used to make him chew tobacco in the clubhouse and he would throw up."

Jacobson would be the object of derision in many columns on many subjects, displaying the disdain Royko felt for television news. But Royko was equally tough on radio talk-show hosts, harpooning conservative Howard Miller on several occasions. He would also go after people he admired greatly, such as Frank Sinatra, if he thought they deserved it and if it meant a good column.

And he remained the voice of the little man whether he wished it or not. He complained about paying fifty dollars for tickets to see the stage musical *Hello, Dolly*, and the $100 tab it cost him to spend New Year's Eve at Chicago's swank Maxim's restaurant. He picked on the number of relatives Richard J. Daley had put on the payroll, and he lampooned Louis Farina, a Daley flunky who had been put in charge of the city's parking-meter maids and designed himself a uniform that rivaled the glitter of any banana republic dictator.

Former U.S. Representative Dan Rostenkowski recalled, "I was riding in a car with Daley the day of the Farina column and I asked the mayor if he had seen it. 'I never read Royko,' he said. Well, I knew he was lying because there was the *Daily News* all over the floor of the mayor's car. Now, we get to City Hall and I walk in with him and there's Louie Farina by an elevator. Daley looked at him and just said, 'Upstairs, now!'

"I went over to Louie and said, 'I don't know if you know, but you're about to get the biggest ass chewing of your life,' and Louie says, 'What'd I do, What'd I do?' Daley read everything Royko wrote."

As mean-spirited as Royko could be about foes, real or imagined, his loyalty to friends and those he admired was ample. A gossip columnist for the *American*, Maggie Daly, once wrote an item about Nelson Algren being arrested in a car where police found marijuana. Royko offered his column to Algren to write a vicious satire attacking Daly.

Royko also admired Mort Sahl. He once wrote a column charging that the Kennedy family had used its influence to have Sahl blacklisted from top clubs and network television shows because he made jokes about President Kennedy and Robert F. Kennedy. Royko never did have much good to say about the Kennedys, although John Kennedy had been killed only a few weeks after Royko began his column and

Royko rarely wrote much negative stuff about the late president other than to join in witticisms about his bedroom habits when they were disclosed in the 1970s. But he generally regarded Bobby and Ted Kennedy as spoiled brats of an indulgent and not very likable father.

One of the pols he did pick on was Ronald Reagan, then governor of California who was "trotting around the country, showing his teeth to voters, doing what his press agents tell him, and hoping that *Death Valley Days* had enough fans to help get him into the White House." Although he later pilloried Reagan for his attitudes toward social issues, Royko's first concern about the one-time actor was that he wasn't tough enough to deal with the Soviet Union. Of course, it was Reagan's tough posture toward the Soviets which would contribute to the ultimate collapse of the Soviet Union. Royko's amazing prescience was taking the day off on that column.

And one of the mobsters he picked on was John "Jackie" Cerone. Royko had earlier written that Cerone had won a golf tournament at Tam O'Shanter that was rigged, and he urged club members to throw Cerone out. The issue became moot after the golf course was sold and the land turned into a housing complex. But Cerone found another club. When Royko wrote that Cerone and Lou Rossanova, "well-known coat holder and flunky," were members of Riverwoods, they were tossed out and the club changed its name. Chicagoans were amazed that Royko could write so flippantly and insultingly of people who were supposedly responsible for some of 1,000 unsolved gangland slayings in the city dating to the 1920s.

Although he never backed off on the mob, Royko sometimes wondered if it might not bend its old rule about only wreaking violence on its own. David Royko remembers one incident in 1967 when his father came home and discovered a dark car parked ominously in front of his house.

"He parked down the street and got out of the car, closed the door real quiet and went around back to the alley and came up to the back door. I remember my mother saying, 'Is that your father at the back door?' when he came bursting in shouting, 'Carol, Carol, lock the front door.' We all went to the front door and suddenly there's this squeaky voice saying, 'Mike, Mike.' It was one of my grandparents' friends who were always around and my father detested this woman. After that night, he despised her. But he was sure it was mobsters who had come to get him."

Royko may not have wanted to risk his life covering a war, but he was willing to anger Chicago mobsters if it meant a good column. The column had become the single driving force in his life by now. It was what he thought about the first thing in the morning and the last thing at night. It had become an obsession. The only few hours of pleasure he had came in the evening when he could enjoy the satisfaction of having just completed one column before the morning came and he would face the dread of doing another.

Still, Royko would occasionally goof off at the office. Ed Gilbreth, who became the *Daily News* political writer and later an editorial page writer with the *Sun-Times*, remembered one of the many occasions on which Royko would feel compelled to resurrect the acting career that began and ended at Central YMCA high school. "I remember the night when Royko jumped on a desk and began imitating the supposed love affair between Adlai Stevenson and Eleanor Roosevelt. He was playing both parts and Royko would imitate Adlai's Ivy League voice saying, 'Oh, Eleanor, your elegance is only matched by the beauty of your soul,' and then he would play Eleanor in a high-pitched voice saying, 'Oh, Adlai, you scamp, you.'

"He loved mimicry and he would often do it at Riccardo's after a few drinks and have everyone laughing. Everyone knew how funny he could be," Gilbreth added. "After too many drinks he could get ugly and everyone knew that, too."

Once, Gilbreth was on the receiving end (literally) of one of Royko's famous practical jokes. Columnist Jack Mabley at the *Ameri-*

can had written a holiday column asking readers to send him their Christmas cards so he could forward them to disabled veterans who would, as part of their therapy, cut them up and make different art items from the various pieces of tinfoil and glitter. But several readers got confused and instead of mailing boxes of cards to Mabley at the *American*, they sent them to Royko at the *Daily News*.

"I was living in Calumet City then, and one day I get a call from the post office saying I had a package with $12.50 due," Gilbreth said. "It was a lot of money then but I thought someone had sent me something so I went to the post office and there were several boxes. I paid the postage and when I got home I opened the boxes and here were all these Christmas cards. I didn't know what the hell was going on. It was several weeks before someone told me Royko had just gathered the boxes and forwarded them to my address," Gilbreth recalled.

Royko also enjoyed being an impresario. In September 1967, after the Westminster Dog Show at New York City's Madison Square Garden, Royko announced in his column that he would sponsor the First Annual Mixed Breed Dog Show. It was good marketing. Newspapers traditionally sponsored events that provided entertainment and tried to endear them to readers. The *Chicago Tribune* , and particularly its sports editor of the 1930s, Arch Ward, were extremely successful with such promotions. In 1933, Ward convinced the owners of major league baseball to hold an all-star game pitting the best of each league in conjunction with the 1933 World's Fair. Ward then matched the best of the college football stars against the professional champions, an event that lasted nearly forty years until the multimillion-dollar contracts offered to the collegians made it too risky for them to play against the professionals.

Royko's mongrel mix was held September 10 at Soldier Field and was an astounding success. Royko created categories for dogs that looked like Charles de Gaulle and dogs that looked like Mike Royko;

for dogs with the shortest tail and the longest tail. Van Gordon Sauter, later a president of CBS News, covered the event for the *Daily News* and called the show "a nightmare of canine genetics that attracted about 750 of the nation's most slothful, homely, inept, endearing dogs." The Royko Cup, awarded for the dog that was most indifferent when given commands, was won by a "huge hound that resembled a cross between a yak and a polar bear." More than 4,000 people attended the event, and the *Daily News* circulation department loved it.

He also sponsored a contest to design a new city seal to go with a new city motto that he provided:

"The old one is 'Urbs in Horto' (City in a Garden)."

"The new motto—'Ubi Est Mea'—means 'Where's Mine?'"

Royko's first collections of columns, *"Up Against It,"* was published in 1967 by Henry Regnery Company, Chicago. Royko got a $2,500 advance, which was a windfall because he was earning only $20,000 a year. His friend and occasional victim of practical jokes, Richard Christiansen, reviewed the collection and praised Royko's "wicked sense of the ridiculous, and above all, his grumpy individuality."

The introduction to the collection was written by another Royko friend, Bill Mauldin, the prize-winning cartoonist who had created the lovable World War II dogfaces, Willie and Joe.

Mauldin was eerily prophetic:

Royko is indeed fast becoming successful, and the poor guy is fighting against it every inch of the way. The same instincts that make him so good at his work warn him that celebrity has a way of alienating a fellow from the very wellsprings of his success....He grumps at admirers in office corridors....The more he glowers at gushy ladies the more thrilled they'll be. Mike is one of the few newspapermen in Chicago who hasn't been afraid to give Mayor Daley hell. Now he'll probably run into the Mayor socially and have to exchange small talk with him.

Royko is like his city. He has sharp elbows, he thinks sulphur and

soot are natural ingredients of the atmosphere, and he has an astonishing capacity for idealism and love devoid of goo. He has written
about Chicago in a way that has never been matched.

Mauldin was among the first to notice that celebrity was changing
Royko, that he was already becoming conflicted by the joy of fame
and the fear of failure that would drive him deeper into daily homage
to the column above all else. He and Christiansen both noted Royko's
"grumps," which became as much a part of his aura as his wonderful
laugh lines and his legendary long nights.

In December, Royko finally got the kinks out of a column he
wanted to run on Christmas. He had written it back in 1963 but didn't
like it. He worked on it again for 1964, but still wasn't satisfied. He
rewrote it for 1965 and again in 1966, but didn't use it. In 1967, he
wasn't entirely satisfied but he put it in the paper on December 19. It
became one of his most popular columns.

The column told the story of a young couple, Joe and Mary, who
arrived in Chicago on Christmas Eve, looking for a place to stay
where Mary could deliver her child. They ultimately encountered
Cook County bureaucrats at every level, from the Chicago Transit
Authority to the Public Aid office to Planned Parenthood and the
Cook County Hospital. When their child was born, three strange
men appeared at the bedside but the cops hauled them off on a nar-
cotics charge for possessing frankincense and myrrh. The humorous
but poignant parody ends with the family fleeing Chicago for the
southern Illinois area known as "little Egypt."

The *Daily News* mailed thousands of copies of the Joe and Mary
parable to readers. The piece was included in several anthologies of
newspaper writing and essays. It was a Christmas classic.

By the end of 1967, Chicagoans were generally accustomed to
greeting one another at the coffee shop or the watercooler with one

of Slats's observations. "Where's Mine?" became part of daily conversation. So did "Ju read Royko?"

In 1968, a lot of America read Royko. There were never so many conflicts tugging at the political, economic, and cultural fabric of society. Blacks were rebelling, and whites were swallowing their ambivalence in huge gulps, confused between guilt and greed. Many whites felt sympathy and sorrow over the centuries of racist oppression, but they also feared that every black man or band of youths was on the prowl for revenge. Meanwhile, teenagers from comfortable suburban enclaves were burning their draft cards and young girls were refusing to wear bras. College students began to toss four-letter words into polite conversation and write them in grease on their foreheads. They didn't get haircuts and they smoked dope and they thought the President of the United States was a baby killer. From January to June of 1968 there were 221 demonstrations involving nearly 40,000 students on more than 100 campuses. Columbia was the angriest, but even staid Northwestern and the University of Chicago had sit-ins and unrest.

It was not a good year for college presidents or for Lyndon Johnson, who was trapped in the White House, forced to listen to daily chants of "Hey, Hey, LBJ, How many kids did you kill today?" But it was a great year for Mike Royko and many other newsmen who reported on the collapse of the prosperous post–World War II complacency and the explosion of dissent that threatened every institution in the nation.

Chicago was the city that ultimately took the rap for the social evils wrought by two centuries of racism and political systems that flourished on patronage, corruption, and deceit. But the tumult and the anger and the uncertainty began on January 30 on the other side of the world when the Vietcong launched their Tet offensive. It was shortly after Tet that Minnesota Senator Eugene McCarthy won 42 percent of the vote in New Hampshire against Lyndon Johnson, proving that the antiwar protest was gaining support from mainstream America. Only a few days later, Robert F. Kennedy announced he was entering the presidential race despite warnings from Richard J. Daley that he could not defeat the sitting president.

Royko opposed the Vietnam War from its start, but he also viewed Bobby Kennedy as a disloyal opportunist who had been doing everything to undermine the Johnson presidency since his brother's death. He wrote a parody about a sly kid from the old neighborhood named Bobby who was jealous of Big Lin and who was always telling his gang, "Boy, someday I'll let him have it good." Bobby's gang always urged him to go ahead but he never did and Big Lin would just sneer at him. Then, Royko wrote, there was a baseball game with Big Lin managing and doing dumb things, but Bobby didn't say anything to Big Lin. It was a third-string pitcher who wore glasses and read books and whose name was Eugene who had the courage to say, "Big Lin, you are stupid." Then, Eugene smacked Big Lin in the nose.

Suddenly, Bobby shouted: "Don't worry, Eugene, I'll protect you." And Bobby socked Big Lin in the back of the head with a catcher's mask. Big Lin just said, "Oww." And he sat down on the ground and held his aching head. Bobby strutted around saying: "See, I told you guys I'd let him have it good some day."

Nobody thought much of Bobby after that. They figured he was a lot more sly than he was brave.

But most of America wasn't in tune with Royko. Kennedy's name, his charisma, and his youth charmed campaign crowds, overwhelming Eugene McCarthy's professorial aloofness and creating a collective amnesia about who had been the first to challenge Johnson and his continuation of the war. The mood of the country was shifting rapidly, so rapidly that not even Royko could keep up with it. CBS sent Walter Cronkite to Vietnam, and when he came home he said he did not think America could win the war.

On Sunday, March 31, 1968, Lyndon Johnson stunned a national television audience by announcing, "I shall not seek, and I will not accept, the nomination of my party for another term as your president."

Richard Daley was not stunned. He had received a call from Johnson a half-hour earlier, and the president had told him what he was going to say. Daley argued that it was a mistake and that the Democratic party convention would draft him anyway.

Immediately after Johnson made his withdrawal announcement he flew to Chicago to make a speech and to visit Daley. The two men, along with Daley's sons, Mike and Bill, and Congressman Rostenkowski, spent more than an hour together on Air Force One. It was a meeting that enthralled the national media. They wondered if Johnson were instructing Daley to go all out for Vice President Humphrey. They speculated that Daley would support Bobby Kennedy.

Royko's readers might have expected Royko to be jubilant that the president who insisted on continuing what Royko saw as a futile, foolish, and deadly war was stepping aside. They might have expected him to praise any likely successor who would immediately end American involvement in what was a Vietnamese civil war. They were wrong. On Tuesday, April 2, he wrote:

> There were those who screamed with a vicious joy when President Johnson in that slow, sad way of his, said he is not running again....The white racists said, "good." The black racists said, "good." The superhawks said good and the doves said good. And most of all the young said good. The young, who are so sure they have the answer in Bobby with the flowing hair.

Royko wrote that to the white racists, Johnson was a "nigger lover," and to the black racists he was a white racist, forgetting that "he launched some of the most ambitious civil rights legislation in the nation's history." He wrote that Johnson offended the youth by failing to pander to them, by not telling them they were the ones who had all the answers.

> Unrestrained hate has become the dominant emotion in this splintered country. Races hate, age groups hate, political extremists hate. And when they aren't hating each other, they have been turning it on LBJ. He more than anyone else has felt it ...
>
> Maybe he wasn't the best President we might have had.
>
> But we sure as hell aren't the best people a President has ever had.

Other columnists, even those sympathetic to Johnson, criticized the president's poor judgment on the war, or his inability to cross the generation gap, or his advisers. Only Royko seemed to understand that a contagious hatred was infecting America. His opinion would be justified tragically only four days later.

His column for Friday, April 5, began:

> FBI agents are looking for the man who pulled the trigger and surely they will find him.
>
> But it doesn't matter if they do or they don't. They can't catch everybody and Martin Luther King was executed by a firing squad that numbered in the millions...
>
> It would be easy to point at the Southern redneck and say he did it. But what of the Northern disc-jockey-turned-commentator with his slippery words of hate every morning? What about the Northern mayor who steps all over every poverty program advancement, thinking only of political expediency, until riots fester, whites react with more hate and the gap between the races grows bigger?
>
> Toss in the congressman with the stupid arguments about busing. And the pathetic women who turn out with eggs in their hands to throw at children.... They all took their place in King's firing squad...

He wanted only that black Americans have their constitutional rights, that they get an equal shot at this country's benefits, the same thing we give to the last guy who jumped off the boat.

So we killed him. Just as we killed Abraham Lincoln and John F. Kennedy. No other country kills so many of its best people...

We have pointed a gun at our head and we are squeezing the trigger. And nobody we elect is going to help us. It is our head and our finger.

There were some ethnic politicians who enjoyed Royko's success, appreciated his admiration for the hardworking veterans of the bungalow belt and even privately, very privately, laughed at his scorching of Richard J. Daley. They just couldn't understand how one of their own could be so obsessed with fair treatment for blacks. Royko often said he was always bothered by discrimination. Perhaps it began when he first saw the black men sitting in their dunking cages at the Dips in Riverview Park.

"I was always getting in arguments when I was 15, 16, 17 years old on the subject of racial discrimination," he told writer Neal Grauer, the author of *Wits and Sages,* in a 1983 interview. "I can't even trace it to any one thing for any particular person, other than perhaps my mother, who never did understand discrimination. But...some chemistry kept me from acquiring the attitudes that were prevalent in the area I grew up in.

"I have very few memories of discussing blacks and Jews and things like that. It just wasn't something that I was really aware of. My father, I'm sure, had more conventional racial and ethnic attitudes, but it wasn't something that he carried around with him. He was too busy," Royko said.

King's murder set off riots everywhere. In Washington, only blocks from the White House, the sadness and rage of black America erupted in flames and looting. A white man was dragged from his car and beaten to death. In Chicago, the Thursday night after King's assassination was calm. Friday was not. Blacks began harassing white students in schools. When schools closed at 3 P.M. thousands of black youths poured over the West Side. Shopkeepers on Madison Avenue wheeled iron gates in front of their stores. A huge crowd gathered in Garfield Park, and the police ignored them. There were too many to arrest. By dark there was sniping and firebombing. Thirty-six major fires broke out on the West Side by midnight. Nine blacks were killed. Saturday was worse. Daley called out the army, and troops from Fort Hood, Texas, arrived to patrol the West Side. On the morning of Palm Sunday, Daley made a helicopter tour of the West Side. When he landed and talked to reporters he was obviously shaken. "I never believed this could happen here. I hope it will not happen again." Then he secluded himself for the rest of the day and night.

On Monday, Royko wrote about a Northwest Side tavern where patrons watched Saturday night's television reports of police chasing looters and National Guardsmen prodding crowds of black youths. They cheered every time a cop swung a club at a black head. They muttered that King got what he deserved. Like Chicago's politicians, some readers were angry that a blue-collar guy like Royko, who grew up using "nigger" as much as any other Polish son of a tavern keeper, blamed them for trying to keep blacks out of their neighborhoods, their jobs, and their families.

On Monday, Mayor Daley issued his infamous order to "shoot to kill any arsonist or anyone with Molotov cocktails in their hand to fire a building because they are potential murderers." Oddly, Royko did not write about it. It may have been his usual inclination to avoid those stories which were covered extensively on the front pages, or it may have been that he halfheartedly sympathized with Daley's predicament. Although he could write brilliantly about a national epidemic of hatred, as he did only days earlier, Royko would not have justified the looting or burning or killing on the West Side.

Of course, Daley had his defenders, beginning with the *Chicago Tribune*, which wondered what the entire ruckus was about. "As the chief executive officer of the city, it is his [Daley's] duty to keep the peace," the *Tribune* editorialized. The *Tribune* was a lonely voice among the nation's press.

Newsmen around the country speculated that Daley, who had turned into a kingmaker when John F. Kennedy gave him credit for his victory in the 1960 presidential election, was now the key to the Democratic nomination. Kennedy's proclamation aside, Daley's reputation as kingmaker was a myth. As Royko often wrote, and was the first to write, the Chicago Democratic organization's tremendous turnout in 1960 did in fact help Kennedy win the state by less than 9,000 votes. But the local pols were drumming up the vote to defeat Daley's dreaded rival, Republican Ben Adamowski, who was running for state's attorney, an office that had the power to indict politicians. That was far more frightening to Daley's cronies than the specter of Richard Nixon. Beyond that, Kennedy could have lost Illinois and still have won the presidency. But myths die hard, and every Washington, New York, Boston, and Los Angeles political writer had come to Chicago to ask about Daley. Royko indulged them:

> The mayor, legend has it, first appeared during the Chicago Fire of 1871. He doused the fire with one hand and milked Mrs. O'Leary's cow with the other. Before the ashes cooled, he hired Frank Lloyd Wright to redesign the city, dug Lake Michigan to cool it, organized the White Sox and set aside land for two airports in case airplanes were ever invented. More restrained admirers say he is simply the greatest mayor Chicago ever had, which is like singling out the best baseball player the N.Y. Mets ever had...
>
> Daley likes to build big things. He likes high-rises, expressways, parking garages, municipal buildings and anything else that requires a ribbon-cutting ceremony and can be financed through federal funds.

He isn't that enthusiastic about small things, such as people.

Robert Kennedy had added to the Daley legend. Asked how important Daley was to obtaining the Democratic nomination, Kennedy said, "Daley's the whole ball game."

After the fires died out, the West Side of Chicago looked like pictures of Berlin after the allied bombings of World War II. There were blocks with nothing but the charred remains of buildings. Many of them would remain vacant for another twenty years.

In June, the year that already suffered enough political shocks and tragedies was hit with another one. Robert Kennedy was shot to death in a Los Angeles hotel shortly after winning the California primary that most likely would have vaulted him to the Democratic presidential nomination over Vice President Hubert Humphrey. Kennedy was shot shortly after midnight on June 5, and doctors offered little hope that he could survive. The shot in the head was tragically similar to the wound that killed his brother.

As the nation waited for word from Los Angeles, Royko wrote his column for June 6:

Maybe it's time to change the words to our song, to bring it up to date and capture the national spirit:

> *Oh, say can you see by the pawn shop's dim light*
> *What a swell .38 with its pearl handle gleaming*
> *In a gun catalog is a telescope sight;*
> *I'll send for it quick, while the sirens are screaming.*
> *And the TV's white glare, the shots gripping in air*
> *Give proof through the night that our guns are still there*
> *Oh, don't you ever try to take my guns away from me.*
> *Because the right to shoot at you is what I mean by liberty.*

His satire was bitter, attacking the National Rifle Association, arguing facetiously for the right to own a fully activated tank and a machine gun, and pointing out that the Health Department was un-American because it would not let him keep pigs, chickens, and goats in his backyard. Keeping livestock, he argued, was an even older tradition than the right to own a gun. He concluded:

> So if you want to fight for your right to own guns, cut this column out and send it to your congressman or your senator.
>
> Quick, while they are still alive.

The chaos continued. By the end of spring, Chicago was a city under siege. Antiwar protesters of all stripes were setting up camp to protest at the Democratic National Convention scheduled for the last week of August at the International Amphitheatre in Mayor Daley's home neighborhood of Bridgeport.

The radical Students for a Democratic Society (SDS), which had led campus takeovers at Columbia and other major schools, came together under the leadership of Tom Hayden. Old line pacifists were being gathered by the Mobilization Committee to End the War (The Mobe), led by David Dellinger, who was jailed for protesting American involvement in World War II. There were thousands of hippies and a handful of yippies led by antiwar activists Abbie Hoffman and Jerry Rubin. It was all called "counterculture," and Chicago cops were detailed to watch the ragtag bunch every time they marched. Other cops were detailed to spy on them and anyone else who might be suspected of having subversive feelings. Studs Terkel and Mike Royko were among them.

Demonstrators filtered into Chicago in July carrying their belongings in U.S. Army knapsacks or canvas airline bags. They found rooms in makeshift hostels, church basements, and cheap apartments. The Mobe had filed applications to use Soldier Field and Grant Park for rallies, but no one at the Chicago Park District responded. Daley had his aides discuss marches with Davis and Dellinger, but his heart wasn't in it. Daley wasn't concerned with the longhaired, foul-mouthed peace protesters. He was worried about another uprising in the black community. Daley rejected all their requests to sleep in the parks after the 11 P.M. curfew.

Whether Royko felt comfortable writing outside the sphere of Chicago political antics no longer mattered. The events of 1968 had unshackled him from the "alderman beat" and thrust him into the national picture. His columns became the leading attraction of the package of syndicate news and features the *Daily News* offered throughout the country. Many newspapers began to run Royko's columns regularly. Now, Royko had a national audience. As Chicago prepared for the Democratic National Convention, thousands of reporters and television newsmen came to the city anxious to read Royko. They crowded into Riccardo's and the Goat hoping to run into Royko. They wanted to be on the scene if he snarled at someone.

By the third week of August the demonstrators were giving Chicago fits. The antiwar crowd predicted there would be 150,000 or 200,000 peace pilgrims in Chicago for convention week. Daley called out the National Guard and flew in 5,000 federal troops to guard the International Amphitheatre. Royko noted there were more troops in Chicago than in Khe Sahn, a Vietnamese hilltop where U.S. Marines were being surrounded by Vietcong. From the moment Daley had uttered his "shoot to kill" order, the David Dellingers, Abbie Hoffmans, and Tom Haydens of the peace movement sensed it would be easy to create a confrontation in Chicago.

Their literature was volatile. "If at all possible, learn first aid measures in your respective cities before coming to Chicago....Bring a supply of Vaseline to be applied against the skin for protection against Mace....Mobile medical teams in white coats will be stationed at strategic areas..."

On August 20, newspapers reported 300 members of the militant Black Panther party had arrived for nonviolent demonstrations, but that other black leaders were leaving Chicago, fearful that the antiwar protests would spill over into their neighborhoods. The *Chicago Tribune* contributed to the hysteria by reporting exclusive stories that the antiwar groups planned to toss LSD into the city's water-filtration plants.

At this point during the crisis of 1968, the role of the media was as much a part of the story as the events themselves. American newspapers were in as much conflict as the nation. Many of the publishers and editors were troubled by the government's role in Vietnam and disgusted by the years of institutional racism in the South, but most of them were hesitant to play as much of an activist and critical role as their young reporters were demanding. Perhaps in no other city did the conflict appear so obvious as in Chicago, where Royko stood clearly as the symbol of change and the *Chicago Tribune* remained the bastion of the status quo.

At that time, the *Tribune* was run by Don Maxwell, one of a trio of disciples that Colonel McCormick had left in charge upon his death in 1955. Maxwell was all that Royko detested about the *Tribune* and the era of coziness between publishers, politicians, and advertisers. Maxwell was known to kill stories that embarrassed city leaders or top businessmen. As the *Tribune*'s sports editor in the 1920s he had been instrumental in building up the fledgling Chicago Bears and the National Football League. Bears' owner George Halas counted Maxwell as one of his closest friends, and Maxwell summarily dumped reporters who treated the Bears unfavorably.

Although Royko and the *Daily News* and the *Sun-Times* under its youthful and talented city editor, Jim Hoge, were trying to give the volatile scene balanced coverage, Maxwell's *Tribune* was hostage to the establishment mentality. The *Tribune* still used the word "Negro," while others newspapers began to use "black." And while the *Tribune* did not support the war in Southeast Asia, it was nowhere near as militantly opposed to the war as it would have been under McCormick, who opposed all U.S. foreign involvement. In fact, the *Tribune*, one of the wealthiest and most resourceful newspapers in the country, never assigned a correspondent to cover Vietnam on a permanent basis.

The *Tribune* mentality was being force-fed to its subsidiary, Chicago's *American*, where the staff was not nearly as subservient. Bill Garrett, who had worked with Royko at City News, had joined the *American* in 1967. He found a gallows humor in the newsroom. "Everyone knew the *Tribune* was going to close the paper down, it was only a matter of when. The first day I went to work there was in October and already the staff was discussing whether they should shop for Christmas since they might be out of work by then.

"The people at the *American* probably hated the *Tribune* more than the people at the other papers," continued Garrett. "Especially when they saw the kinds of things Royko was doing and we couldn't do. He was attacking Daley and the establishment and our bosses just went along with them. All the young people envied Royko and wanted to copy him."

Hundreds of young reporters from all over America and the world were in Chicago when the convention opened August 26 with a welcoming speech by Daley and a huge confrontation in Lincoln Park.

The protest groups had been trying to stay overnight in the park for a week but always gave up at the 11 P.M. curfew when police moved in, gently prodding them with nightsticks. But on August 26, the demonstrators refused to leave Lincoln Park and the police grew impatient. At 11 P.M. the cops mounted a full running charge into the park, clubs aloft. Forced out of the park, the demonstrators began to cluster at the triangular intersection of North Clark and LaSalle Streets, clogging traffic for several blocks. About 2,000 policemen surrounded them, tried to disperse them, and began clubbing randomly.

The victims included twenty newsmen, photographers, and cameramen. The hostility between the police and the media was sealed for the remainder of the week.

The *Tribune* noted the incident with an inside story and downplayed the beating of newsmen. Royko did not:

> The puzzling thing about the police attacks on newsmen is that we do not want to sleep on the grass in Lincoln Park. Everybody knows that it is perfectly proper for the police to club the young protesters because they want to sleep on the grass in Lincoln Park. The grass in Lincoln Park, as well as the Lake, the sky, the moon and sun, belongs to Mayor Daley. And he doesn't want anybody sleeping on his grass in Lincoln Park.
>
> But I repeat: We, the newsmen, do not want to sleep on the grass in Lincoln Park. We just want to watch the police beat up those who do want to sleep on the grass...
>
> The least he [Daley] can do is impose a bag limit. No policeman would be permitted to take more than one newsman in one night. If it keeps up, we shall become extinct and who will be left to describe the glories of the mayor's reign?

Tuesday night was a repeat of Monday. A group of clergymen erected a cross in the park and sang the most popular refrains of that year, "We Shall Overcome" and the "Battle Hymn of the Republic." At 11 P.M., an electronic megaphone boomed, "Move out now!" The familiar reply: "Hell, no, we won't go." Tear gas canisters exploded. Obscenities, rocks, and bottles were thrown. "Hit me, pig!" "Shoot me, pig!" The chase began again. That night, many demonstrators raced south to Grant Park, where, oddly enough, police allowed them to remain all night. The cops ringed the area as the demonstrators lit tiny fires, fired off an occasional obscenity, and continued their vigil until dawn. Yippie founder Abbie Hoffman was arrested while eating in a restaurant for having the word, "Fuck" painted on his forehead.

On Wednesday, Royko changed gears. He did not write a humorous satire. He wrote a straight condemnation of U.S. Attorney Thomas Foran, a Daley disciple, for his defense of the police conduct.

"Foran," Royko wrote, "is either stupid or a liar."

No one had ever written that way about a top federal lawyer in Chicago. While the *Tribune* was printing stories leaked by the Chicago Police Department's "Red Squad," an undercover unit that spied on black militants and suspected communists, Royko was damning the whole establishment.

Royko also wrote:

> The only time I've run to save my hide was Monday night. A group of Chicago police were after me. My crime was watching when they beat somebody who didn't seem to deserve it. When Foran talks about "wonderful discipline," he sounds like a boob. He's not. It's just that, like anyone else on the public payroll in Chicago, he is a flunky for the mayor...
>
> But our mayor, the architect of the grand plan for head bashing, is wandering around loose and making predictable statements. The great dumpling says newspapermen ought to move faster when cops come at them. That way they won't be banged about.

On Wednesday, as the convention began to debate the war in Vietnam, a crowd of 8,000 filled Grant Park. One young man climbed a flagpole to pull down the American flag. Police stopped him. Bottles and rocks then flew. Police charged a small group. David Dellinger spoke to the crowd and said it was time to march to the amphitheatre. They moved a block before police stopped them. National guardsmen blocked the southern and eastern exits to the park. The demonstrators moved north, turned onto Michigan Avenue, and wheeled south toward the Conrad Hilton. Inside the convention, the

roll call for presidential nominations started and Congressman Rostenkowski, a Daley loyalist, was handed the gavel to introduce the
speakers. A fight broke out in the New York delegation. Rostenkowski called on the sergeant at arms to clear the aisles. Suddenly
the blue helmets of Chicago police were scurrying among delegates.
The police took three New York delegates into custody. CBS newsman Mike Wallace, who was trying to figure out what was happening, was knocked to the floor.

Meanwhile, outside the Hilton, the swelling crowd faced off
against the police. The obscenities continued. Someone threw a beer
can. The police charged. Nightsticks thudded on bones. Dozens of
persons were shoved or dragged into patrol wagons. The television
cameras were turned on. The violence spread along the sidewalks,
and police went into the hotel lobbies, chasing demonstrators, newsmen, spectators, and anyone who stumbled in their path. It was all
over in eighteen minutes.

At the amphitheatre, Senator Abraham Ribicoff of Connecticut,
an old ally of Daley from the 1960 Kennedy convention, rose to nom-
inate George McGovern. He taunted Daley for supporting Vice President Hubert Humphrey, who had refused to defy his President and
denounce the war. "With George McGovern, we wouldn't have
Gestapo tactics on the streets of Chicago," Senator Ribicoff said.

Royko ushered everyone out of town on Friday by writing a column with hints on what visitors to Chicago should do and see. He
said he had planned to run it before the convention, but the rush of
big news prevented that.

There will be time to cross Michigan Avenue to beautiful Grant
Park. Don't hesitate to walk on the grass or to spread your coat and
take a nap. Chicago likes its parks to be used by people, so you won't
see any "Keep Off the Grass" signs. And you can feel safe in the park.
It is well patrolled by policemen and many Chicagoans sleep there

on hot nights....If you should become lost in Chicago, don't hesitate
to ask for directions. Any policeman will be glad to point you in the
right direction....Just looking out your hotel window into the street
can be fun. There's always something happening....When it's over,
you'll have memories that will last a lifetime.

In September, Henry Regnery published a second Royko column
collection, *"I May Be Wrong, But I Doubt It,"* for which Royko received
a $5,000 advance. Herb Caen of the *San Francisco Chronicle* reviewed
the collection and declared, "He [Royko] is at the top of his form
when he is indignant, a mood that is frequently upon him as he pon-
ders the hard lot of minorities in Chicago, the excesses of police and
FBI, the foolishness of the Mayor and other politicians. Along with
being gutsy, he is a hard-digging reporter—another quality that sets
him apart from most columnists. He can also be very, very funny."

In February 1969, Royko took the train to Washington to receive
the coveted Heywood Braun Award, presented by the American
Newspaper Guild, which Braun helped establish and presided over as
its first president. The citation was based on fourteen columns "scoring police mistreatment of Negroes, racism among Chicago whites
and police attacks on newsmen during the Democratic National Convention." The award committee added, "The work of Mike Royko—
sardonic, bold, courageous, always stressing the little man and what
our society sometimes does to him—seemed to us to be in special
harmony with Braun's concept of journalism."

Marshall Field wrote Royko, "I know I am not the first—and I certainly won't be the last—to offer congratulations on your receiving
the Heywood Braun Award, but you have none more sincere! This is
just another reason we are glad to have you as a member of our team!
Heartiest congratulations! P.S. The $1,000.00 also sounds good!"

It seems clear Field already knew that while Royko was basking in
his celebrity, the son of the tough Ukrainian immigrant always kept his

priorities aligned with his bankbook. That's why Terry Shaffer was stunned when Royko returned from Washington, shuffled into his cubicle, and uttered his usual morning grunt followed by "come over here."

"He gave me a check for $500 and said it was half the Broun award and he was giving it to me because he figured I had done at least half as much to earn it. Then he grunted and turned around. He knew what that meant. Hell, it was more than two weeks salary for me. He would just do that kind of thing and then never talk about it again," Shaffer said.

In the spring of 1969, the first Pulitzer Prize for commentary was awarded and Royko didn't get it. He was disappointed and angry, suffering again from the old insecurities that he fretted about when he couldn't get hired at the *Daily News*. He felt the eastern establishment media wouldn't honor him because he was a Chicago writer and didn't cover Washington.

The first commentary award went instead to Marquis Childs, the longtime *St. Louis Post-Dispatch* correspondent in Washington whose columns dealt with Johnson's decision to step down, the Vietnam War, Bobby Kennedy's entrance into the presidential race, and the eventual election battle that put Richard M. Nixon into the White House. In fairness, there was so much news in 1968 that Childs or any number of distinguished writers deserved the award. But Royko could have won just as easily. Along with his columns on the death of King and Kennedy, on Johnson's abdication, and on Daley and the Democratic convention, he wrote wonderful asides that writers like Childs couldn't touch. They didn't have his knack for finding ironic stories. They didn't have his voice or his humor.

When Royko observed on the praise and popularity surrounding the movie *Bonnie and Clyde*, which received rave reviews for its realism, he interviewed Jim Campbell, a pipefitter who was a twenty-year-old junior college student in Oklahoma when his father was killed on April 6, 1934. Campbell's widowed father had raised him on his fifteen dollars-a-week salary as a town constable. Campbell told Royko:

> A farmer came to town that day and told my father some people he passed were in trouble. Their car had gone off the road. My father and another policeman—they were the whole force—drove out there. When they stepped from the car, my father was killed instantly. I'm sure my father didn't know who killed him. He was just going out to help someone.
>
> I never went back to my classes. I guess I became…oh…bitter, you might say. I didn't see much point in anything. I just brooded. As far as that movie goes, I guess Bonnie and Clyde seem glamorous.…They may have had reasons for doing what they did. But they weren't glamorous.

Royko also wrote about the burial of Phillip Craig Skinner, who had hoped to join the Chicago Police Department after finishing his tour as a marine in Vietnam. Skinner came home in a casket, and the first thing the owner of a funeral chapel told his mother on the telephone was that the government didn't provide good caskets and that he was sure she would want to look over his line. When Skinner's father visited the chapel the owner explained, "While we are willing to handle this body, you understand, there are many excellent Negro undertakers on the South Side."

Skinner's mother told Royko, "So we did not inconvenience the white chapel by asking them to handle the body of a young man who gave the most there is to give. But I'll never understand what he gave it for. For that undertaker? Did he and other black boys give their lives in the hope that a few dozen black children could walk into a school without being spectacles? What did he give his life for?"

Royko ended the column: "If anyone has a good answer to that question, I'll be glad to pass it on to Phillip's mother."

As Royko's popularity grew, he began to make frequent appearances on Chicago television stations. He appeared on talk shows and newscasts, interpreting the secretive machinations of Daley's City Hall and discussing political election strategies. In 1969, he earned more than $4,000 from NBC affiliate WMAQ-TV for his television work.

That money along with the income from his two books and his $25,000 salary was enough for the family to move from the second and third floors of the Duckman home on Central Avenue to a three-bedroom brick home at 6657 N. Sioux Avenue in Edgebrook, a comfortable neighborhood of lawyers, police commanders, and judges on the northwestern edge of the city.

In the summer of 1969, Hal Scharlatt, an editor from E. P. Dutton and Company, a New York publisher, approached Royko about writing a book. Royko was wary about taking time from his column. He also wasn't sure it was worth the effort. The two books of columns he had published had not involved any work, but they didn't really earn a lot of money either. He discussed the idea with Studs Terkel, Nelson Algren, and Saul Alinsky, activist community organizer. They urged him to try.

chapter 8

By the summer of 1969, Royko and his column were inseparable. The column was a god that demanded sacrifice from Royko every day, and he made it; it left little room for anything or anyone else.

"I had no intention of writing a book of any kind because I just didn't think I could do one while I was writing a column," Royko told the *Daily News* in a 1971 interview.

But Dutton's Hal Scharlatt persisted, telling Royko that a book about Daley and Chicago could be a national best-seller.

"That scared me," Royko said. "That applied even more pressure. So I made another excuse. I said, 'Daley won't cooperate.' They said, 'Will he ever cooperate with anyone?' So I thought about it and went over my files, the columns I've written on him, the material I've developed on the city and its government and I realized I'd already done a tremendous amount of research. If a guy came out from New York to try to do it, he'd have to start basically where I started 15 years ago.

Everyday I worked as a police reporter, every hour I spent covering a political beat, was part of my research for this job. And I've got the contacts.

"And that decided me, that and Studs Terkel nagging and nagging me to do it. 'Ya gotta do it Mike, only you can do it, you gotta do it.'"

Nelson Algren and Saul Alinsky were also encouraging Royko to write the book. He finally told Scharlatt he would do it.

In July, Scharlatt wrote, "I'm enclosing a copy of a contract between you and E. P. Dutton for the book on Mayor Daley and Chicago. The excitement in the house over publishing this book is tremendous. Please take a look at the terms of the contract and especially the division of the advance. If it all meets with your approval, just sign the original and the enclosed card and return it to me and I'll send you by return mail our check for $3,000."

Royko told his editor, Roy Fisher, that he would be working on a book but that unlike many journalists who do books he did not want a long sabbatical. He said he would continue writing five columns a week. He planned on researching and writing on the book on weekends.

When he began, the book project was like everything else. Royko declared war on the subject and mapped his battle plans. He spent most of August lining up interviews. "The way I organized them was like a target. Mentally, I visualized this target, and Daley was the bull's eye. The circles closest to him were composed of men who are closest to him as friends and associates. The more distant circles were the people who were safest to interview, people who were not as apt to alert him to what I was doing."

In September he took four weeks vacation and a two-week leave of absence. He spent the entire time on the interviews.

"Then, after I felt I had everything I could get from every other source and I had checked it out three times, I took the last logical step: I wrote to Daley and asked for an interview. I don't know how I phrased it, something like 'I'm doing a book on your era in Chicago and I'd like to talk to you about it.'

As Royko expected, he received no reply. He called Earl Bush, the mayor's press secretary, and asked about the interview. "What is it about?" Bush asked. Royko laughed at the question. He knew that after all his interviews Bush and Daley knew exactly what he was doing. Apparently, Bush finally told Royko the answer was "No interview."

During the winter and spring of 1970, Royko spent his weekends at his cubicle at the *Daily News*. He took one of the huge rolls of paper that was used by the Associated Press and United Press International teletype machines and placed it on the floor beneath his typewriter stand. It was a psychological device that forced him to write until he reached an appropriate place to stop without worrying about the end of a page or how much work he had done.

One of his assistants said, "I remember coming in on Mondays and seeing this big roll on the floor and ten or fifteen feet of paper hanging over the back of the typewriter. I'd read some of it upside down but I had no idea he was writing a book."

No one had any idea because Royko was trying to keep it secret, partly because he didn't want to be distracted by well-meaning friends who would try to give him information he didn't need, and partly because he did not want to raise expectations. He also didn't want anyone to think he was not devoting all his attention and energy to the column.

One Saturday afternoon in the summer of 1970 Royko finished the book and went to Riccardo's, which was typically deserted on weekends, when only skeleton staffs manned the newsrooms that provided the restaurant's clientele. Royko went to the empty bar and told the bartender he wanted a martini because he was celebrating. He had just finished a book.

The bartender fixed the drink and then reached under the bar and proudly produced a Mickey Spillane paperback. "I'm almost finished with this one, too, Mike. I'll give it to you when I'm done."

It was a story that Royko loved telling, perhaps because it reminded him that no matter what success he enjoyed or what triumphs lay ahead, to the bartender he was just a good customer, nothing special.

Boss, which was published in January 1971, was an instant success in the publishing industry. In a matter of a few weeks it had risen to number four on the *New York Times* Best Sellers list and would eventually sell more than 125,000 hardback copies and more than 1 million copies in paperback.

In Chicago, the sale was phenomenal. Kroch & Brentano's bookstore called it "the fastest moving book we've ever seen." Sales at Marshall Field's and Carson Pirie Scott book departments were "fantastic," the stores said. One of Chicago's premier booksellsers, Stuart Brent said, "We've never had anything like it."

Royko received $62,000 for the paperback rights, another $10,000 when *Boss* was picked as an alternate Book-of-the-Month Club selection, and more than $50,000 in foreign rights from European and Asian publishers.

When he was told the book was going to be published in England, Royko cracked, "Who's going to do the translation?" In Japan, his name was translated as Maiku Loyko and for a while he called himself Loyko-Royko.

Sam Sianis, nephew of "Billy Goat," remembered Royko bringing one of his first royalty checks to the tavern. "This is the first time in my life I've seen a check like this."

Mike and Carol were euphoric. David Royko described their glee:

I remember the first royalty check. It came not long after we were in the house on Sioux Avenue. I remember my father sitting in the kitchen, and he and my mother were just staring at each other and my father was just beaming. He had this strange look and my mother had this smile and was shaking her head like, "I can't believe this." I remember saying, "What, what is it?" My father said, "You want to see something?" I said, "Sure," and he said, "You can't tell anybody about this. You can't tell your friends." He showed me this check for $35,000 and, of course, I couldn't appreciate what that much money meant except I knew it was a lot. My reaction, I was like eleven or twelve, was that it doesn't change my life.

Later that night, we were up in the study, which was the spare bedroom, and my dad said, "Do you know what it's like to get this check?" I said, "No," and he said, "I'm going to show you what it's like." My father got two five-dollar bills from his wallet, and he handed Robbie and me each a five-dollar bill. We got like fifty cents a week for an allowance, and I looked at him and said, "Can we keep this?" And he said, 'Yeah," with that big grin. I looked at Mom and she said, 'Yeah," and my dad said, "That's what this check is like." At the time I thought, great, but what I really thought was, "Wow, I've got five dollars."

For Royko, the success must have brought back memories of hustling pop bottles in the alley, setting pins for seven cents a line, and those days spent agonizing over whether he would have enough money to get married.

Long before the royalty checks were the reviews. The two that meant the most to Royko came from two peers, Jimmy Breslin of the *New York Daily News* and Russell Baker of the *New York Times*.

Breslin wrote, "The best book ever written about a city of this country. And perhaps it will stand as the best book ever written about the American condition at this time."

Russell Baker, after receiving an advance copy of the book, wrote to Scharlatt.

"Yes, you are lucky to be publishing Mike Royko. He is the best thing to come out of Chicago journalism since Ben Hecht. He has wit, style, high intelligence and marvelous ability to set public matters in perspective, and there's nobody better writing politics anywhere in the country today. You may quote me. About the book? It's Daley; Royko's got him to the life. And it's Chicago. Even if you've never been there you know it's Chicago. A fine job."

Baker was more prescient than perhaps he knew. While other writers knew as much about Chicago as Royko, they were never as convincing in its depiction. Royko's conversational writing style engaged readers personally, and they nodded and chuckled knowingly as they read *Boss*. They thought they knew as much about Chicago and the machine and mob as he did. The reality was that they didn't, but Royko reinforced their opinions, assuring and humoring the readers. And Royko's writing also worked for those who had never seen Chicago. Somewhere, from other stories or from movies, or television, or from Algren or Studs Lonigan, people knew he had got it right.

Algren praised the book in the *New Republic*: "Mike Royko not only owns the most richly endowed comic sense of any Chicago writer since Ring Lardner, but also possesses as lucid an understanding of how political machines survive corruption as does any political reporter in any big American city."

Another old colleague, Nicholas Von Hoffman, reviewed the book for *Life* magazine. "Excellent, first rate…the best thing ever written about Mayor Daley and fully worthy of Mike Royko, the best newspaper writer in the country."

Almost every review was a rave, but there were exceptions. The *Chicago Tribune*, expectedly, said Royko had taken a one-dimensional view of the mayor and had deliberately overlooked his great contributions to the city.

But even those who nitpicked couldn't deny the power of *Boss*. Christopher Lehmann-Haupt wrote in the *New York Times*, "One might wish that Mr. Royko's descriptions of the Mayor Daley were a little less second-hand. One might wish that he had put matters in perspective with a more sympathetic treatment of the conditions that produced the Daleys.... One might wish, in sum, that Mr. Royko had written a book that one could read with emotions other than rage. But all in all, this is muckraking at its best, a remorseless book that bites and tears."

Boss was not a traditional biography. It did not have balance; it did not have a variety of viewpoints. Royko did not deal with Daley as a personality; he dealt with him as a force of power that ruled too harshly and too unfairly. The author made no pretense of trying to explain why Daley might not be sympathetic to Negroes; why he put so much effort into the central city and shrugged off the plight of decaying neighborhoods; why he repeatedly gave important jobs to incompetent cronies and crushed political foes ruthlessly. He merely showed how he did it. He ignored any justification for the building boom that made contractors rich, or the razing of an old ethnic neighborhood to build an urban campus for the University of Illinois. He was relentless in creating a caricature that was perhaps more banal than the reality.

Most Chicagoans mistakenly thought *Boss* was an indictment of Daley and the political system that had flourished for forty years. But it also was an indictment of the voters who allowed Daley's system to reign unchecked, benignly following George Orwell's aphorism that people don't mind tyranny as long as it's imposed on them in an acceptable way. Royko was delivering the same lecture he gave when voters in black wards continued to sheepishly elect white aldermen. But *Boss* was different in style than the daily feeding Royko gave his readers. The humor was not as sharp as the kickers he used to end each column. A great deal of the book was straightforward reporting. And, despite his many interviews with supposed insiders, *Boss* provided little about Daley that people who watched him closely did not know.

It was how Royko put it all together that mattered. His uncanny eye and ear made it work. For example, Royko began his book with the terse, bland responses that Mayor Daley gave when he was called as a witness in the federal trial stemming from the 1968 convention rioting. Daley's words—only brief responses as to his name, address, title—meant almost nothing to the many reporters who heard them and the readers who read them in the newspapers, but Royko used them to capture the essence of the man who ruled Chicago.

The trial was a great legal circus. In its haste to blame someone other than Daley and his policemen for the chaos during the convention week, the U.S. Justice Department and U.S. Attorney Thomas Foran (whom Royko had labeled a "liar or stupid") indicted eight men for conspiring to cause riots in Chicago. They included Bobby Seale of the Black Panthers, David Dellinger of the Mobe, Thomas Hayden and Rennie Davis of Students for a Democratic Society, Abbie Hoffman and Jerry Rubin of the yippies, and two obscure associates, Lee Weiner and John Froines.

The U.S. District Court lottery system picked Julius J. Hoffman as the presiding judge. Hoffman was an elderly conservative, sure to clash with the defendants and their chief counsel, activist lawyer William Kuntsler.

The five-month trial from September 1969 to February 1970 provided almost daily flashbacks of the 1968 violence, reigniting the bitter divisions between cultures and generations. Supporters of Mayor Daley—mostly ethnic and older Chicagoans, including most of Royko's neighbors—rallied to his defense. Young people, including many newspapermen, blamed Daley for Richard Nixon's election and the continuation of the war. The press coverage of the Conspiracy Eight trial revealed these biases. The *Chicago Tribune* rarely treated the trial as a political controversy, but depicted it as a criminal case with all the implicit seriousness of a murder trial. The *Daily News* and *Sun-Times* described the carnival atmosphere and quoted supporters of the defendants freely. Alice Hoge, the wife of *Sun-Times* city editor Jim Hoge, was a daily spectator and supporter of the defendants.

The decorum of the federal courtroom was destroyed on the first day when Bobby Seale accused Judge Hoffman of being a racist and refused to remain silent. Hoffman had him shackled and bound and gagged and carried into court strapped to a chair. Sympathetic reporters wrote of inhuman cruelty. Older reporters wrote of anarchist defendants.

Seale's case was eventually severed from the other defendants, which only changed the name for history's purposes to the Conspiracy Seven.

Kunstler clearly wanted to be the star, baiting the aging judge continuously with groans and remarks that brought laughter from the defense table, which provided its own antics such as appearing in court wearing judicial robes. Kunstler once failed to get an appropriate reply from the judge and said, "You're like a child saying, 'Because, because,' and the tiny, bald jurist leaped from his seat and shouted at his bailiff, "Let the record show that Mr. Kunstler compared me to a child!"

Kunstler called all sorts of witnesses; beat poet Allan Ginsberg, who chanted from the witness stand until Judge Hoffman asked what language he was speaking. "Sanskrit, your honor," the poet replied. There were writers Norman Mailer, William Styron, and Terry Southern. Julian Bond and the Reverend Jesse Jackson were defense witnesses, and so were singers Pete Seeger, Judy Collins, and Phil Ochs.

The trial turned into a legal farce. Eventually all eight men were acquitted of everything, including a series of severe contempt sentences that Judge Hoffman had meted out for their daily insults. They had been charged with a new law that made it a federal crime to cross state lines with the intent to incite a riot. Kunstler called it "crossing state lines with a state of mind."

The late Abbie Hoffman summed up the entire case at the beginning of the trial when he told reporters, "We could never agree on where to go to lunch; how could we conspire to do anything?"

Royko attended the trial occasionally and wrote a few pieces about it. But he was in the front row reserved for the press on January 6, 1970, when Kunstler called Richard J. Daley to the stand.

And when Daley took the stand, Royko knew how his book was going to begin.

"When I was covering the Conspiracy Seven trial the day Daley testified," he said in the 1971 *Daily News* interview, Kunstler asked him, 'What is your name?' and Daley said, 'My name is Richard Joseph Daley,' and, 'What is your occupation?' And he said, 'I am Mayor of the city of Chicago.' Not lawyer, not politician, but 'I am the mayor'.... That's not an occupation, that's an elective office. But that's how he sees himself. Each morning, he gets up and goes to work to do his job, to be mayor, and this is his life. So it was the natural beginning, to write about his day and, weaving through it, to bring in the city as it is today."

The opening words of *Boss* are Daley's testimony, and each chapter of the book began with excerpts of that testimony. Since almost all of Kunstler's questions were challenged by prosecutor Foran and most of the objections were sustained by Judge Hoffman there wasn't a lot of material, but Royko made the most of it and the clipped testimony gave Daley an austere, chilling presence in the book.

The opening of the book follows Daley as he leaves his home in the Back of the Yards neighborhood of Bridgeport and travels four miles in his official limousine to the Loop, where he attends mass at St. Peter's Church. "Regardless of what he may do in the afternoon, and to whom, he will always pray in the morning," writes Royko.

Royko's description of Daley's morning trip with notations on all the expressways that Daley built and named and all the buildings he constructed and all the people he placed in them conveys a sense of empire, and indeed the cover of *Boss* depicted Daley in full armor as a Roman emperor.

When Royko describes Daley as the spoiled, only child of a sheet-metal worker whose parents came from County Waterford and a suffragette whose parents came from Limerick, he evokes the texture of a Chicago that had hardly changed from Daley's birth in 1902 to Royko's birth thirty years later:

> The neighborhood-towns were part of larger ethnic states. To the north of the Loop was Germany. To the northwest Poland. To the west were Italy and Israel. To the southwest were Bohemia and Lithuania. And to the south was Ireland....
>
> But you could always tell, even with your eyes closed, which state you were in by the odors of the food stores and the open kitchen windows, the sound of the foreign or familiar language, and by whether a stranger hit you in the head with a rock....
>
> The ethnic states got along just about as pleasantly as did the nations of Europe. With their tote bags, the immigrants brought along all their old prejudices, and immediately picked up some new ones. An Irishman who came here hating only the Englishmen and Irish Protestants soon hated Poles, Italians, and blacks. A Pole who was free arrived here hating only Jews and Russians, but soon learned to hate the Irish, the Italians and blacks.
>
> That was another good reason to stay close to home and in your own neighborhood-town and ethnic state. Go that way, past the viaduct, and the wops will jump you, or chase you into Jew town. Go the other way, beyond the park, and the Polacks would stomp on you. Cross those streetcar tracks, and the Micks will shower you with Irish confetti from the brickyards. And who can tell what the niggers might do?

Royko understood how and where and when Daley had grown up

and with what prejudices and loyalties. He could surely forgive Chisel and Dutch Louie and the other patrons of his father's saloons for their prejudices. He even chalked up his own father's casual anti-Semitism to old-world prejudices. But he would not let Daley off the hook for being a turn-of-the-century, Back of the Yards kid who was raised at a time when blacks were hardly considered Americans, and grew up to see them as an economic and social threat. Royko felt compelled to make an argument, unconvincingly, that Daley had been an active participant in the deadly 1919 race riot in which twenty-three blacks and fifteen whites were killed. Much of the violence took place in Bridgeport and was attributed to the Hamburg Athletic Club. Daley was a member and would later serve for fifteen years as the club's president.

There never was any evidence of Daley's involvement in the riot, but Royko repeatedly noted that Daley never discussed the incident and refused to answer questions about it. It was the kind of overkill that Royko regularly employed in his column to spark laughter. But the issue of Daley participating in the slugging and maiming of blacks was too serious for exaggeration. It was the only discordant note in the book.

Royko also attacked Daley in a characteristically provincial fashion when he concluded that the clash between police and demonstrators during the 1968 Democratic convention had immense historical significance: "This is what may have determined the election and altered the course of world history—the decision that nobody would be in Lincoln Park after 11 o'clock."

While Royko did not write about Daley's "shoot to kill" declaration after the Martin Luther King assassination rioting, he used the incident to demonstrate the arrogance of Daley's City Hall, quoting press secretary Earl Bush's locally famous defense of the mayor's words: "They should have printed what he meant, not what he said."

The supporting cast was as entertaining as Daley: "The Hawk got his nickname because in his younger days he was the outside lookout man at a bookie joint. Then his eyes got weak, and he had to wear thick glasses, so he entered politics as a precinct captain."

When Royko traced the machine's nepotism he collected more than fifty names of families that had passed on political power from one generation to another, and he did it in Old Testament fashion, noting who begat who from the various Kerner, Stevenson, Burke, Clark, Keane, Hartigan, Dunne, and Touhy clans. He jumped to the New Testament to add, "They are their brothers' keepers, too." And he listed all the pols who made certain their brothers had cushy spots at the hall or the courts. As some critics observed, none of this was new. Chicagoans were aware that political families moved from office to office, that judgeships traveled from father to son. But the way Royko packaged it, filling two pages with the genealogy of clout, was a profound statement about the incestuous and exclusive nature of the machine.

He also showed a rare moment of Daley kindness to a political foe. John Waner, the 1967 Republican sacrificial candidate against Daley, told Royko the morning after he was trounced that Daley called and said, "John, this is Mayor Daley. I'm not calling to gloat. You ran a decent campaign, John, and I know how it feels to lose. I lost when I ran for sheriff in 1946."

But there weren't many instances of Daley's kindness to outsiders, because Royko didn't believe that Daley allowed anyone except for his family and his tiny core of Bridgeport friends inside his inner circle.

As Daley forgave no transgressions, Royko forgave Daley nothing. He took a shot at Daley's public support of President Johnson's war by recalling that Daley never volunteered to serve in World War II. Daley was thirty-nine at the time, and Royko could have pointed out that few middle-aged men with families served. He also chided the mayor because all four of his sons who had reached draft age during the Vietnam conflict tucked themselves away in reserve units that were never called to active duty. Their decision was not unlike Royko choosing the air force in the futile hope that he would not be sent to Korea.

Royko never claimed he was trying to be fair. "It is a subjective study of the man, the machine and the city. It is one man's view of Daley. I am not one of his great admirers. Anyone who expected my book to be a warm-hearted, sympathetic study of the man is just foolish. But, just because I didn't portray Daley as a sweet grandfather figure, I didn't do a hatchet job. I described Daley as he is. I am confident that ten years from now, my book will have held up. The city's history is going to show that," he told a high school interviewer.

He was more than right. *Boss* remains a stylistic marvel and a political science road map that, not ten but thirty years after its publication, functions as the best history of a political organization that ruled a city and influenced a nation for four decades.

Boss hit Chicago bookstores two weeks before its national publication because Royko hoped it would have some small impact on Daley's bid for a fifth term. Some Daley admirers noted in their reviews that no matter how well received the book was, it obviously did nothing to pierce the swell of the mayor's popularity. Daley won with 70 percent of the vote.

But if Royko couldn't hurt Daley, Daley could help Royko. Or at least Mrs. Daley could. In September, Eleanor "Sis" Daley strolled into her National supermarket at 1918 W. Thirty-fifth Street and saw *Boss* on display. Sis Daley had told the *Chicago Sun-Times* in June that "the book is trash, it's shallow, secondhand hogwash at best. I advised the mayor it wasn't worth his time reading." She flipped the books at the store so their covers were upside down and then told the manager that she wouldn't shop there anymore unless the book was removed.

Royko gleefully wrote a column about the incident, and reported that a day after Mrs. Daley's tirade, a directive went out from the National Tea Company ordering *Boss* removed from the shelves of all

Chicago stores. Royko wrote, "I have issued a directive to my wife that all National food products are to be removed from our shelves immediately and fed to a goat."

National Tea reversed its ban a few days later, but Royko also learned the city's airports, O'Hare, Midway, and Meigs, had removed the books. Royko replied loftily, "I travel by train."

Earl Bush said, "The banning of the book came as a complete surprise to the mayor."

Royko was thrilled. "Every writer wants to be banned. Good grief, the publicity! Sis probably did it impulsively. But Mrs. Daley did me a tremendous favor. I'm probably going to have a book leather-bound, embossed in gold and sent to her for Christmas because she put a couple of dollars in my pocket."

In the same month, the school board at Ridgefield, Connecticut, about sixty miles from New York City, banned *Boss* because it slandered law-enforcement officials, presumably the Chicago police, who were described as burglars (which they were) and head-bashing storm troopers (which they were).

Royko was again delighted. "I have been banned from the children's library. I'm going to ban *A Connecticut Yankee in King Arthur's Court.*"

The entire year of 1971 was a heady one. Royko appeared on the Dick Cavett and David Frost network television shows, an ordeal he always claimed he hated but that he usually accepted. He and Carol went to New York on a writer's tour in March. A book party was being held for Royko at the toney 21 Club. Jimmy Breslin was bringing New York City mayor John Lindsay, who wanted to meet the author of *Boss*.

Studs Terkel recalls what happened: "Well, Mike must have had a drink or two and when he showed up at '21' he wasn't sure where to go, or he didn't say who he was, and they told him to sit at the bar. So that's where he was when Breslin showed up with Lindsay. By then, people are apologizing and trying to get him to go the room where they were having the party, but Mike refused. He said, 'You put me at the bar, that's where I'm staying.' He could get like that."

Even in the midst of living every writer's dream, with a best-seller and national celebrity, Royko had to be contrarian. In a letter to Hal Sharlatt, he wrote, "The *New York Times* has asked me to write 700 words for the op-ed page on Daley and his current style and whether he'll be a factor in '72. They promise to mention the book. Jeez, the *New York Times*. I think I'll start wearing a hamburg [*sic*].

"I gave the editor of the *Daily News* a copy of the book. He said he learned more from it than he did in his three years on this paper. The city editor, a life long Chicagoan, said he sat up all night. Dummies. I made the whole thing up.

"In fact, I'm getting so sick of hearing how good it is I'm going to do a column blasting it."

The letter displayed all those mixed emotions that constantly tumbled inside Royko. Although he topped it off with his usual sarcasm, Royko clearly was awed that the lofty *New York Times* would ask him —the tavern keeper's kid—to write an op-ed page piece. And to make sure Sharlatt didn't think the tough, taciturn Royko was sailing on an ego trip because of the book's success, he teased about trashing the book in a column. Inside, he was immensely proud. But he would only share those feelings with his wife, never with an outsider.

Royko said repeatedly in interviews that he had worked harder on the column in 1970 than at any other time because he didn't want anyone, especially the *Daily News* brass, to think that writing *Boss* diluted his daily effort. To produce five columns a week for thirty years, he had to be more than a great reporter and a gifted writer; he had to be a planner and he had to be efficient. Usually his letters columns— those wacky responses to real and imagined complaints—were sporadic. His son Robbie described them as the "hangover columns." But in late 1969 and through most of 1970, the columns appeared with regularity every other Monday. Royko wrote them either late Thursday or early Friday, and that gave him the rest of the weekend to work on his book without worrying about the column until Monday. His 1970 output was just as intense and energetic, angry and funny as it had been in previous years. There was no discernible letup in his attacks on bigotry and bureaucratic idiocy.

Just as remarkable was that during 1971, when he spent a lot of time promoting *Boss*, traveling to other cities, doing book signings, and appearing on television, his columns were as good, and maybe even better, than anything he had done previously. On January 22, he wrote a classic piece about a Chicago payroller, the pervasive system of petty graft, and an innocent, naive fellow who gets taken for a ride. He turns the tables on the reader with his trademark Royko ending. The column was about Jerry Parrott, a graduate student who went to take his test for a driver's license and ran into a driving examiner named Sam.

> Over the years he has had about six different political sponsors, he has manned precincts in several different wards, and he has had jobs with the city, the county, and the state. In fact, if you include two years working for the U.S. Post Office in the 1920s, Sam is a superpayroller.
>
> But Parrott couldn't know he was sitting in the car with so remarkable a creature, a man who knew more about ticket-fixing, vote-hustling, pushing tables for political banquets, ward heeling and other political sciences than all the city's civics teachers put together.
>
> The test finished, Sam shook his head and sighed. "You flunked it, Jerry. You'll have to come back next week."
>
> Disgusted, Parrott said: "OK, OK, I'll come back next week."
>
> Sam looked at him in surprise. "Hold on, hold on. Maybe if you pop for lunch, I'll give you a break."
>
> "Gee, sure. What time do you go out to eat?" Parrott said. Sam stared at Parrott. Some people catch on slowly.
>
> "I eat later," Sam said, "so why don't you give me lunch money now?"

"How much do you need?" Parrott asked.

"How much you got?" Sam asked. Parrott looked in his wallet. "All I have is $4."

"That'll do fine," said Sam, deftly pocketing the cash.

Later, Parrott pondered the incident and discussed it with a native Chicagoan. "Did I do the right thing?"

The truthful answer was: No, you did the wrong thing. You should have left $2 on the seat when you first got in. That way you would have saved yourself $2. Everybody knows that, dummy.

Royko also wrote about a police gambling raid that netted eleven old Jewish men and twenty-one dollars in cash. When the judge dismissed the case, the defendants asked for their money back.

The $21 was the amount of money that was confiscated from the game when the two vice detectives raided the club where the old gaffers gather to play cards. It comes out to about $1.90 a man. It must have been a wild game, with nickels flying all over the place.

The judge agreed that they should get all their money back. "But not the cards," he chuckled. "That is gambling material. It wouldn't be ethical."

"What did he say?" one of them whispered.

"He said the cards are not ethical," someone else whispered.

"What does that mean? Are they marked?"

"I think he means that we might use them for gambling."

"Dat's silly. We'll just play nickel-dime poker with them."

In March, Royko wrote about the travails of Cliff Parrilli, a truck driver who lived in Daley's Bridgeport neighborhood. Parrilli was mugged and beaten twice. On Christmas Eve his house was robbed of all holiday gifts. His motorcycle was later stolen from in front of the house and his stereo was robbed from his house.

It may be that Parrilli is being unfair to the venerable Bridgeport neighborhood. Bridgeport has the distinction of having produced every mayor we have had since 1933. It has given us at least one county assessor, a fire commissioner, a police chief and more top officials and employees in City Hall and the County Building than any other neighborhood in the city.

It is inconceivable that a neighborhood of thieves and head-crackers would achieve such a remarkable record in politics. Then, again, it might be inconceivable that it wouldn't.

Royko also liked to write about stupidity in other places. Over the course of his career, he insulted Florida, California, Indiana, the entire South, part of Wisconsin, and New York whenever he got the chance.

On Sept. 1, 1971, his target was St. Louis, which, according to the excerpt below, "must have the most dangerous barstools in the country." He wrote about Lee Wilson of Herrin, Illinois, who went to St. Louis to visit his mother and was sleeping off a hangover in the backyard when his mother, worried because he had a bleeding ulcer, asked police to come and take him to a hospital. But Wilson, in a stupor, refused to leave, and the police kicked him for a while. Five hours later Wilson died in jail, and the case was sent to the coroner, a Mrs. Helen Taylor.

After hearing evidence, Mrs. Taylor came to the conclusion that an accident caused Wilson to suffer seven fractured ribs, a punctured lung, a ruptured pancreas and lacerations of the liver. Mrs. Taylor said that earlier in the evening, before Wilson fell asleep in his mother's yard, he had fallen off a barstool. And that she said is what probably gave him seven fractured ribs, a punctured lung, a ruptured pancreas and lacerations of the liver.

That's what I mean about barstools in St. Louis....

The worst barstool injury I had previously heard about was the case of a man who lost the tip of his ear when he fell off a stool and landed on a big, mean, sleeping dog, which bit him.

When Royko wanted to provide expert advice on the subject of saloons and drinking, he turned to the owner of the Billy Goat: "Sam Sianis, who owns a tavern, said, 'Maybe if you fall off a barstool about 100 times in a row, it would be possible. But I don't think any tavern owner would let you keep doing that all evening. It wouldn't look right.'"

One of Royko's most memorable columns of 1971 was an allegory about a visit to a Wisconsin farm. Royko talked with an immigrant who had farmed the land for sixty years until he grew too old to work and sold his land. The man now took care of bees and sold honey. He described his boyhood in Europe, his arrival in America, and his work in the Pennsylvania coal mines. He spoke a lot like Royko's father and the other older men who frequented the Blue Sky lounge. He worried about jobs.

> "You got regular work in Chicago?" he asked. "You got steady job? Good. That's good. What you do?"
> I told him I worked for a paper.
> He nodded. "Good. Every day you work, huh? Regular work. Good. Is that hard work on newspaper? Hard work?"
> I told him that I used to think it was. But not anymore.

In November, Royko told a tale of two buildings, one owned by a divorce lawyer named Ted Korshak. Ted's building had holes in the roof, and the water was shut off for days because he didn't pay the bill. But each time Ted was hauled into court the judge continued the case. Royko wondered why.

> Maybe it is because he's a cousin of Marshall Korshak, the City Hall powerhouse. Cousin Marshall says he doesn't even know cousin Ted.

That may be so. But as Ted says: "There is only one Korshak family
in Chicago and we're all part of it." And when your name is Korshak,
it is unlikely the building department will take chances with the family toes.

The other building was a neat, clean four-flat on Sacramento
Street in Royko's old neighborhood. The building was owned by
Angie Pieroni, who had been a Democratic precinct captain but had
turned against the mayor and worked for Daley's opponent, Richard
Friedman, in the 1971 mayoral election. Suddenly, Royko related,
building inspectors were all over Angie's building. They couldn't find
much wrong, but they finally decided Angie needed to build more
entranceways and staircases, which would cost about $4,000.

Royko quoted Angie: "When we're done, this place would make a
great whorehouse. There will be so many doors, the police wouldn't
be able to catch them all. They wouldn't know which one to break
into if there was a raid."

"These pols could teach Mao Tse-tung a lesson on how to crucify
people."

Royko concluded: "People like Ted Korshak exploit a neighborhood. The Angie Pieronis are the only hope for neighborhoods. You
can see which one City Hall appreciates most."

Royko also took advantage of his fame from *Boss* to do freelance
magazine pieces, including one for *Sport* magazine on Cubs star Ernie
Banks. He hired a literary agent, Sterling Lord, and got $1,000 for the
Banks piece.

He was hardly ever home. David Royko said, "I remember him
working seven days a week. That was the irony of the dedication in
Boss, that he was never home on Sundays or any other day. I remember for a while we had a tradition of going downtown on Sunday
mornings to the Little Corporal restaurant across the river from the
Daily News Building. We would have breakfast—I always had a

Monte Cristo sandwich—and then he would walk across the bridge and go to his office, and Mom would drive us back home.

"I remember when he was home, he would play the guitar and my mother would sing on the back porch on the second floor of the house on Central Avenue. But our family was not the *Donna Reed Show*. There were never family dinners. My mother would tell me to make up lists of things that my dad could do with me. He would ask if I wanted to go to a hockey game or a baseball game. I didn't. I think it disappointed him that I didn't like those things but he accepted it, never made a big deal of it. A lot of times I was the little kid standing at the top of the stairs on the third floor, listening to the yelling and wanting to go down and tell my dad to stop yelling at my mother."

The shouting was about drinking, and undoubtedly Carol wondered what other attractions were keeping Royko out past midnight night after night.

When he was home, he was often cranky and tired. "He could be real easy but then sometimes he'd lose his temper, especially if there was noise," David Royko said. "I remember it seemed my mother was always telling us to be quiet because my dad was sleeping, usually because he hadn't come home until late."

Royko did what most fathers do with sons. He tried to teach them sports.

"He was not very patient," David Royko said. "He tried to teach Robbie or me how to hit a baseball or softball, how to bowl. He'd start out real patient. He was good at explaining things, but then it would go downhill. He expected kids to learn at the same rate as an adult."

Some of the best times the family shared were at Bohner's Lake, where the Duckmans had a tiny cottage a few miles from the lake. "We would all go up, my mom, grandparents, uncle, and cousins. And Dad would come up late at night or the next morning. I don't remember us doing much with him, but until my grandparents sold the cabin sometime in the late sixties we had fun," David Royko said.

Bohner's Lake sits outside the small town of Burlington, Wisconsin, just outside the Lake Geneva tourist sprawl. In the fall of 1971, with the royalties from *Boss,* the freelance money and a new, five-year, $35,000 a year contract from the *Daily News,* Royko and Carol went looking for a summer home. They canvassed Lake Geneva, but waterfront property was expensive and the area was cluttered. They expanded their search to Bohner's Lake and found a cottage with glass sliding doors opening to the yard that sloped to the lake. For the next eight summers they both enjoyed the place, although on too many occasions they visited it separately.

Terry Shaffer was invited to the cottage many times. "My wife and Carol had gotten to be good friends, and they would go up there and Mike and I would stay in town drinking. Then we would drive up but we didn't spend much time with them. One time we got there and decided we would try to hit every bar in Burlington. I think there were seventeen of them and we made it through about eleven or twelve before we passed out. When we got to the cottage we would grab the beer and jump in the boat and go out in the middle of the lake to continue drinking. Once in a while we caught a fish."

Boss was nominated for the National Book Award but didn't win. It hardly mattered, because Royko was collecting awards from every Chicago civic organization and university. To go along with the Heywood Hale Braun Award he won in 1969, he picked up the Headliner's Club award in 1971.

On May 1, 1972, he was visiting his brother in Madison, Wisconsin, when he got a message to call his secretary at the *Daily News.* Before making the call, he used the bathroom. "So, I'm standing there with my schwanz in my hand wondering why the paper was calling me and then it comes to me, I won the Pulitzer."

He added that when he returned to his home in Chicago he was greeted by his cleaning lady, who whispered to him in awe, "Your office called. You won the Nobel Prize."

Royko was the third recipient of the Pulitzer for commentary. His prize-winning entry included the stories of Angie Pieroni's fight with City Hall, Cliff Parrilli's ordeal in Bridgeport, the St. Louis barstool killing, and the nickel-and-dime poker gang of eleven old Jews.

Royko celebrated at Riccardo's, where he accepted congratulations and said, "Sure, I like to win but newspaper awards are meaningless now, and they will be as long as only stars are recognized. An award will have meaning to me when it seeks out and recognizes the sparkle that most people never see, the rewrite men, the copy editors, and the reporters who don't get many bylines. Those are the guys who make a daily newspaper work." But Royko's ego was equal to his modesty. "Yeah, I write a good column and I deserve the Pulitzer." He thought he deserved it every year.

Royko was just shy of his fortieth birthday, and he was acclaimed as the best columnist in the country. Perhaps he wasn't read as widely as the humorous Art Buchwald, or taken as seriously as the Washington crowd that searched the entrails of Richard Nixon's White House, but he was the best. And now he was one of the most prominent celebrities in Chicago, a town that admittedly didn't have many heroes and was as likely to become enamored with a hit man or a skinny shortstop as with a newspaper writer.

By 1971 the newspaper conditions in Chicago were changing rapidly. At the *Daily News*, the benign and benevolent Roy Fisher, who received a gracious acknowledgment in *Boss,* was gone and Daryle Feldmeir was named the new editor in January 1971. Feldmeir got the same old mandate: Stop the bleeding. Both the *Daily News* and its afternoon rival, *Chicago Today,* were losing circulation faster than ever before.

A native of Montana, Feldmeir had joined the *Daily News* as managing editor in 1968 after holding the same post for twelve years at the

Minneapolis Tribune. He was immensely popular at the *Daily News,* where he was heavily involved in the newsroom and not nearly as aloof as Fisher had been. A master sergeant in World War II, Feldmeir was the kind of guy that Royko instinctively liked and became one of the columnist's favorite editors.

Meanwhile, the other side of the newsroom was flush with success. Under Jim Hoge, the *Sun-Times* was making circulation gains at the expense of the *Tribune*, whose dogged determination to represent law and order, higher authority, and just about anything the police department said or did was costing it readers. Its credibility was battered by its lackluster coverage of the 1968 Democratic convention disturbances and its defense of State's Attorney Hanrahan's role in the December 1969 raid on a West Side apartment in which two Black Panther leaders were killed. Hanrahan's police staged the raid; afterward, he claimed that the police had been fired on from an apartment filled with weapons. But reporters determined that it was unlikely that the occupants of the apartment, Fred Hampton and Mark Clark, did much firing. The door to the apartment was riddled by bullets that were on their way in, not out.

The *Tribune* printed Hanrahan's argument that the holes in the door were nail holes. The paper looked ridiculous.

While older *Tribune* executives, handpicked by Colonel McCormick, seemed oblivious to the social and cultural changes erupting around them, one new editor did not. Clayton Kirkpatrick had taken over the editorship of the *Tribune* in 1969 from the caustic Don Maxwell. Kirkpatrick, who had also been a McCormick protégé, had risen steadily from reporter, City Hall reporter, city editor, and managing editor to the paper's top editorial position. He was not unaware that the *Tribune*'s stodginess and its reluctance to accept new mores and realities were driving away readers. He knew something needed to be done to recruit those readers driven away by McCormick's ultraconservative positions. Jews hated the *Tribune*. The city's growing number of African Americans, who were rarely featured outside the *Tribune*'s crime blotter, were becoming loyal *Sun-Times* readers.

Kirkpatrick made immediate changes in the *Tribune*. Maxwell had been bombastic and provincial. He had acted as though the *Tribune* were a partner with various Chicago institutions—City Hall, the Bears, the Cubs, and major retailers. Kirkpatrick was more dignified. He wanted to break those ties that undermined the *Tribune*'s credibility, especially with younger readers. He hired younger reporters, including women. The *Tribune* actively began to recruit African Americans, although in the early 1970s black newsmen were a scarce breed.

Tribune circulation was suffering most in the city. The newspaper's strongholds were North Shore communities stringing from Evanston to Lake Forest, and in DuPage County where Kirkpatrick lived and where McCormick had established his baronial estate, Cantigny. But in ethnic neighborhoods on the city's Northwest and Southwest Sides, the *Sun-Times* was dominant. The *Sun-Times* popularity was also driven by the economic and demographic patterns of the 1970s. The metropolitan workforce had been commuting to Chicago. Rush-hour traffic was one way, to the city in the morning, to the suburbs in the evening. Blacks and other minorities were still trapped in service-industries jobs, working in restaurants, hotels, or minor clerking positions at banks and other large companies whose offices were downtown. These city residents took the El or bus to work. Suburban commuters rode the trains; while the trains were spacious enough for the cumbersome folding and refolding of the *Tribune*, the El and bus passengers were cramped and the *Sun-Times* tabloid form was much easier to read. Sports fanatics could simply begin on the back of the tabloid and work their way forward. In hopes of cutting into *Sun-Times* readership, the *Tribune* remade its afternoon paper, the *American*, into a tabloid and renamed it *Chicago Today*. But the strategy failed, and *Today* continued its circulation decline. The oil embargo of 1973 boosted public transit riders, and this, in turn, boosted *Sun-Times* street sales.

Tribune executives knew the struggle to keep *Chicago Today* alive was a lost cause. They also knew that the *Daily News* would take what little was left of the afternoon audience. The *Daily News* had little chance of long-term survival, but it did have Mike Royko. Kirkpatrick could achieve two goals if he could lure the columnist to the *Tribune*: He would prop up the ailing *Today* and, more important, change the image of the austere, aloof, and arrogant *Tribune*.

At the time, the *Tribune*'s lawyer was Don Reuben, one of the most powerful men in Chicago. Reuben also represented the Roman Catholic archdiocese and the Chicago Bears. While Reuben served mainly as a First Amendment lawyer, he also filled many other roles. One of his jobs was to recruit Mike Royko as a columnist for the *Tribune*. In the summer of 1971 Reuben made several overtures to Royko, taking him to dinners at the Tavern Club and to the Vernon Park Tap, one of the few Italian restaurants in the old West Side neighborhood. Terry Shaffer recalled Royko telling him the story: "Don Reuben came by and picked me up and we drove up on Michigan Avenue. He stopped his car right on the bridge and said, 'Look at that,' pointing at Tribune Tower. 'It's been there forever; it will be there forever. You belong there.'"

Reuben convinced Royko to get together with Kirkpatrick. Royko was driven by limousine to a clandestine meeting in west suburban Oak Brook. He and Kirkpatrick discussed newspaper philosophy and some specifics. Kirkpatrick promised Royko total editorial freedom and five columns a week on page three and a three-year contract at $50,000 a year, about $15,000 more than the *Daily News* was paying him.

Royko tested the *Tribune* editor by suggesting that his first column could be on the McCormick Place convention hall that the *Tribune* had lobbied mightily to plop down on the sacred lakefront. His second column, he offered, could rip Daley's proposal for a new, publicly financed sports stadium, a project the *Tribune* was supporting editorially. Kirkpatrick grimaced. The dinner ended with an agreement to continue the negotiations.

But with Royko there was always the kicker. "When I went to leave I notice that the limousine driver is a guy I know from the old neighborhood so I got in the front seat to talk to him and somebody said, 'Mike, at the *Tribune*, you ride in the back,' and I said, 'That's why I'm in the front.'"

A week later, Royko went to Reuben's office and studied the proposed contract. "I had the contract right in front of me and I was going to sign," Royko told Shaffer. "Then I put the pen down. I thought, 'I've hated the *Tribune* all my life. I hate all the people who work there. I hate what it stands for, I can't do it.' I told Kirkpatrick, 'Every time I think of walking into the *Tribune* Tower, my neck muscles tighten up.' He said, 'You can have an office in another building.' I got up and said, 'I'm sorry I put you through this' and I left."

Reuben told a reporter, "I'm just as disappointed as I think Mike will be. He really wanted to come, but he's just sentimentally attached to the *Daily News*. He remembers it when it was great. He's still burning a torch for Fanning." Royko scoffed at the notion that the spirit of the former editor was that powerful, declaring that his new boss, Daryle Feldmeir, "is probably the best editor the *Daily News* has had since I've been there." There is no question that Royko thought highly of Feldmeir, but it was also typical of him to pay homage to whomever his current boss was. Fanning, the editor who had given him his chance, had died earlier that year.

Kirkpatrick did not give up. "Every time my contract expired," Royko said, "the *Tribune* would make me an offer. I just couldn't do it."

In 1972, Royko got a new $50,000-a-year contract from the *Daily*

News, and the royalties from *Boss* were still arriving regularly. He decided he could afford to treat his family to his version of a grand tour. Planning as diligently as the Allies in 1944, he took his family to see the lands of Shakespeare, Balzac, Beethoven, and Bach.

"It was six weeks and two days and the reason I know that is that Robbie and I were counting every day for the trip to be over so we could go back home and be with grandma and grandpa," David Royko said. "But for my father it was another one of those things. It was a sign of his success, that he was able to take his family to Europe like rich people did. On the ship it was okay because there was lots of stuff for kids to do, but in Europe, all we did was traipse around and look at ruins and art and architecture," David added.

Royko booked his family first class on the SS *France* in August and then dragged them to St. Paul's Cathedral and Westminster Abbey, to the Louvre, the Rhine, and Monte Carlo. His idea of touring was to see everything important he had read about.

"If it wasn't significant, he had no interest in it," David said. "If my mother stopped to talk to a sidewalk vendor or somebody who made scarves, my father would mutter, 'God damn it, we have to go.' He had something else he wanted us to see even if we didn't want to see it. He had planned to see everything in Europe that was significant for 2,000 years and he wanted to do it thirty days."

Royko often told a story about meeting a man in the ship's lounge who was pontificating about Chicago politics. Royko finally told the man he didn't know what he was talking about. The man replied, "There's a new book out by a man named Mike Royko and if you read it, then you'll know."

David Royko can't verify that story, but he did remember a similar episode with a customs agent in London. "The guy looked at my dad's passport and said, 'Oh, Mike Royko. Chicago. Terrible place.' My father loved that. We'd been on a boat for five days and there's this guy with a British accent who knows who my dad is. That really impressed me. That's when I realized what a difference *Boss* made. It hadn't meant much in Chicago because everybody already knew who my dad was even before *Boss,* but the idea of going to another country and they knew who he was, wow. He loved it."

When he returned, Royko did not write of Shakespeare or Balzac or Beethoven. He wrote of Monte Carlo.

"I stepped up to the chemin de fer table and said, 'Banco,' just the way James Bond did."

chapter 9

Mike Royko loved movies. He loved John Wayne movies and James Bond movies and Fred Astaire movies. He dreamed up fantasy alter egos from his favorite movies. He could quote dialogue from hundreds of movies. He virtually memorized *The Godfather*. He watched Steve Martin in *The Jerk* whenever he was depressed, tired, or happy.

Royko wrote about movies because they enhanced his "just another guy" image. His columns on movies were done in that conversational tone that Studs Terkel said "made the reader think he was talking just to you."

When John Wayne died, Royko wrote that a liberal friend of his had criticized him for liking the star. The friend argued that Wayne glorified violence, male chauvinism, and racism in the way he casually killed Indians and openly supported the Vietnam War. Royko replied:

His movies make me feel good. Other movie cowboys were more popular when I was a kid. But there was something unreal about them. Roy Rogers, for example, never shot anyone, except in the wrist, and seemed to be in love with his horse, Trigger. Gene Autry never shot anybody, except in the wrist, and he played a guitar and sang in an adenoidal voice. Then John Wayne came along and he shot people in the heart, and drank whisky, and treated his horse like a horse. In fact, he treated women like he treated his horse. He seemed real because he reminded me of the men in my neighborhood....

We knew he would not become bogged down in red tape, or fret about losing his pension rights, or cringe when his boss looked at him....

He would do exactly as he did in True Grit, my choice as his greatest movie, when he rode out to bring in Dirty Ned Pepper, whom he had once shot in the lip....And Dirty Ned sneered and said something like: "That's mighty bold talk for a one-eyed old fat man."

Ah, it was a wonderful moment. And it got even better when Wayne, in a voice choking with anger, snarled: "Fill yer hand, you sonofabitch!"

Now that he's gone, I don't know what we'll do. I just can't see somebody like Johnny Travolta confronting Ned Pepper. He'd probably ask him to dance.

The column said as much about Royko as it did about Wayne. It evoked those boyhood memories of Saturday afternoons when kids of Royko's generation escaped the mundane life of Milwaukee Avenue to find role models for their futures—a new one every Saturday. Movies were also a marvelous kind of storytelling, different from the books in which Royko made his daily trips to fantasy. In the vast, high-ceiling darkness, a kid with a big nose and sloping forehead could imagine himself as dapper Cary Grant sipping a martini from a cocktail glass and always winning the loveliest girl. It was a place where a boy could imagine that he would look and sound just as sophisticated as David Niven if he had a tweed jacket, or that he would draw down on some polecat if he tossed back a shot of red-eye and lit up a cigarette like the Duke.

If the entertainment police of the politically correct 1990s were around in the 1940s they would have worried that a young boy would become an alcoholic from watching movie heroes drink their way to love and glory. They might have been right. Hollywood's glamorous depiction of drinking and masculinity reinforced the belief in most first- and second-generation immigrant households that liquor was part of the high life. Booze accompanied both the good and sad times; the holidays, the weddings, the baptisms, and the funerals. It was certainly a major part of Royko's boyhood, whether in saloons or on the silver screen.

When a television network broadcast Wayne's *True Grit*, and cut out the "sonofabitch" line, Royko spewed his outrage in a column condemning the network for censorship and insensitivity to great art. In another column, he confessed that he had hated it as a child when the nearby Congress Theater played Astaire films "with their sappy stories, mushy love songs, and dance after dance after dance." But by the time Royko was shaving regularly, he decided Astaire was the sharpest, coolest guy around. "He could stroll across a room with more style than most dancers can dance," he said. When Astaire died in 1987, Royko eulogized him in his column:

He went privately and quietly—a class act right up to the end. So when I finish writing this, I'll go home, have dinner, then get out my video-cassette of "That's Entertainment." I'll fast forward to the part where Gene Kelly tells us about Fred Astaire and his remarkable talents.

For about the twentieth time since I've had that cassette, my wife is going to have to sit and listen to me say: "Will you look at that? He's dancing with a coat rack...on the ceiling...look at that move…look at that timing…you know, he's an incredible athlete…fantastic."

But when Astaire finishes gliding through "Dancing in the Dark," with Cyd Charisse in Central Park and they almost float into a carriage, I won't say a word. I never can.

One of Royko's closest friends, John Sciakitano, who spent a lot of evenings with Royko at the Goat, said, "Whenever women came around, he loved to play his James Bond role, you know, talking about whether they would like to fly on his private plane to Florida and board his yacht."

One Royko column was inspired by the announcement that the literary executors of Bond author Ian Fleming had decided to bring the spy back to life. Royko concocted an imaginary dinner date with the spy. He asked Bond if he was still driving the "last of the great four-and-one-half-liter Bentleys," to which the resurrected spy replied, "Good lord, no. Couldn't afford the gas. I'm driving a VW Rabbit these days."

When the imaginary waiter arrived at the imaginary dinner, Royko ordered Bond's favorite, a dry martini with the usual instructions: "Shake it very well until it is ice cold, then add a large, thin slice of lemon peel. Do I have it right, James?"

But Bond opted for Perrier, explaining to Royko that the reason he killed all the Smersh agents was "because I was loaded to the gills most of the time." The satire went on until the new James Bond admitted he had no idea why he had come back. There was a sly inference, however, that the old Royko was still the old Royko. No Perrier.

But westerns were Royko's favorite. The western captured the spirit of the lonely hero. It appealed immensely to teen-aged boys who, in the throes of adolescence, almost always suffered some imagined or real unrequited love and self-pitying solitude. Young Mickey Royko, with his hidden passion for Carol Duckman, was no exception. To such lost souls, Duke Wayne's aging cattle baron in *Red River* was the perfect sad and heroic role model. And it was easy to imagine being Alan Ladd's silent *Shane*, or, best of all, the conflicted sheriff in *High Noon*, torn between love and duty.

Royko eventually tossed the Wayne classic, *Red River*, off his personal top-ten list of westerns, declaring, "The movie's many fine points were bogged down by the female lead, Joanne Dru, I believe, who was so clinging, manipulative, gooey and intrusive that she should have been tossed to the Indians. And by the ridiculous notion that delicate Montgomery Clift would get in a knock down fistfight with the Duke and hold his own."

Royko's top ten always featured a pair of obscure Burt Lancaster westerns. He claimed to have discovered them since both were received with lukewarm reviews, and then praised by revisionist critics: *Ulzana's Raid* and *Valdez Is Coming*.

His eyes would dance brightly when he described his favorite scene in *Valdez Is Coming*, the story of a peace-loving constable named Valdez who has decided he must revert to his warlike ways after a bullying ranch owner refuses to help the peasant widow of a man he unjustly killed.

"When Lancaster goes to his little room and kneels on the floor and opens that trunk and pulls out his old pistol and buffalo gun, it was just like Arnold Palmer in his prime, hitching up his pants on the back nine. It was saying, 'The game is on.'

But *Shane* was Royko's favorite, and he was incensed after reading and hearing glowing reviews of a Clint Eastwood movie called *Pale Rider*:

It was obvious that "Pale Rider," was a brazen rip-off of "Shane." The plot, almost scene by scene, was nothing but a remake.

I asked the city's—maybe the nation's—two best known film critics why they hadn't blasted the makers of "Pale Rider," for the blatant theft. I got blank stares. Through pitiless interrogation, I learned that neither of these film scholars had ever seen "Shane."

The critics were lucky that a nice guy such as me rather than someone from the National Enquirer, had uncovered this potential scandal.

The critics, "Thumbs Up" pair Roger Ebert and the late Gene Siskel, were not lucky. Royko made sure every other film fan knew the two "film scholars" had never seen *Shane*.

It was not the first time Royko had tweaked Ebert. Ebert was working part-time for the *Sun-Times* on January 1, 1967, when he met Royko at the office coffee machine. The beginnings of a blizzard were raging on Michigan Avenue, and Royko offered him a lift home. Ebert recalled, "He was in his thirties then, but looked older. He was one of those guys like Robert Mitchum who always looked about fifty, no matter what age he was."

As they drove north, Royko got thirsty and he stopped at a small bar under the El tracks at North and Milwaukee. The bartender was listening to the Chicago Black Hawks hockey game on the radio. "What a game," Ebert said, "the Black Hawks seem to be scoring every thirty seconds!"

"You jerk!" Royko replied, "That's the highlights."

A few years later, after Ebert had begun his prize-winning critical career, he wrote the screenplay for a 1970 movie, *Beyond the Valley of the Dolls*, and invited various friends, including Royko, to a private showing on West Jackson Street.

Dorothy Collin, then a reporter for *Chicago Today* and later a Washington correspondent and columnist for the *Chicago Tribune*, was with Royko during the screening. "He offered me a ride to O'Rourke's bar where everyone was going to meet. All through the ride he kept saying, 'Geez, what am I going to say to Roger. The movie is terrible, it's awful, but Roger's my friend. How can I tell him? I can't, I can't tell him how bad it really is.'

He did. He did it at O'Rourke's, and then he wrote a column about the screening and his anguish and his opinion that the movie was terrible. "I believe," he wrote, "that every young man is entitled to one big mistake, despite what the alimony court judges may say. And this movie is Ebert's, and I urge you to avoid it. Someday he will write another movie, and I'm confident that it will be excellent. Even if it is dirty, it will be better. I'll be his technical advisor." Friends are friends but the column was always the column.

It was the column which led to a movie about Royko. *Continental Divide* was a 1981 flick starring John Belushi as a chain-smoking, alderman-hating Chicago newspaper columnist. Royko was pleased.

> The move's publicists have said that he is supposed to be a "Roykolike character." So, obviously, part of my reaction is that I'm flattered....as much as I like Belushi personally, I think the producers made a mistake in casting my part. I think Paul Newman would have been a better choice, although he's older than I am. And in appearance we're different between he has blue eyes and mine are brownish-green. And I would have been satisfied with Clint Eastwood even though he's taller than I am.

Royko had fun ridiculing the movie's plot, which has the columnist being beaten by Chicago cops and sent by his editor to a remote mountain wilderness to interview a female expert on birds.

> That's unrealistic. Chicago cops haven't beaten up a newsman since the 1968 Democratic convention and most of those newsmen were from New York and Washington, so they had it coming....And I'd never climb a mountain to interview a bird-watcher. I'd ask her to

come down the mountain for dinner so we could study such birds as coq au vin, or Rock Cornish hen. On the way up the mountain the columnist loses his supply of liquor and cigarettes and is heartbroken. That's stunningly realistic. In fact, I wept during that part of the film.

Continental Divide meant a great deal. After all, Hollywood hadn't done any movies based on the life of Jimmy Breslin, or on Art Buchwald, or William Safire or any other columnist. The movie was another reminder of how far Royko had come.

In the early days of his career and marriage, when Mike and Carol were like most young couples—full of hopes and dreams and each other—Royko would finish work at the weekly and go to a Logan Square grill not far from the doctor's office where Carol worked as a receptionist. He would drink coffee until she arrived. Carol ate her lunch at the grill most days. The owner was Pete, a young Albanian immigrant married to a lovely Greek-American girl named Marion who worked the cash register and waited on the people who sat at the counter. Mike and Carol became close friends of Pete and Marion, and Pete's brother, Adam, who had also come from Albania.

The two couples didn't have much, so they traded visits to each other's apartments for dinner and cards. Pete and Marion introduced the Roykos to pastichio and dolmades and baklava, and then Pete would pull out a bottle of Metaxa, the fiery Greek brandy which Royko thought was a menace to a drinking man. They attended each other's family gatherings. When Robbie was born, Pete was his godfather.

When Royko finally succeeded in getting hired by the *Daily News*, Pete and Adam opened a steak house on North Avenue near Harlem. It was a success, and soon the brothers wanted an even bigger, ritzier place. They opened a popular restaurant called Fair Oaks on Dempster Street near the Edens Expressway in Morton Grove. Pete got involved in other businesses, but the economy suddenly went into a tailspin as interest rates flared during the Carter presidency. By 1980, the restaurant was closed and Pete was ill.

Royko wrote: "He died at an age when most businessmen are just hitting their stride. It seemed unfair—all those eighteen-hour workdays, all the sacrifices. And he had his great American dream for so brief a time."

In one of many boxes of papers, letters, magazines, and assorted odd material that Royko saved in his office at home was a restaurant guide from 1975. It had advertisements from north suburban businesses. There was a small ad for the Fair Oaks restaurant that Royko had circled in pen and stored away. The ad carried the names of the owners, Pete and Adam Belushi.

At Pete's funeral, Royko talked with the whole family, including Pete's nephew, Adam's son, John, whom he had known as a kid running around the yard at family parties. John had gone into show business instead of the restaurant business.

Royko had run into John Belushi years earlier at a political rally, when a young man approached him tentatively and said, "Uncle Mike? You don't remember me?"

Royko said, "I know you're one of the Belushi kids by your goofy face, but I'm not sure which one." Royko later went to see John perform at Second City and brought him to the Billy Goat for a "chizbooga, chizbooga," and "Pepsi, no Coke," which Belushi later made famous on *Saturday Night Live*.

When Belushi made the *Blues Brothers* movie he called Royko. "He sounded shy when he asked if I liked his movie. I was touched that he'd want my opinion. I told him, yes, I had seen 'The Blues Brothers,' and I thought it was hilarious. And how was he doing? And his family? After we said good bye, I thought about my friend, Pete Belushi and his American dream, part of which was: 'Maybe somebody in my family will be famous, uh?'"

The last time Royko saw John Belushi was at a party after *Continental Divide* opened in Chicago. A reporter wrote that the evening ended with Royko and Belushi hugging one another. That was in 1981. A year later, on March 7, 1982, Royko recalled the moment, and his friends, Pete and Marion and Adam, in another column.

> When you feel like a proud uncle, and see the kid up there on a movie screen, you ought to give him a hug. This column seems to have rambled. I'm sorry, but I just heard about John's death a few hours ago, and I have difficulty writing when I feel the way I do now.
>
> He was only thirty-three. I learned a long time ago that life isn't always fair. But it shouldn't cheat so much.

During the 1970s, especially in the summer, movies were not Royko's main passion. That was reserved for sixteen-inch softball. Mike Royko loved sixteen-inch softball. He loved being part of the playground game that was unique to Chicago. He insisted on being the general manager, manager, star pitcher, and promoter of the *Chicago Daily News* softball team, which played in a media league against other newspaper teams and some teams fielded by television stations. They played at Grant Park, where four or five diamonds would be filled with softball teams by 6 o'clock. They played at Thillens Stadium on the Northwest Side, and they played at Jackson Park on the South Side, and they played any place else that would let them.

During the depression, sixteen-inch softball was Chicago's most popular pastime. Every neighborhood had two or three teams. The Wildcats were Royko's favorites. They usually played for beer paid for by the tavern that sponsored the losing team. The tradition continued through Royko's playing days. The Windy City Softball League, a professional circuit, played from 1932 through 1948 and drew crowds of several thousand. Several Windy City players went on to the major leagues, including Phil Cavaretta, Royko's favorite Cub; Lou Boudreau of Cleveland; and Bill Skowron of the New York Yankees. Nat "Sweetwater" Clifton, later a member of the world famous Harlem Globetrotters, was a big home-run hitter in softball. Unlike softball, in which a twelve-inch ball that was almost as hard as a baseball could be pitched fast and hit far, sixteen-inch softball was designed so that anyone could hit and anyone could catch. The game almost disappeared in the 1950s with the advent of television, the flight to the suburbs, and Little League baseball which took over sandlots. But it was revived by all the Little Leaguers who couldn't make the major leagues. All sixteen-inch softball required was the same inexpensive equipment kids had used in the depression, one ball, a bat, and twenty players. (Unlike baseball, sixteen-inch softball had one more defender, a short center fielder.)

Royko began playing when the *Daily News* team was organized in the 1960s. He soon took over as manager. As with everything else, he became the game's historian, chief student, and most enthusiastic player.

Tim Weigel, a Chicago television sportscaster for more than twenty years, was a young reporter when Royko recruited him for the softball team. Royko always prided himself on being a shrewd judge of talent, whether in softball or journalism, and he sized Weigel up as somebody who could help the team. Probably because someone told him Weigel had been a starting halfback at Yale.

"I'd been at the paper for four or five days in July '71 and the word was out that I'd played football in college and they needed a right fielder for the big match at Thillens Stadium with *Chicago Today*. They were big rivals, fighting for survival, both professionally and on the softball field. Royko approached me—I hadn't met him—and said he needed somebody to play right field. I had never played softball which was strictly a city game and I was a farm boy. I was smart enough not to ask where the gloves were because everybody's throwing the ball around with bare hands and I figured out that there's no gloves involved.

"In the first inning, Max Saxinger, a great hitter, comes up and Royko pitched him outside, and he lined one down the right-field line. I ran to the line, dived for it, made the catch, rolled over, and threw the ball to second base for a double play. And 'til the day he died, Royko always said it was the greatest catch he'd ever seen in a sixteen-inch-softball game. And it really kind of forged our bond, even though I did swing and a miss at the first ball I saw as a batter. 'How the fuck could you miss a sixteen-inch softball?' he said."

Royko always envisioned himself as the star, but in reality he played the least important position—pitcher. "All a sixteen-inch pitcher does is lob up the ball and get pounded. I always thought it was very odd that Royko, of all people, picked that for his final active athletic expression outside of golf, which of course doomed him to frustration. Golf will frustrate anybody, but here's Mike, who spent his whole career both as a journalist and as a person, a friend, dishing it out. He's the writer. He was always in control. He's the raconteur. He was always in control. If he's sitting across a bar from you, he's got the last word. He's the zinger," Weigel said.

"So to say he was a good pitcher, there's just no such thing as a good sixteen-inch-softball pitcher. I don't care all about the jumping and fake pitches and different windups, it's basically lobbing the ball up over the plate and saying here, asshole, you either hit it or you ground out," Weigel said.

"The best athlete on a 16-inch softball team plays shortstop or short centerfield. Mike couldn't run so the only place he could play was pitcher. But he loved the game so passionately, so deeply. I think he loved it even more than golf, which is saying something. He was as competitive as Michael Jordan. He surely wanted to win as much and it really is amazing. I played in front of 80,000 people at the Yale Bowl and in high school won 38 straight football games at Lake Forest High

School, and yet one of my most—maybe because it's more recent—
one of my most vibrant memories is after we beat *Chicago Today* in a
championship. He ran out and he hugged me in left field because I
made the last catch."

Royko's competitive nature could also be brutal. Mike Argirion,
features editor at *Chicago Today* and later at the *Chicago Tribune*,
remembered one of the fiercer championship games: "I had hit a
double or triple in the first inning and Royko just stared at me. Later,
I was covering third base and Royko was on when somebody hit a
double and he came crashing into me and knocked me over and
wrecked my knee, tore all the cartilage. I had to have surgery. It was
deliberate. He didn't even try to slide past me, he wanted to nail me."

Years later, when Royko joined the *Tribune*, Argirion stopped by
his office to say hello. "He remembered me, I know, but he never said
a word about the softball game. He never said he was sorry when it
happened and he didn't say it ten years later."

Bill Garrett also played for the *Today* team. "We were always better than the *Daily News*, most of the time anyway, and Royko hated
that. He and I were friends but for the three hours of the game, he
was your worst enemy."

Sometimes, Royko was his team's worst enemy, too. In 1974, Don
DeBat, a *Daily News* real estate writer who had organized the team
before Royko took over as manager, and some other players who
were disgruntled that Royko kept adding his personal friends to the
team, staged a mutiny. Royko grudgingly agreed that Weigel could
take over as manager.

"But Mike was still pulling all the strings, and I was merely his
mouthpiece," Weigel said, "and it all boiled down to one moment in
Thillens Stadium, and we were playing the Chicago Bears in a charity
game, and the Bears had these big bumpers and the fences were really
short at Thillens, and Mike gave up six home runs in a row. It was the
most amazing thing. Pow! Pow! Pow! Pow! Six homers in a row go

soaring over the fence. Finally, DeBat and others kept looking out at me in left field to make a change. After the sixth home run, I yell, 'Time out!' and I come trotting in from left field, I said, 'Mike, give me the ball. I'm taking you out.' He said, 'Get the fuck back there in left field.' I said, 'Hey! I'm the manager.' And, of course, he yelled back, 'I'm the manager. Get the fuck back there or I'll kick you off the team.' Here we are, arguing like Billy Martin and Reggie Jackson right in the middle of the field and he's yelling 'Fuck you,' and I'm yelling, 'Fuck you.' Finally I just turned away and he finished the inning, gave up another home run, finally got a guy out, and then he took himself out and left the game. The next week he announced he was reinstating himself as manager. He was always in charge."

In 1976, the biggest moment of Royko's athletic career occurred. He broke a leg in a softball game and never left the game. In fact, he didn't know the leg was broken until he had it x-rayed the next day. But he immediately cast himself in the mold of the toughest professional athletes who learn to play with pain. He often referred to this personal experience when he damned the greedy multimillionaire athletes of the 1990s, particularly those who seemed to be injured all the time.

In 1977, Royko spotted another great promotional opportunity for him and for softball. The Chicago Park District decided to allow players to use gloves. Royko was outraged and filed suit, charging that the ruling had been issued after he had paid a $240 membership fee for the *Daily News* team to play in the Grant Park league, and it was impossible for the team to withdraw and find a league "in which men played like men."

Royko also claimed the rule "runs contrary to the spirit of 16-inch softball and unfairly penalizes those with talent and calloused hands and gives an unfair advantage to those with tender and well-manicured hands."

Royko also got very lucky because during a conference in a judge's chambers—an episode that Royko described as if his civil suit could result in a prison life sentence—a surprise witness showed up on his behalf. It resulted in a delightful column.

> Suddenly, it happened. The turning point. The door opened and a long, black cigar entered. Attached to the cigar was a short, squat man, wearing a cashmere overcoat, a gray fedora, a huge pinky ring, and a face like that of the late Rocky Marciano....The uninvited, unexpected visitor was Bernie Neistein, former state senator, former boss of the West Side's 29th Ward, and one of the Machine's most efficient vote hunters.
>
> It was Neistein who once walked into Daley's headquarters after an election and said: "One of my precinct captains brought in his precinct 350 to zero. I told him, 'Hey, dummy, that's the kind of stuff that brings in investigators. You go back and give the Republicans three votes.'
>
> "I'll tell ya what," said Neistein, pointing his cigar at me. "I heard about this case, and I happened to be in the building, so I came up here to testify for him."
>
> "You are my witness?" I asked, trying to keep from falling out of my chair.
>
> "Yeah, I'll testify that there ain't never been no gloves in 16-inch softball..."
>
> In a few minutes the agreement was drawn up and I claimed a legal victory....And as Neistein left the courtroom, he said:
>
> "Gloves? Why the only time anybody on our old team ever wore any kind of gloves was when they didn't want to leave fingerprints."

The *Daily News*, which had happily run a picture of Royko on crutches when he broke his ankle, now ran *Daily News* coverage of his lawsuit. In fact, many of the *Daily News* front pages of the 1970s referred to Royko columns and stories and followed up on his reports of bureaucratic idiocy.

Two of Royko's leg persons, Ellen Warren, who worked for him in 1971 and 1972, and Paul O'Connor, who succeeded her, remembered how important softball was to Royko. "He would take hours working over the column, sweating out each word," O'Connor said, "unless there was a softball game. Then he'd rip it off in about twenty minutes, grab his uniform, and take off."

Softball also became a family affair, as Carol would take David and Robbie and drive to Grant Park to watch the games, then join the team and the assorted hangers-on for an evening of triumphant toasting or gloomy drinking at the Goat.

"I remember it as one of the best times we had," said David Royko. "What kid didn't want to sit around the Goat drinking Cokes, eating cheeseburgers...it was a really happy time. Mom was always happy and it was one of the few times we were together as a family."

Royko once said, "I play softball because, if I hit a game-winning home run and I'm hoofing around the bases, I can kid myself into thinking I'm 17 again."

Tim Weigel said, "I have a thirty-year-old son now who's an actor, who's doing quite well, and he tells me that the thing he remembers most fondly in his life and in his childhood is sitting around the Billy Goat at five- and six- and seven-years-old listening to Mike Royko. He said he's been around a lot of these so-called big names of Hollywood and whatever, and his fondest memories are as a little kid just sitting there watching Mike hold court. And so you can imagine how it was for us as adults, that it was unspeakably exciting, really. I felt that I was able to sit there, and it was like having a drink with Hemingway or playing cards with Mozart or smoking cigars with Churchill. Mike was one of the greatest men of our time, and we got a chance to just listen to him talk. It could get abusive and it could get really weird— anybody gets weird after they get drunk, but there were those windows of extreme brilliance that will live with me forever."

John Schackitano was there for most of the moments of brilliance and many of the weird times as well. Schackitano was a composing room supervisor at the *Sun-Times* who grew up on Kedzie Avenue in Royko's boyhood neighborhood. They became friends when "Shack" was prevailed on by a neighborhood buddy, Dave Reilly, to drop in on Royko and mention he had seen Reilly, who had played with Royko as a child.

"I went to his office one day in 1972 and asked if he had a minute," Schackitano recalled. "He looked up at me with that stare over his glasses and said, 'Yeah.' I told him Dave Reilly wanted to me to say hello. 'Sonny Reilly? You know Sonny Reilly?' Well, Reilly had been this skinny little kid that everybody picked on, and then when he was a teenager he began lifting weights and became Mr. Atlas. Royko hadn't seen him in years but he told me a story that once when some neighborhood bully was threatening to beat up Reilly, Reilly ran into the Royko house through the back door and Mike was there with his mom. The bully followed Reilly, and Mike's mom cracked him with a frying pan.

"Then he asked me where I grew up and what I did and all that and the more we talked the more we had things in common," Schack-itano said.

Schackitano's grandparents had emigrated from Sicily, and, like many of the turn-of-the-century Mediterranean island natives, had opted to enter America through New Orleans rather than New York. Schackitano's father was born in Texas, and his mother was born in Louisiana, but both moved to Chicago at an early age. They got married and lived in a basement apartment of a six-flat on Kedzie near Armitage. Schackitano was born in 1940. He was eight years younger than Royko, but they shared a similar childhood.

"I was kicked out of Lane Tech—I didn't get anything higher than a D—and enrolled in Tuley but I did even worse. I tried to get into Weber (a Catholic school) but they wouldn't take me. Finally, my mother begged Lane Tech to take me back and they did. This time, I really did well and I was on the football team, baseball team and wrestling team. But I was having problems at home—my dad was drinking, there were fights—and I had a counselor at Lane who had retired from the Navy and he encouraged me to join. So I did."

Shack's grandparents were immigrants. He was from a blue-collar family where both parents worked. He lived on Kedzie Avenue and had served in the military. Plus, Shack went through radio-operator school, just as Royko did in the air force.

"In those days, newspapers had pneumatic tubes to send copy from the newsroom to the composing room, and Mike and I would tap out messages in radio code," Shack said. Schackitano went to a trade school after the navy to become a linotype operator and got a job at the *Sun-Times* in 1968.

Schackitano was a brawny six-footer with deep-brown eyes and a mop of black curly hair. Through Royko, he became close friends with Bob Billings, and like Billings, he was an imposing figure. When the three of them were huddled at a table, they did not invite interruptions. But while Billings usually wore a scowl, Schackitano had a quick smile and a big laugh. Their conversations ranged from the mundane to the sophisticated.

"Billings was the smartest guy I ever met. He was an absolutist. With him, everything was black or white, no shades of gray. He was a natural athlete who had gone to the University of Illinois and played football for the army in Germany. He was an authority on James Joyce and threw a party every June 16 to celebrate Bloomsday. There would be a bunch of us guys all reading different parts of *Ulysses*. Bob and Mike both loved classical music, but Billings was a true audiophile. He knew just how many grams of weight a turntable arm needed, and he only had records. He turned his nose down at tapes and compact discs. He would often fly to New York to hear a certain symphony at the New York Philharmonic. But he just couldn't get along with people. Most people that intelligent can get along with anyone. Mike could when he wanted to, but Billings couldn't. He and Mike were really close but Billings wasn't a regular at the Goat," Schackitano said.

Almost before Schackitano had mentioned that he played ball at Lane, Royko recruited him for the *Daily News* softball team, conveniently ignoring the fact that Shack worked at the *Sun-Times*.

"Those were the best times, going to the Goat after the games, sitting around with a bunch of guys, replaying the game. It was really great if we left just when Mike had a buzz on and was feeling good, and it wasn't one of those nights that lasted forever," he said.

"There were so many great times at the Goat. Once someone planted an M–80 firecracker in a mop in the men's room and just as some guy was going to open the door it blew and pieces of mop were flying all over the place. Sam Sianis went running into the men's room and he came out holding this smoking mop and says, 'Ees joost one of those exploding mops.' Mike loved telling that story. Another time we were sitting there and here's a guy at the end of the bar with two steins of beer in front of him. He's got his eyes closed, even when he's taking sips of beer. I said to Mike, 'Never trust a man who drinks with his eyes closed,' and a few minutes later this guy throws a stein of beer at us, just the beer, not the glass. Sam goes over and throws him out of the place but for years Mike would always say, 'Shack says never trust a guy who drinks with his eyes closed.'"

Royko always embellished and exaggerated the brilliance and feats of loyal friends. For those who wondered how Chicago's journalism and literary giant could have a linotype operator as a best friend, he would declare, "Shack is the brightest man I know. He could run the newspapers. He's tough. He's strong as a bull. He could play for the Bears."

Shack's strength had more than passing appeal to Mike, who was glad to have it around on some occasions.

"We were drinking at a bar in Niles one night and a young guy came over with his wife and give Mike the usual 'I've been an admirer for years and could I have an autograph,' and he gives Mike a pen but I notice the wife's kind of giving Mike the fish eye and practically is rubbing up against him. The guy's going on about the columns he liked, and I make a quick trip to the men's room. When I came back, the guy's got Mike bent over the bar and he's choking him because I guess Mike put a move on the wife. I had to peel his fingers off Mike's neck while the owner—he was a cousin of Carol's—calls the cops. When the cops get there and take the guy out, Mike, who was still rubbing his neck, had to get in the last word. 'Hey, asshole, I've still got your pen.'

"The best times were what he called Night of the Assholes, when we'd sit at the bar and for some reason an unusual number of people would start bothering him and he would say, 'This is like John Wayne fighting them off,' and he would trade insults and cut these people down. But I got nervous when I heard him say, 'Chum,' because that was the last thing you wanted to hear. That meant he was about to go after somebody, take a swing at him."

Often, Royko and Schackitano drove into work together, since Schackitano lived with his wife, Linda, and three children at Milwaukee and Foster, which was only a few miles from Royko's home in Edgebrook.

"One night we had stayed out late and on the way up the Kennedy Expressway, Mike started to complain about a corn on his foot. He took his shoe off and suddenly tossed it out the window and it bounces on the express lane. The next morning, I get a call about 7 A.M. and this real deep voice says, 'You got my shoe?' He drove to pick me up and I swear he only had one pair of shoes because he had a sneaker on one foot. When we got downtown, he started limping so people would think he was wearing a sneaker because his foot hurt. He went to Marshall Field's that afternoon and bought another pair of shoes."

In addition to watching movies, playing softball, and drinking at the Goat, Mike Royko loved fishing. Ironically, so did his arch rival,

Richard J. Daley. Both Royko and Daley were given their first bamboo rod and worm by their fathers. Daley's father, Martin, a sheet-metal worker, often took his only child on exotic trips to Wisconsin to hunt muskie and brought him to Lake Michigan to net smelt. Daley's rare moments of cheerfulness at news conferences usually occurred after his annual trip to Florida to fish for tarpon. He would chortle when reporters asked him if he had had any luck. He would occasionally dream that someday the Chicago River would be stocked with trout and that workers in the Loop could cast for them during their lunch hour. Those utterances were greeted with ridicule by the press corps, much as they had snickered when the mayor talked of putting Venetian gondolas in the Chicago River.

Royko never snickered at Daley's fish tales. He, too, had been taken by his dad to the Humboldt Park lagoon to hunt blue gills and sunfish. He, too, had been taught to net smelt. But it was Schackitano who got Royko back into fishing. Royko had not fished much as an adult, devoting any free time in the early years of his marriage to golf, and then taking up softball after work.

"I had been fishing Lake Geneva for twenty years so I brought him with me." Schackitano recalls. "The next thing I know he's buying all this expensive equipment, pricing bass boats, and looking for houses on a lake. That's when he bought the place at Bohner's Lake. He became a good fisherman. Whatever he did, he had to do it better than you. The first time I took him to Lake Geneva he caught an eight-pound walleye. I had been fishing it twenty years, and the biggest I ever caught was a three-pounder. He might not always catch the most fish—I usually did—but somehow he always caught the biggest one."

Royko told the *Washington Journalism Review*, "I don't fish to catch the fish so much as I really like dawn on a lake. I like mists over a lake in the Ozarks. I like the way the air smells. I like the feeling of tiredness after a long day of fishing."

Since everything he thought, everything he did, and everything he saw or heard eventually went into a column, it was not surprising that Royko began to write about fishing. He could have written with some expertise about going after the great American game fish, the largemouth bass, which he chased in Wisconsin, Kentucky Lake, and Bull Shoals. But he knew instinctively that a serious column about bass fishing written by a kid from a Milwaukee Avenue tavern wouldn't have the right tone. He knew what kind of fish his readers would expect from Mike Royko:

> All my life, I've gone after only one fish—the mighty bullhead. I have pursued it from the Calumet River to the Chain of Lakes. I have tempted it with hooks baited with Feta cheese, Swedish meatballs and Slotkowski's sausages. I have hauled it up out of the mud and respected it for its courage as it leaped up and shook its head and splattered mud on my glasses.

In ridiculing a *Harper's* magazine piece on the joys of trout fishing, Royko wrote:

> Why do snobs like trout fishing? Probably because it is safe, since most of it is done in shallow, narrow streams. The trout fisherman has no stomach for venturing out into the vastness of the Chain of Lakes and facing the treacherous water-skier and the dreaded drunken speedboater.
>
> Why do I prefer the bullhead above all others? Because he is one of the few fish that actually exist.
>
> The legendary muskie does not exist....The bass is another myth....I know the bullhead exists because I have caught him. I have fished for muskie, bass, walleyes, bream, crappies, gar, tarpon, snook, coho, steelheads. You name it, I've never seen it. But whatever I have fished for I have caught bullheads.

So I prefer the bullhead above all others, and especially above the overrated trout. Try catching a trout on a piece of salami. They bite only at gnats, fleas and flies....Common sense should have told Harper's that anything that prefers eating gnats to salami is really dumb.

No less an authority on trout fishing—and writing—than Red Smith took offense at the column. In his *New York Times* column of August 26, 1974, Smith accused "the gifted columnist of the *Chicago Daily News*" of succumbing to Second City Syndrome.

Royko's trouble is that he is a bullhead fisherman. This is a Class D catfish, a poor relation of the creature Mark Twain describes in his "Life on the Mississippi."

The fish Royko tries to entice with a worm or piece of salami is a poor thing compared to Twain's but Mike enjoys sinking a barb into him, perhaps because of his facial resemblance to Mayor Richard J. Daley in pensive moments.

Royko's fishing mania was encouraged by his boss, Marshall Field, an avid fisherman who invited Royko on many trips to Coleman Lake, Field's private preserve sixty miles north of Green Bay, Wisconsin. The fishing party often included Tom Tallarico, a Field's executive, and Al VonEntress, circulation chief of the Field newspapers and a natural enemy of editorial staffers.

VonEntress was the scapegoat of one of Royko's great pranks. In collusion with the others, Royko changed the clocks so that when the alarm went off at 6 A.M., the scheduled start of the fishing day, it was in reality only 3 A.M. VonEntress protested that it couldn't be 6 A.M., that he was still sleepy, but the others persisted, and he dressed and went out to the campfire in pitch darkness. One by one the others slipped away and crawled back into their beds, leaving VonEntress wondering why he was alone in the dark on the shore of a lake. Of course, the incident was turned into a column.

On another occasion, Schackitano said, "Royko was in the car with Field and two other guys and they were speeding like mad down some Wisconsin road. Royko turned to the scion of Chicago's greatest family and said, 'You know, if we have an accident the newspaper headlines will read, Royko and Three Others Killed.'

Rob Royko remembered, "All the fish I catch when no one else is getting any, I learned from him. He used to concoct stories for me about what the fish was thinking down there under the water. He had one story he loved about a guy who was starving who sees a sandwich by the lake. He picks it up and rips off the wax paper—and my dad would make all the facial expressions—and the guy takes a bite out of the sandwich and suddenly he's pulled into the water by a big fish holding a rod. Another fish says, 'Is it a white one or a black one?'"

Terry Shaffer, another fishing companion, thought many of Royko's trips were nothing more than beer-drinking excursions to the middle of Bohner's Lake.

"Once we went on a trip to Kentucky Lake, Mike, Shack, and Larry Green, who was working in the Springfield office [of the *Daily News*] and who Mike really respected as a newsman. We picked Green up on the way and Mike is bitching about Green not being ready when we got to his house, and Larry came out with this fishing tackle that looked like it came from K-Mart. Mike was really pissed. He always had to do everything perfect. If you were going to fish you had to have the best stuff, or at least, what he said was the best. If you didn't do his way, he'd let you know. So all the way to Kentucky Lake he was ragging Larry about his fishing gear and swore he wouldn't fish with him," Shaffer said.

"The next day we have guides and two men to each boat, so Mike says he'll fish with me and puts Shack and Larry in the other boat. We fish all day and we get skunked, nothing. Except for Larry who comes in with a six-pound bass. From that point on it went downhill. Mike was on Larry all night. We were doing some drinking and playing poker but he wouldn't let up. Larry finally gets mad and goes to his room and starts packing. He's going home. I begged Mike to shut up because we didn't want Larry to leave because it's no fun fishing one man to a boat and you couldn't do three in boat.

"Mike took every competition seriously. He just had to win. It didn't matter to him if Shack and I caught no fish, but if he had no fish, he was pissed. There was another trip with Larry where Mike accidentally hooked him with a lure. Larry kidded that he was going to sue, and Mike said, 'See, never take a Jew fishing.' He kept that up for quite a while," Shaffer said.

"It wasn't always easy being his friend," Schackitano said. "He could be the kindest, most considerate, most generous man and then...I've seen him be so gracious to people who came up and asked for an autograph. He would ask where they were from and what they did, but if someone yelled, 'Hey, Royko, your column today stunk,' that could set him off.

"He only turned on me once," Schackitano recalled. "I had some friends whose son worked on a suburban weekly, and they asked if they came down to Billy Goat's would I introduce them to Mike. I said sure, and the night they dropped in he was at the corner table— the wise guys table—and he was holding court. I went over and said, 'Mike, I got some friends here who want to meet you.' He looked at me and growled, 'I'm not some fucking trained bear.'

"So I went back to these people and said, 'He's kind of busy right now.' Then I took them over to Riccardo's and I got a drink and now I'm steaming. I'm getting madder by the minute so I storm back to the Goat's and back to his table and I read him the riot act. 'Who the fuck do you think you are pulling that shit on me. I'm not some ass-hole.' The next day he called—we talked everyday—and said, 'What did you get so hot about?' as if he couldn't remember what happened.

That was the first and last time he ever pulled that shit with me. He did it to a lot of friends, but never again with me.

"There was no way in the world that anyone could be the center of attention when he was there. If you were with him alone, that was fine, you could say anything, but when there was a crowd, he always had to be the focal point. He always was. He wanted to dominate every conversation, and he could. He turned every event to himself. He could walk away from you if the spotlight was on you, not him.

"But if you were his friend, he'd do almost anything for you and praise you to the sky. His friends were always the best at anything and the brightest people." Royko's obsession with his friends and with fish was not that much different than his obsession with politicians. He explained it in a November 30, 1977, column supporting an early parole for Alderman Thomas Keane, the most powerful member of the City Council, who had been sent to federal prison for fraud.

> I was against putting him away in the first place. That may surprise those who assume that I dislike politicians because I occasionally try to catch them with the cash still in their hands. Anyone who believes that is mistaken. I enjoy chasing and sometimes catching politicians for the same reason I enjoy fishing.
>
> I don't have anything against the fish. I surely don't dislike them. So when I catch one, I usually put it back in the water. That way, I might catch it again someday.
>
> That's how I feel about Chicago politicians. I don't dislike them. How can anyone dislike people who have brought so much entertainment into our lives?
>
> There are laws in most places limiting the size and quantity of fish and game that can't be taken, and setting the season. We should consider preserving our political wildlife in the same way. Or some day City Hall will be depleted. That's why I favor catching them, letting them wiggle and flop around and breathe hard, but tossing them back for another day.

chapter 10

In fact, Royko liked politicians. He liked the game of politics and admired the way the best of them played it. Although he was a great writer, great storyteller, and great investigative journalist, Royko was basically a political reporter. He grew up in an era when politicians affected people's lives. He was fascinated as well as outraged over the power that politicians wielded, from the precinct captain to the president of the United States.

What he detested as a young reporter were journalists who were so beholden to politicians that they never wrote anything negative about them. He often warned young journalists not to become compromised by their close relationships to politicians. His own reason for keeping his distance was not that he feared the temptation to ignore a potentially disparaging story about a friendly political figure; on the contrary, his problem was, he said, "If you get too close, then you're going to feel uncomfortable when you have to stick it to them."

One way to infuriate Royko was to suggest that he was not as influential a national columnist as some of his peers, who wrote lofty essays from the shadows of the White House, or that he was not as politically sophisticated as those who dealt daily with the intricacies and intrigues of Capitol Hill.

No other columnist in America had been so influential in so many elections. Neither Royko's indignation, satire, humor, nor outrage could have much effect on Chicago's idolatry of Richard J. Daley, but he destroyed or salvaged more politicians than any one else, and he did it at a time when American newspapers had all but lost their once unshakable grip on American opinion.

By the 1970s, the cherished role of newspapers as the primary source of election information had dwindled. One obvious reason was television, which candidates were using to portray themselves in the best possible light and to send subliminal messages to viewers. Another was the disintegration of party loyalty. As more and more voters strayed from their traditional Democratic or Republican ideologies, they began to make political choices based on personality rather than philosophy. Television was far better suited to reveal those personalities than newspapers, which were only beginning to look into private lives and not yet peeking in windows. Beginning in 1972, there was also a dramatic shift in how American presidential candidates were chosen. Following the 1968 Democratic National Convention debacle in Chicago, the parties opened up their systems and made the primaries the battleground for the nomination. In a few short years, the political conventions that had produced presidents for more than 100 years—with more good results than bad—became nothing more than a dreary summer television special, devoid of the mystery, excitement, and exultation of party unity that once commanded American attention.

Before the primaries took over, political bosses met in those storied smoke-filled rooms, often emerging with candidates who were virtually unknown to the American public, or, in many cases, known only in their regions. It was left to the newspapers to begin educating the public about the stewardship of a William McKinley in Ohio, or the internationalist leanings of a Woodrow Wilson, or Franklin Delano Roosevelt's dry attitude toward Prohibition, or even the civil rights positions of John F. Kennedy. What few primaries mattered before 1972 were almost gentlemanly. Opposing candidates left their chances in the hands of state party professionals. There were no debates, no commercials, and no public or private bashing of opponents. If there was anything bad to be revealed about a presidential contender, it was up to the press to find it.

Moreover, even when the political bosses were hesitant about their choices, as they usually were up to and including John F. Kennedy, they always presented a united front to the public. No Democrat in 1960 whispered about possible philandering, and no Senate colleague pointed out that Kennedy was rarely around and rarely involved.

It was newspapers that charted voting records, collected negative information, and printed what was deemed proper for the public to know. Those were the years when the Walter Lippmanns and Drew Pearsons and even such scolds as Westbrook Pegler were relied on for inside information.

The primary system introduced a new element to nomination battles, and negative campaigning, which newspapers only reported, rarely instigated. George Bush called Ronald Reagan's 1980 campaign theme "voodoo economics," then squirmed when Reagan selected him as his running mate. Veteran Democrats squealed on Gary Hart's woman problem in 1984. It was the staff of the 1988 Democratic nominee Michael Dukakis that crushed Joseph Biden's primary campaign by leaking that Biden had delivered a plagiarized speech. In the same year, Robert Dole called Bush a liar. And most of this happened live on television. Newspapers were around to document and disseminate the exposures, but they weren't that important in discovering them.

Yet from 1972 to 1982, Royko alone changed the course of several local and state elections, and meddled happily in many others. His first victim was State's Attorney Edward V. Hanrahan, the pugnacious leader and defender of the 1969 raid on the Black Panther apartment where two men were killed by police. In the fall of 1971, Hanrahan went dutifully before Daley's slate-making committee to ask for another term as state's attorney. Slate-making was a hallowed tradition in Chicago politics. A group of seemingly powerfully ward and state committeemen would gather in private at either the Bismarck, LaSalle, or Sherman House Hotel, depending where Daley had placed his office as chairman of the Cook County Democratic party. They would interview candidates for a few days and then emerge and announce whoever Daley had picked to run for state and local offices.

Royko wrote a column in 1968 explaining who the slate makers were:

> John D'Arco, the well-known 1st Ward boss, was a slate maker. It has been said that D'Arco takes orders from people like Sam Giancana. That's a terrible thing to say about anyone, but D'Arco has not sued anyone for saying it.
>
> Bernie Neistein, like D'Arco, has been known to sip tea with the kind of people who hold hats over their faces when they see a camera....
>
> In the days before he got to be alderman and a party slate maker, Joe Burke was a singing waiter in the South Side's Canaryville neighborhood. Burke is now acknowledged by most serious students of local government as having attended more wakes than any other alderman. Alderman Thomas Keane was on the slate-making committee. He is famous for many things, including making the city's alleys safe for real estate developers.

The slate makers did what Daley told them, but the fallout from the Panther raid was too much even for Daley, who usually defended his cronies until the cell door closed. Hanrahan was dumped, and a faceless judge named Raymond Berg was installed as the party's endorsed candidate. But Hanrahan did the unthinkable. He defied Daley and ran in the March primary against Berg. He won. It was the same primary in which Dan Walker, a former general counsel of Montgomery Ward & Company, who authored a 1968 report on the convention violence calling it a "police riot," defeated Daley's candidate for the gubernatorial nomination, Paul Simon, who would reemerge years later as a U.S. senator. Royko chortled over the machine's disaster:

> I believe the machine is beginning to collapse because it has been run in a very old-fashioned, unbusinesslike manner. It can't rely on old-timers and their sons and cronies. The Parky Cullertons and the Vito Marzullos would be laughed off the debating platform by the new 18-year-old voters. Little wonder that Dan Walker came on like a member of the Kennedy clan.
>
> On Tuesday night, Mayor Daley was surrounded by people such as Charlie Swibel, the oily real estate opportunist; Matt Danaher, who has been a protégé for about 30 years and looks it; and the mayor's sons, Curly, Larry and Moe.

Hanrahan was the favorite in the November general election against Bernard Carey, a cherubic Republican who was undermanned and underfinanced. On the Wednesday before the election, with the polls showing Hanrahan leading, Royko wrote a column about Hanrahan meeting privately with the same John D'Arco that Royko had labeled a stooge for crime-syndicate boss Sam Giancana. The column was devastating, inferring that Hanrahan, chief law-enforcement officer of Cook County, was cutting a deal with the crime syndicate to ensure his reelection.

F. Richard Ciccone | 218

Hanrahan's campaign was over. On Election Day, underdog
Bernie Carey was elected, as was Governor Dan Walker, defeating the
politician that Royko had known longer and admired more than anyone, Governor Richard B. Ogilvie. Ogilvie had been elected Cook
County sheriff in 1962 when Royko was covering the county beat.
The two became good friends: golfing, lunching, and dining together.
Ogilvie was a good source, and Royko was a hungry reporter, one
who either violated his own creed about becoming close to politicians or was perhaps too young to have made it.

Ogilvie had been elected governor in 1968, and Royko never
picked on him in four years, which said more about Ogilvie's
integrity than Royko's favoritism. Royko avoided the gubernatorial
campaign in his columns, perhaps fearful that his admiration for
Ogilvie would reveal itself. He never swiped at Walker, whom he
considered a "goo-goo," his favorite tag for reform liberals, and who
won the primary after walking around Illinois with a red bandanna
wrapped around his neck, a strange wardrobe for a former corporate
lawyer who had never been wrapped in anything but pinstripes.

If ever it was clear that Royko could like politicians, it was in the
column he wrote about Ogilvie's postelection news conference.

It would have been understandable, had he been late. But that's not
the way Richard B. Ogilvie handles unpleasant chores. When they
come up, he faces them. He's a firm believer that the buck has to stop
somewhere, so make it here. So he came in and it was a poignant
thing to see. He doesn't have that big toothy smile of today's
upward-bound politician. By his nature, he is reserved. Also, he
caught some shrapnel in his face while running a tank in World War
II, and that didn't help his smile....

A few months ago, he and I talked about this thing, charisma. He
said a man needed personal magnetism to keep moving up in politics. I argued that today's voters are smart enough to look beyond a

man's style, his looks. He was right. He knew that flashy tactics, such as walking the length of the state, can be more effective than a creative public health program....

So the calm Scotsman with the wry smile tried to make it with an honest approach and excellent performance. It wasn't quite enough. The voters weren't buying it, and this time the voters were wrong.

But Richard Ogilvie wasn't Royko's favorite politician. Phil Krone was:

a clean-cut sort, is not the kind of person who could ever be mistaken for a lst Ward committeeman. He used to be a school teacher. Most lst Ward politicians used to be bookies. He doesn't spit olive pits. He doesn't smoke, drink or swear. I'm not sure how we got to be friends.

Krone has been active in politics since he was 12. He became a Republican because he doubted that Parky, Paddy, Tom Keane and the rest of them would be interested in the view of a teen-aged reformer...He once ran for the legislature, his slogan being: "There is no substitute for integrity and ability." At that time, I wrote: "Oh, yeah? How about money, clout or a famous name? And I predicted he would lose, which he did.

Phil Krone was never personally successful at electoral politics, but he was the person who most influenced Royko's take on Chicago politics and provided him with all the facts and figures that helped make Royko's political columns and election-night television forecasts so accurate. Krone was probably responsible for more column ideas than any of Royko's individual legpersons.

Krone first ran for office as a twenty-one-year-old seeking a seat in the City Council in 1963. He lost to a Daley disciple, but not for lack of campaign funds. "Paddy Bauler gave me thousands of dollars for my campaign. He loved reformers," Krone recalled. It was in Bauler's saloon that he first ran into Royko.

When Royko got his column, Krone became a frequent visitor to his office, and a more frequent caller. "I usually called around 4:30 in the afternoon and he'd growl, 'Whadda ya want? If you don't have an idea for me, hang up.'

"I'm probably responsible for maybe 100 columns that came up at 4:30 in the afternoon. When I first met him he was an irascible reporter. He became the most important person in my life. He was my 'Chinaman,' he was always there for me. I wasn't a golfer or a softball player or a drinker so I was a different kind of friend. We'd have dinner every couple of weeks in the early 70s and talk politics. Sometimes, for some of his columns, I'd play Daley and he would interview me and I would say Daley kind of things and then he would write a column. It was hilarious.

"I used to tell him, 'I hate trading on your name, but I do it.' He would laugh. We would just sit and come up with ideas and he would make you feel like you were the brightest guy he ever knew, even if you knew you weren't. He was incredibly loyal to friends," Krone said. "He was a major figure in my life, one of the few people who could really depress me. When he was down, he was really down and when he was up, he was really up."

Over the years, Krone's access to Royko made him a highly influential media consultant. Although he aligned himself with such reform groups as the Independent Voters of Illinois early in his career, he moved among regular Democrats and Republicans with ease and his clients included Neil Hartigan, who served as lieutenant governor, attorney general, and narrowly lost a 1982 bid for governor; Thomas Hynes, Cook County assessor; and Richard M. Daley, who Krone cultivated during the younger Daley's apprenticeship as a state senator, and whose eventual ascent to the mayor's office made Krone one of the city's insiders. But Krone is an indefatigable gadfly, and his acquaintances also ran the political scale down to John D'Arco, the lst Ward Committeemen that Royko loved to ridicule—and filmmakers such as Oliver Stone.

"On election nights, Royko always had my numbers and he would call from Channel 5 or whatever station he was on and I would give him the numbers from the ward committeemen, the numbers that nobody had. It was funny, these old committeemen got a kick out of helping Royko," Krone said. In 1972, Krone and Royko wound up on the same side of an issue that found Royko at his most contrarian, stunning some of his most fervent admirers.

The Democrats had routinely elected fifty-nine delegates to the Democratic National Convention scheduled in August in Miami. The delegates were elected under new rules that had been approved by a committee headed by Senator George McGovern of South Dakota, who, coincidentally, had become the darling of the liberals and the front-runner for the party's nomination. McGovern moved to the front after one of the first causalities of the new dynamic of primaries and media. Senator Edmund Muskie fell out of the 1972 campaign when, during a campaign appearance prior to the New Hampshire primary, he broke into tears when discussing an insult to his wife printed by a conservative publisher. The incident was televised. That bit of humanity allowed McGovern forces to characterize Muskie as a weak candidate, undeserving of the presidency.

The rules also were intended to eliminate the party bosses, and about the only boss left was Richard J. Daley. In Chicago, a group led by the Reverend Jesse Jackson and independent alderman William Singer handpicked their own slate of delegates in church basements. Their delegate choices violated party rules far more than the regular party slates. Yet the McGovern forces insisted that the Jackson-Singer group would be the one seated at the convention and Richard J. Daley, the man who elected Jack Kennedy in many minds, would stay home.

The decision to keep Richard J. Daley out of a Democratic

convention was greeted with euphoria by the liberals, who regarded
Royko as their champion spokesman. But Royko called it stupid. He
rose to the defense of the man he caricaturized in *Boss,* the man who
he personally blamed for failing to improve the condition of minorities in Chicago. It was typical of Royko to cut through the rhetoric
and argue that the wishes of the hundreds of thousands of people
who actually voted in the primary, and had elected the Daley delegates, were far more valid than the ideological wishes of the dissi-
dents. The Democratic party was losing its mind, if not its soul, in
usurping the wishes of the Chicago voters.

That August in Miami, Royko's columns were passed out to delegates by some of the regular Chicago Democrats who usually found
themselves embarrassed by his writing. When Royko went to Miami
he was verbally assaulted and abused by many liberals, who called
him a turncoat. He did what he usually did when confronted by dis-
satisfied readers. He ignored them.

Royko's other main gripe was with politically correct reporting
and the solicitous treatment of black political leaders by the establishment press. In October 1972, he wrote:

> It is time to add another chapter to the exciting political saga of Jesse
> (The Jetstream) Jackson, leader of men.
>
> Let us go back to the spring when the Jetstream saga begins. It is
> primary day, Democrats are going to the polls to decide who the del-
> egates to the 1972 convention will be. But not the Great Jetstream.
> He is too busy being a leader of men to bother to cast his vote for
> someone. The voting ends and they add up the ballots. Nowhere
> among the winners can be found the name of the Great Jetstream.
> Or even among the losers. That's because he was too busy being a
> leader of men to run as a delegate....
>
> Now we jump ahead to early summer. There we see the Great
> Jetstream, testifying that he should be a delegate. The Great Jet-

stream is saying that the mayor violated the new party guidelines by not having enough Chicanos, blacks, women, youths, students, blah, blah....

Now we leap forward to the new chapter and some recent events. The Great Jetstream is speaking again. He is saying that he will vote for McGovern, but he won't campaign for him. The Great Jetstream says that since the convention, McGovern has not called him, come around to see him, shown him the respect due a leader of men. That's no way to treat the Great Jetstream.

The column was a brilliant example of why Royko couldn't be called the voice of anyone. In his columns from Selma, Alabama, he certainly was not the voice of Chicago's racist ethnics, but in the Jackson column he made them roar with laughter. After the column was published, newsmen and television celebrities all over America suddenly began referring to Jackson as Jetstream. It was typical of Royko to choose such a wacky, wildly appropriate nickname to prick pomposity.

Royko loved handing out nicknames. Perhaps his most famous was his 1972 labeling of eccentric California governor Jerry Brown as "Moonbeam," a sobriquet that was still being used when Brown ran for mayor of Oakland, California, nearly thirty years later. He dubbed Daley "the Great Dumpling," and renamed Daley's sons after the Three Stooges. Jane Byrne had hardly been in office long enough to fire a couple of police chiefs and department heads before he, and all Chicago, were calling her "Mayor Bossy."

The other remarkable thing about the Jackson column was that Royko liked and admired Jackson. He even used a Jackson anecdote in *Boss*. Jackson told Royko how, as a newcomer to Chicago, he had been sent to see Daley, who told him he could have a bright future in the Democratic party and offered to help him get a job as a toll-booth collector.

Similarly, in 1984, when Jackson made the first legitimate bid by an African American for the Democratic presidential nomination, Royko praised his efforts and again turned the tables on his readers, this time appalling the ethnics and delighting the liberals.

As for Daley, he said, "The fact is, after I did *Boss*, I was kind of all Daleyed-out. After years of writing about him and doing a book on him, I just found it painfully hard to sit down and write about the guy. I used to turn down magazines that wanted me to write about him. When *Newsweek* started that contributing column they carry, "My Turn," they started out with great hopes of getting all the best known journalists in America to do it, and the editor contacted us and I said, 'Yeah, I'll do one for you if you suggest the idea for me, because I've got to think of five columns a week. I haven't got time to think of something for you.' And what do they think of? Daley. I said the hell with it. I am capable of writing about other things."

But, in February 1973, Daley gave Royko one of his most laughed-about columns. It was reported that Daley had turned hundreds of thousands of dollars of city insurance business over to an agency that employed one of his sons. The mayor held a news conference, and in his typically apoplectic manner, he invited anyone who saw anything wrong with his actions to kiss the mistletoe hanging from his jacket.

Several theories have arisen as to what Mayor Daley really meant a few days ago when he said: "If they don't like it they can kiss my ass." On the surface, it appeared that the mayor was merely admonishing some of those who would dare question the royal favors he has bestowed upon his fine sons, Prince Curly, Prince Larry and Prince Moe. But it can be a mistake to accept the superficial meaning of anything the mayor says.

One theory is that he would like to become sort of the Blarney Stone of Chicago. As the stone's legend goes, if a person kisses Ireland's famous Blarney Stone, which actually exists, he will be endowed with the gift of oratory. People from all over the world visit Blarney Castle so they can kiss the chunk of old limestone and thus become glib, convincing talkers.

So, too, might people flock to Chicago in hopes that kissing "The Daley" might bring them unearned wealth. Daley, or at least his bottom, might become one of the great tourist attractions of the nation.

After that column, Royko's phone rang all day with calls from irate Irish readers. One woman said the column offended her. Royko said, "Did the mayor saying 'kiss my ass,' offend you?"

"No," the woman replied.

"Well then, kiss my ass," he said as he hung up.

But the day of December 21, 1976, was not a day to be contrary or triumphant or funny. It was a day most Chicagoans were saddened by the death of Richard J. Daley, for better or worse, the man who had shaped and symbolized their city for twenty-one years. And Royko's farewell was neither sentimental, revisionist, nor spiteful. His clarity caught the mood of those who adored or despised the mayor:

> If a man ever reflected his city, it was Richard J. Daley of Chicago. In some ways, he was this town at its best—strong, hard-driving, working feverishly, pushing, building, driven by ambitions so big they seemed Texas-boastful.
>
> In other ways, he was this city at its worst—arrogant, crude, conniving, ruthless, suspicious, intolerant. He wasn't graceful, suave, witty or smooth. But, then, this is not Paris or San Francisco. He was raucous, sentimental, hot-tempered, practical, simple, devious, big, and powerful. This is, after all, Chicago....
>
> Daley was not an articulate man, most English teachers would agree. People from other parts of the country sometimes marveled

that a politician who fractured the language so thoroughly could be taken so seriously.

So when Daley slid sideways into a sentence, or didn't exit from the same paragraph he entered, it amused us. But it didn't sound that different from the way most of us talk.

For the next year, Royko would flare up when people suggested that his own performance would suffer now that Daley was gone. Royko and Daley had, without doubt, a symbiotic relationship that Royko never acknowledged. There might have been the same political misdeeds, the same avaricious characters, and the same large and small corruption that Royko uncovered throughout his career, but Richard J. Daley was the face for it. He was the personification of ruthlessness, suspicion, and intolerance. Yet from 1963 until Daley's death in 1976, less than 100 of Royko's nearly 3,000 columns mentioned Daley directly. For anyone to suggest that Daley was the only subject Royko could tackle was insult enough to have Royko lean close and say, "Chum."

Royko wrote dozens of columns about Watergate, most of them filled with humorous satire. On May 1, 1973, he described a Nixon television appearance: "I want to be the first person to congratulate President Nixon for his bold new effort to achieve burglary with honor. If anybody can do it, he can. And he is off to a rousing start."

He invented conversations between Nixon and the Founding Fathers to celebrate July 4, 1974:

JEFFERSON: "I have never been able to conceive how any rational being could propose happiness in himself from the exercise of power over others."

NIXON: I want the most comprehensive notes on those who tried to do us in.

FRANKLIN: "Half the truth is often a great lie."

NIXON: (To Ron Ziegler) Just get out there and act like your usual cocky, confident self.

HANCOCK: "Some boast of being friends to government! I am a friend to righteous government, to a government founded on the principles of reason and justice."

NIXON: Expletive deleted.

Royko argued against any criminal punishment for Nixon, raising as he had in his farewell column to Lyndon Johnson and his piece on the murder of Martin Luther King the collective responsibility of the American public, a theme he never tired of using:

> My personal reason for not wanting Mr. Nixon prosecuted is that he really didn't betray the nation's trust all that badly.
>
> The country knew what it was getting when it made him president. He was elected by the darker side of the American conscience. His job was to put the brakes on the changes of the 1960s—the growing belief in the individual liberties, the push forward by minority groups. He campaigned by appealing to prejudice and suspicion. What he and his followers meant by law and order was "shut up."
>
> He did his job. Let him quit and go.

Watergate did more than give Royko column material for more than a year, it was a watershed in American journalism. Carl Bernstein and Robert Woodward's reporting in the *Washington Post* inspired college students in journalism schools everywhere. Royko watched as newsrooms suddenly filled with young people who wanted to be stars. If some of the reporters of Royko's era were lapdog journalists, a new breed, the junkyard-dog journalist, was emerging. Adversarial reporting began to replace access journalism. Public figures became targets whose downfall could elevate a reporter into a television personality.

While Woodward and Bernstein had done as much as any American journalists in history to uncover the biggest constitutional crisis of the presidency, they also sowed the seeds of a new era in which politicians would defend themselves from journalistic excesses with a battalion of consultants, media experts, and spin doctors. The off-the-record conversations that allowed reporters to explain the mechanics of government vanished for two reasons. Politicians in the 1980s and 1990s could no longer trust journalists to keep anything off the record, and journalists lost interest in the mechanics of government. Now journalists wanted flashy stories. Important, serious stories, such as the scandal in the Department of Housing and Urban Development or the $500-billion disaster in the savings-and-loan industry, were routinely buried by editors and networks who thought them boring compared to the travails of Donald Trump.

But even in the midst of Watergate, Royko made national headlines. In May 1973, Vice President Spiro Agnew, who had become a darling of conservatives for his media-bashing, gave an especially nasty speech about welfare recipients. Royko tore him apart. "I just peeled his goddamn hide," he later said.

Agnew asked for a meeting with Royko. "We sat there for half an hour calling each other assholes. I told him exactly what kind of a shit I thought he was." But Royko knew the value of having an exclusive session with the vice president. He knew that an interview with the vice president, who had turned down similar requests from everyone, could stand on its own. Agnew's responses were far from antagonistic. The vice president was empathetic with draft resisters and war protesters and less doctrinaire than his national image. The Associated Press picked up both of Royko's Agnew columns, and they were carried in hundreds of newspapers that normally did not have access to Royko though the *Daily News* syndicate.

The legman who handled the arrangements for the Royko-Agnew sit-down was Paul O'Connor. O'Connor was twenty-five at the time, and had been working for Royko for less than a year. "He found me where he found everyone else, City News Bureau," O'Connor recalled. O'Connor was the son of Len O'Connor, the WMAQ-TV political commentator who Royko knew from the days they both frequented Paddy Bauler's saloon. "He may have hired me because of my dad. He just told people he hired me because he couldn't find anybody smarter. That was his story and he always built up whoever worked for him." Like most of the other early legpersons, O'Connor was in awe of Royko.

"I sat in front of his cubicle, which to me was the tabernacle of fear. I never knew what he might say or do. When I saw him enter the newsroom at the far corner I would put on my headset and call the weather number or something so he would think I was talking on the telephone and ignore me," O'Connor said.

"The issue was: never let him down. I had two primary functions: check everything and then reduce it all to a double-spaced memo. Then he would take you to graduate school. He had a City News mentality. He'd ask questions until he got one you couldn't answer and then he'd make you go back and get the answer. After you got better at it and he began to trust you, he'd leave notes like, 'deputy coroner so-and-so—asshole.' That was it. So I'd try to find out everything I could about the asshole without knowing why he was an asshole. It was great education.

"He wanted to educate me because I grew up in Winnetka and I wasn't a Chicago kid. He'd send me to the housing projects. Once a rich oilman read one of his columns about life in the projects and sent a big check for the family Royko had written about. He sent me to deliver it and I found the family, supposedly on welfare, living in luxury. When I told him, he said he wanted me to learn that everybody was on the make," O'Connor said.

"What really distinguished him as a reporter was that he believed you could get Mother Teresa. She's a human being, so you can get her."

In early December 1973, O'Connor got a telephone call from an accountant in the western suburbs who thought Royko could do something about an unbelievable bureaucratic mess involving the denial of veterans' benefits to a young man who had been wounded in Vietnam. It was O'Connor's double spaced memo that turned into Royko's most celebrated column. It ran on December 10, 1973.

> Leroy Bailey had just turned twenty-one. He was one of seven kids from a broken family in Connecticut. He had been in the infantry in Vietnam for only one month. Then the rocket tore through the roof of his tent while he was sleeping and exploded in his face.
>
> He was alive when the medics pulled him out. But he was blind. And his face was gone. It's the simplest way to describe it: He no longer had a face.

Royko described Bailey's horrible condition, including his inability to eat solid food. He had to use a large syringe to squirt liquid food down his throat. He sat mostly in the basement of a home in suburban LaGrange, knitting hats that a friend sold for him. The home belonged to his brother, who moved from Connecticut to be near Bailey while he was in Hines Veterans Hospital west of Chicago. A surgeon at Chicago's Mercy Hospital told Bailey he could reconstruct his face so that he could eat solid food, but it would require six operations. Bailey eagerly agreed, and after the first was performed, the bill was sent to the Veterans Administration, which replied, "Any expense involved for this condition must be a personal transaction between you and your doctor."

Royko wrote:

> Until he was hit by a rocket, Bailey had teeth. Now he has none. He had eyes. Now he has none. He had a nose. Now he has none. People could look at him. Now most of them turn away. Bailey believes the

VA thinks he wants the surgery just to look better, that it is "cosmetic surgery."

Even if that were so, then why the hell not? If we can afford $5,000,000 to make Richard Nixon's San Clemente property prettier, we can do whatever is humanly possibly for this man's face....If his appeal is turned down...not even once will he be able to sit down and eat at the dinner table with his brother's family, before going back down to the basement to knit hats.

The headline of the next day's *Chicago Daily News* read: NIXON READS ROYKO'S COLUMN; ORDERS VA TO AID FACELESS VET.

Nixon also wrote to Leroy Bailey and said, "I have been personally concerned with your situation since I first learned of it through a column in the Chicago *Daily News*..."

The story was front page all over the country, and Royko was a finalist for the Robert F. Kennedy Award for journalism. Paul O'Connor was besieged with requests from the other Chicago newspapers, television networks, the *New York Times,* and just about everyone else in the media. They wanted to know where to find Leroy Bailey.

"We had the source locked up. It was an accountant who had gone to Leroy's brother's house and saw him in the basement knitting hats. He gathered up all the paperwork. He was meticulous, and he gave us all the army records and all the VA medical reports. We told him, 'You button up,' and we'll do the story. Mike could have milked the story for weeks but after the third column—he had a rule to never do more than three columns on any subject—we turned the accountant loose and the rest of the media finally got a piece of the story, but it was over by then.

"He owned that story, but it's hard to say how much satisfaction he got out of it. I never heard him talk about that column again," O'Connor said. "He was always worried about the next one."

As far as elections go, the next one where Royko had a major impact was in 1978, when U.S. Senator Charles Percy's bid for a third term was stumbling toward defeat. Historically, third terms are the most difficult for U.S. senators; they've been around for twelve years and voters are usually willing to make a change. In Percy's case, he hadn't been around enough to suit the suburban Republicans who formed his political base. Many of the GOP leaders were griping they hadn't seen Percy since his last election. And his Democratic opponent, Alex Seith, was running a hard campaign.

Royko wasn't particularly fond of Percy, a former business wunderkind who headed Bell & Howell at age twenty-nine, but he liked his brand of Republicanism, which was moderate, socially responsible, and not hawkish on foreign excursions such as Vietnam. The fact that Percy was one Republican the *Tribune* continually grumbled about made him even more attractive to Royko. Seith, on the other hand, seemed to Royko an opportunist, a pinstripe lawyer with solid intellectual credentials who was rolling about with the remnants of the Daley machine in exchange for their support.

A week before the 1978 election Royko wrote a metaphoric column which compared Nice Norbert, who was Percy, and Bad Russell, who was Seith. He said, "On the surface, Seith, too, seems to be a gentleman. He went to an Ivy League college, is wealthy, dabbles in foreign affairs and has a good tailor. But there is a definite streak of Bad Russell in Seith. As a political campaigner, he is an alley fighter. His TV and radio commercials—and that's all the campaign consists of—are among the nastiest and most deceptive that I've ever seen."

The desperate Percy campaign got permission to run the column as a paid advertisement in the *Sun-Times* and *Tribune*. A few days before the election, Royko struck again with an old theme. He reported that Seith had been a character witness at the trial of Paul Marcy—the brother of Pat Marcy—who was the First Ward Committeeman and mob chief Sam Giancana's link to city hall. Marcy was convicted of taking a $55,000 bribe from a builder who needed a zoning case fixed. When he took the money, Marcy was secretary of the Cook County Zoning Board of Appeals.

"The chairman of that board," Royko wrote, "is Alex Seith." Royko reported that Seith had testified that he had always found Marcy to be truthful.

"I'm sure Seith thinks it is not sporting of me to mention that he was a character witness for Marcy. But there is an old saying in politics: If you lie down with dogs, you get up with fleas. Start scratching, Mr. Seith."

A few days later, Percy was reelected.

While Krone and Royko took credit for salvaging Percy's senate seat, the final margin was sufficiently safe to conclude that some of the Republicans vowing to get even for Percy's aloofness stayed in the GOP column. But there is no question that Royko gave the impetus to whatever last-moment dynamics sent Percy back to Washington.

Krone's influence on Royko was never greater than when Royko resurrected the mayor's oldest son from the churlish Prince Curly to Richard M. Daley, statesman. Royko had mocked the younger Daleys for everything from ducking the Vietnam War to profiteering and mopery. Only two days after Daley's death, Royko noted, "Young Richard isn't much of a charmer. He is considered something of a bully and doesn't make much effort to hide his arrogance....Had he been a pleasant, humble, respectful young prince, he might now be able to exploit the magic of his name and become a fixture in Chicago politics. But that's unlikely."

Yet the person who most made it likely was Royko. After Daley, the machine chugged on. Daley's own Eleventh Ward alderman, Michael Bilandic, became interim mayor, then filled out the remainder of Daley's term until 1979. But as Royko had predicted after the 1972 elections, the machine was wobbling on its last legs. Individual committeemen were balking at taking orders; reformers were winning elections; and black politicians were making sounds about representing their constituents. New faces, such as Alderman Edward R. Vrdolyak of the Tenth Ward, were moving toward the centers of power. Feeling left out, Daley loyalist Jane Byrne charged that City Hall was run by a "cabal of evil men" that included Vrdolyak and Alderman Edward Burke, the son of the Canaryville singing waiter. Byrne declared she would run against Bilandic in 1979, and Royko cheered her on, begging Chicagoans to finally break the machine's hold on the city. Royko's columns wouldn't have helped without the blizzard that crippled Chicago in January. Mayor Bilandic delivered a number of silly public announcements, utterances that Daley could have made with political immunity, but which sounded ludicrous from the virtually unknown Bilandic.

Chicago was also a changed city. Although its voters had habitually handed Daley his landslide elections, and still routinely elected Democrats in the city and county, the sanctity of the machine was dissolving. With most young Americans heading for college, it was no longer necessary to stay on the good side of the ward committeeman in case Junior needed a job at the Hall. In the aftermath of Vietnam and Watergate, government was seen as an evil necessity, not an opportunity. The petty corruption of traffic cops taking a few bucks to give a speeder a pass was viewed as a crime, not as a benevolent civic perk. Lyndon Johnson's Great Society had federalized every form of government welfare, and no one needed City Hall for his Christmas turkey. Television had long ago replaced precinct captains as the regular visitor in Chicago's living rooms. While Daley was alive, it mattered little which of his opponents ran a few commercials or got free time on the evening news. With Daley gone, Chicago paid attention to the shrill blonde lady they called the Snow Queen.

On February 28, 1979, Jane Byrne was elected mayor. Royko didn't take the credit, he gave it:

It was the most stunning upset in the long, wild history of Chicago politics and this column is about the single most important person involved in that incredible upset—the remarkable individual who made it happen.

And who would that be?

No, I'm not talking about some brilliant campaign manager, or media manipulator, or generous back-room financier, or any of the other political operatives who usually get top billing in day-after-election stories.

And, no, it isn't about Jane Byrne, although little Ms. Sourpuss finally has something to smile about.

This column is about you. That's right—YOU there, on the L train or bus, or in your kitchen reading this over morning coffee. You, at your punch press, or in your firehouse, or hospital cafeteria. You, behind the counter in the department store, or jockeying the cab or unloading that truck.

You did it, you wild and crazy Chicagoans.

Royko's euphoria was real but short-lived. In a matter of days, Byrne began firing department heads and surrounding herself with her own cabal, including many of those she had once denounced as "evil." She began acting more like a boss than Daley ever did. With Byrne, Royko got an even more intolerant machine. Perhaps it was that betrayal that encouraged Royko to support young Richard M. Daley. Regardless, Byrne vowed that neither Daley nor his family would regain their privilege or power.

In 1980, Byrne nudged Alderman Burke, now a member of her inner circle, into the race for state's attorney against Republican Bernard Carey, who was seeking his third term. Although 1980 was looking very much like a big Republican year with an embattled Jimmy Carter hiding in the Rose Garden, Carey also seemed vulnerable, so Byrne was planning on recouping the state's attorney's office before it caused her or her friends any embarrassments, not to mention indictments.

Just before the December filing date in 1979, a new challenger appeared. Richard M. Daley was going to run in the primary against Burke. Daley was anathema to Byrne and all the other ambitious Democrats who had assumed he was no longer a roadblock to their schemes for higher office. Daley was the underdog. But Daley had Royko. Royko was relentless in attacking Byrne's heavy-handed tactics, her coterie of advisers, whom he labeled "the Sleaziest," and her reluctance to cut away the mayor's ties to the notorious First Ward crime-syndicate characters. He ripped Burke and praised Daley. Daley won.

In the fall general-election campaign, Royko suddenly discovered the guy he helped put in the state's attorney's office in 1972, Bernie Carey, had made a mess of the job. In a succession of columns he blamed Carey for allowing mass killer John Wayne Gacy to roam the streets for a year. Carey's office had declined to prosecute Gacy, who had been charged with rape. Day after day, Royko disclosed new cases where Carey's office dropped charges against child molesters and worse. He ripped Carey's television commercials that claimed he was "the best prosecutor in the country."

On the Sunday before the election, Carey said that "Royko has attacked me viciously about five or six times and I think he has two more columns to go before Election Day so we can expect two more columns of that variety."

Carey was wrong. Royko bashed him on Monday and Tuesday, then wrote a column for the tiny edition, less than a 10,000 press run, that went to the train and El stations. He wanted one last chance at the commuters heading home to vote before the 7 P.M. poll closing.

Daley won. "I think," Krone said, "that altogether he did seventeen columns knocking Bernie Carey. Not only did he write a special column just for the commuters, he even ripped him the day after the election.

"He certainly was the most influential political writer of his time. He toppled Bilandic. I told him to stop fooling around or that Jane Byrne would get elected. He didn't care and he got her elected. I don't think he realized what she was going to be. He really didn't care. He loved to play."

But Royko couldn't get Mario Cuomo to play when he tried to convince the then-New York governor to enter the 1992 presidential race. Royko enlisted Krone and another savvy Chicago political operative, Don Rose, who was the guru behind both Bernie Carey and Jane Byrne when Royko supported them.

Royko sent them off to the snows of New Hampshire in January 1992 to form a draft-Cuomo committee and to provide fodder for several columns, not that he needed any material, because he had long been convinced that Cuomo was not only the nation's best orator, but the kind of liberal who remembered when liberals cared about little people.

In November 1991, Royko leaped into the fray when Senator Phil Gramm, a Republican conservative from Texas, downplayed Cuomo's presidential chances by declaring, "We don't have too many Marios in Texas."

Royko asked, "If Cuomo is a candidate, would Republicans be able to chant, 'Mario, Mario, Mario, and frighten millions of American voters into believing that with a president named Mario they will be forced to listen to grand opera and eat garlic?"

In December, Royko lashed out at Ron Brown, the chairman of the Democratic National Committee, who said that if Cuomo wasn't going to be a candidate he should drop out of the race.

"He is dumb," Royko wrote, and in the same column announced that Krone was starting a write-in campaign for Cuomo in New Hampshire. Royko pretended to be dubious, asking, "Who outside of

Chicago has ever heard of Phil Krone?" That let him put the following words in Krone's mouth. "Who the hell has ever heard of Ron
Brown?"

By the end of 1991, the Royko-Krone conspiracy produced a column that included the new address and a 1–800 number for the
Cuomo write-in campaign, with the promise that for ten dollars anyone could get a Cuomo bumper sticker. Krone explained that the
money would go for a mailing of all registered Democrats in New
Hampshire. "But the mailing will cost about $15,000 and right now
I'm $15,000 short of that goal." Royko used the rest of the column to
brazenly solicit funds for the Cuomo campaign, something he had
never done before.

The *Chicago Tribune* sent reporter Charles Madigan to the Krone-
Rose office in Manchester, New Hampshire. Madigan wrote, "There
is a whole pack of Chicagoans, led by veteran political operatives Don
Rose and Phil Krone, working in an office on the second floor above
a luggage store on Main Street. The essence of Chicago columnist
Mike Royko is here, too, but not the guy himself. There are lots of
copies of Royko columns in which he suggests, among other things,
that Democratic National Chairman Ron Brown should stick his
head in an oven, or at least resign.

"There have been many responses, from Illinois and elsewhere,
with $10 contributions attached to notes and letters. Rose says the
campaign is looking at two possible scenarios. 'One is that he finds
some resolution to his budget problem by April and he decides to get
in, and then his own people carry it from there, figuring out how to
get him into the primaries,' Rose said.

"'The other is that he doesn't do that and we go along primary to
primary with uncommitted delegates. We pick up a couple hundred
of them, and then we go to the convention. Nobody would have it
(the nomination) at that point, and so we would have, in effect, a brokered convention, but one brokered from the bottom on up.'"

Krone read letters to Madigan. "From Peoria: 'Mario! Mario! Mario! Mario! It's time the Democrats took back the White House, and every Democrat knows Mario Cuomo is the only one who can do it. You must not let us down because of the sleazy, sly...tactics.'"

Even those who disliked Royko were for Cuomo: One Chicagoan wrote, "Let's get a few things straight from the beginning. Royko is a sexist pig exceeded only by Bush....Cuomo has what it takes to be president."

But the air was taken out of the Cuomo boomlet by the revelations regarding Arkansas governor William Clinton's sex life and Clinton's subsequent *60 Minutes* television show mea culpa that boosted him into second place in New Hampshire and set the stage for his presidential nomination.

Royko's fling with king-making at the national level was over. And he had no intention of dining with kings, either.

On December 10, 1991, Janan Hanna, Royko's fourteenth legman, rushed into Royko's office and gasped, "Do you know who's downstairs in the Billy Goat? The President! President Bush is down there and he's asking for you. Sam Sianis just called. He said that that was the first thing Bush asked—if you were there. And he wanted to know at which part of the bar you usually sit."

Royko was watching the William Kennedy Smith rape trial on television. He sent Hanna to the Billy Goat, and she reported back, "The people Bush ate with said that they talked about the economy, the Soviet Union, the homeless, and AIDS. They said that Bush told them the economy was getting better, but it would take some time."

Royko explained in the next day's column why he preferred watching the Willie Smith trial to chatting with Bush: "I wanted to explain in advance, in case someone rapped me for snubbing the prez. It had nothing to do with politics. It was news judgment. I already know about the economy but I didn't know about the pantyhose."

On one wall of Royko's office was a sign: The Eagle Does Not

Hunt Flies. His acolytes gazed on it approvingly, interpreting it as a creed of the omnipotent journalist. Maybe it was, but it was also a reminder to Royko, who did hunt flies early in his career. He wrote columns that cost lowly city inspectors or park district truck drivers their jobs. He later reformed, admitting that he realized he wasn't changing the system but only hurting the families of people making a couple of hundred bucks a week.

He even confessed that he was not opposed to bribery, provided the guy who takes one supplies good service. "The advantage of a bribe is that it cuts through red tape." He often recounted the case in which he gave a city inspector a pass after his legman watched the man take $200. "You've got two hours to put the money in an envelope and give it back. And if I ever hear of you taking one cent again, I'll see that you go to prison," Royko told the inspector.

"If I was the city editor and a reporter came to me with a story like that, I would tell the reporter to get the facts and write the story. That's the way it should be. But I am not the city editor. This is a personal column. I make my own rules. My rules are that I can play God if I want to."

Playing God, snubbing presidents, or trying to elect Chicago mayors wasn't the soul of Royko's political coverage. Royko's brilliance was in the word pictures he painted of a breed that he grew up with and watched disappear in a wave of reform and cultural change. None was better than his piece on Twenty-seventhth Ward committeeman Edward Quigley:

> The first time I met Big Eddie Quigley was 21 years ago. He had just been given the lofty job of chief of the city's sewers. I found that sewer chiefs dressed well. He had on a silk suit, a silk shirt, a silk tie, a diamond pinky ring, diamond cufflinks, and a jewel-studded wrist watch.
>
> So I asked him: "Did you ever actually work down in the sewers?"
>
> His answer was a classic. It summed up the public service careers

of the many ward bosses, aldermen and other forms of political wildlife who sat atop the Chicago Machine.

"No," he said. "I never was in the sewers. But many a time I lifted a lid to see if they was flowin'...."

Hardly an election has gone by without some of Quigley's precinct captains being put on trial for vote fraud. It's almost become a tradition. They hand out money to the poor, wine to the shaky. If they don't have enough warm bodies in the precinct, they just sit down and start yanking the vote lever like a slot machine.

Quigley hustled votes with the best of them. No Skid Row wino was too drunk to not vote at least twice. No dearly departed was too deceased to not vote at least one more time.

"When it came to bringing back the dead," one of Quigley's old cronies said, "that Dr. Frankenstein was a piker compared to Big Eddie."

11

One of the most inventive and humorous passages in *Boss* came at the beginning when Royko used biblical terminology to describe nepotism in the machine. If he had adapted the trick to his leg creatures it would have read: Terry Shaffer begat John Foreman who begat Ellen Warren who begat Paul O'Connor who begat Wade Nelson who begat Hanke Gratteau who begat John Fennell who begat Patty Wingert who begat Helene McEntee who begat Lynn Terman who begat Paul Sullivan who begat Nancy Ryan who begat Paul Sullivan who begat Suzy Kuczcka who begat Janan Hanna who begat Pam Cytrnbaum who begat Suzy Fritsch.

Hanke Gratteau worked for Royko three different times at three different papers. Ellen Warren was the first woman he hired. These two women would become more than legpersons, and his affection for them led him to hire women almost exclusively in the ensuing years.

"He felt more at ease with women. I think women were his closest friends," said Lois Wille. "Hanke, Ellen, Gee Gee (Geyer)...He seemed much more able to confide in his women friends. Once Ellen worked for him he never wanted men again, although he did hire a few."

Warren, who later covered Washington and the war in Lebanon for Knight-Ridder newspapers, hounded Royko for her job. She was a native of Washington, D.C., who had scrounged a job at City News Bureau in 1969.

"My first day as a reporter I was sent out with a more experienced person who was going to show me the ropes. My first assignment was to cover Royko in traffic court where he was on trial for giving the bird to a police officer who had stopped him for running a light or a stop sign. He was most accommodating to the press which was mainly me and my handler. I remember being so nervous that while he was talking I fell backward over a chair in the pressroom," Warren said.

She later got a job at a suburban weekly in Wilmette on the North Shore and became a friend of Larry Green, Royko's fishing and handball buddy.

"Larry, Mike, and I were at Riccardo's one night, and I was making a pitch to work for him. Mike said, 'I would never hire a woman because I couldn't yell at her.' That was a Friday night. On Sunday, I dropped off my resume at his office with a note that said, 'I can stand the yelling.' I got an interview, and Mike loved to tell the story that I had a cup of coffee and was shaking so bad that I spilled it all over. I don't remember that but he loved to tell it."

Royko hired Ellen Warren in August 1970, while he was engrossed in the final words of *Boss*.

"Legman responsibilities and the relationship with him were very different than they were later. I was not Mike's friend. I didn't pal around or go drinking with him. He was my boss. I gave him a memo on everything. I never used any independent judgment on what might be column. I gave him everything. In later years he made fun of me for bothering him so much. As soon as he came in I'd be asking what I could do, and he said I made his head ache. I was too young to figure out that when his mouth was rimmed with Maalox it wasn't a good morning. I never wanted to be idle a moment, because I wanted to produce for this guy," Warren said.

"He was a terrible boss in the sense he never gave any direction or any feedback. Never said, 'Call so and so or so and so,' No way. So I used to give him long rambling memos with everything in there. I never exercised any judgment. The one interview I got which he loved was when Alderman Vito Marzullo told me Mike Royko had shit in his blood. He loved that.

"After a while I figured out what kind of questions to ask. Many times it was just as good if the guy on the other end was evasive or refused to talk. That worked just as well for Mike in a lot of cases."

One of Warren's most productive assignments was to bribe a driving examiner with a fifty-dollar bill to get a license. It stirred up one the many investigations of corruption in the Illinois secretary of state's office, and Warren recalls, "I drove on that license until I went to Washington."

Paul O'Connor succeeded Warren when she left Royko's world to become a general assignment writer. After two years, during which the Leroy Bailey story was published, he was followed by Wade Nelson, Royko's fifth legman.

"I was working at the *Aurora Beacon News,* and I had run out of hope of getting hired at any of the Chicago papers, which were looking for women and minorities. Royko told people he hired me because I was a twenty-eight-year-old white guy who would never have gotten hired at the *Daily News* unless he hired me. I only stayed nine months when he told me I was moving to the education beat. There was no career counseling, no discussion of what you might do. He'd just call and say you were going to the city desk."

In September 1975, Royko hired Hanke Gratteau. It was the first of three stints Gratteau did with Royko. She became the closest of his many "Dr. Watsons, Tontos, and Igors," as he called the legmen in the dedication to a 1983 anthology of columns, *Sez You, Sez Me.*

Gratteau had been a copy girl at the *Daily News* while attending classes at the University of Illinois–Chicago Circle. One of her assignments was to do a paper on Mayor Daley's old nemesis, Ben Adamowski, the former state's attorney who had given Daley his toughest mayoral election run in 1963. She gingerly walked to Royko's desk and asked if she could interview him. He growled and then graciously told her everything he knew.

"Later, he asked Lois Wille to help me get some assignments at the paper and I eventually got a job at City News Bureau. I was working four to midnight there and I'll always remember the day I got a telephone call from Mike. I had gone to the racetrack and bet on a horse named Rich Passions. I won $7.50 which was a lot of money. I was feeling flush and I was home in Des Plaines when the telephone rang."

"Gratteau?"

"Yeah."

"Royko. You going to stay in this business?"

"Yeah."

"You want to work for me?"

"Yeah."

"Can you start in two weeks?"

"Yeah."

When Gratteau went to work for Royko she was twenty-two, and he was forty-three. "I called him the old geezer. He was only twenty years older than me, but he really looked old. The column was taking its toll. He put in an incredible number of hours on it. And he was having problems at home.

"He was an amazing taskmaster. When you worked for Mike, you weren't just a reporter, you became many things. One of them was a news editor. You had to make decisions on every story that came in, not only what would make a great story, but what would work for him. Then you had to develop a thirst for quotes. The main thing you learned was to work incredibly hard. In by nine and go through the memos that Mike had left from the day before. He always left 'forget this,' 'work on this for later,' 'check this out for tomorrow,' memos. Before you could get started there were the calls from readers and then Mike would wander in and the first thing out of his mouth was, 'What do you got for me?' And, 'You better have something.'"

Like her predecessors, Gratteau went through a postgraduate course in the basics.

"You didn't tell him something unless you had pulled all the clips and checked the information. You didn't dare put a name in the column without checking the spelling in the phone book.

"The toughest part was you never knew who was going to be walking in (the office) in the morning. Was it the expansive Mike who would pull up a chair, put his feet up on the desk, light a cigarette, and say, 'Let's talk about the old man' or was he going to bark and throw you out of the office saying he didn't want to hear anything from you?

"There were bad days when he was really hung over, and as the years went by there were days when I knew he didn't have anything and was scared. But he wasn't the kind of person who would come out and say, 'I'm really worried I won't find something for tomorrow,' or 'I'm getting insecure over the column.' All he would say is, 'Get the hell out of here and don't come back until I tell you.'"

Royko didn't need his legmen as friends. The entire newspaper was his friend, or at least he thought so, and he continued his habits of his early column years, strolling the newsroom, trying a scenario out on various colleagues: Wille, Gilbreth, Bob Schultz, who was now the city editor, or younger pals like John Hahn or Larry Green. His closest friend on the paper, Bob Billings, had left the *Daily News* and had gone to work at *Playboy* magazine, took a turn at public relations work at city hall, and then started a holographic museum on the near West Side. Billings was somewhat of an urban pioneer, since the area had been blighted by its longtime proximity to skid row on the east and the ravaged areas of the Martin Luther King assassination rioting to the west. But skid row was changing, and Billings was among the first to recognize the possibilities of the area. It was to Billings's home on Washington Street that young Robbie Royko was sent on weekends. Undoubtedly the Roykos hoped that the stern visage of the hulking Billings would provide some discipline.

"Bob Billings showed me how to throw a fastball," Rob Royko recalled. "When I first threw him a baseball he said, 'You throw like a little girl.' Then he told me I had to get stronger so he made me push his Volkswagen up and down the block. I always liked him but he could be an arrogant person. He helped me later get a job with the city."

Unlike her predecessors, Gratteau was permitted to edge her way into Royko's personal life. "My husband was a drummer, and David wanted to learn how to play drums so we went over to Mike's house a couple of times. I met Carol but I didn't really get to know her. I saw her sometimes at the Goat after softball games, but I never knew her that well. All I knew was that Mike never quite got over the fact that Carol had married him; this incredible, lovely, prettiest girl in the neighborhood. He ended up marrying her and always thought it was astonishing.

"At the same time, here was this homely guy who had the power and stature as the most powerful columnist in town, which meant all sorts of women who wouldn't have been interested in him if he were a pipe fitter found him wonderful and that was pretty hard to turn down. He used to call them 'the young lovelies' and they were beating a path to his door."

On the day that Mayor Daley died, the *Daily News* city desk called Royko and asked if it could use Gratteau on Chicago's biggest story in years. "I was sent to Northwestern Hospital to do a ticktock, what happened minute by minute after Daley was rushed to the hospital. Mike also showed up at the hospital. A doctor recognized him and took him aside from all the rest of us and shook his head, so Mike was really the first reporter who knew Daley wasn't going to make it.

"I went back to the paper to write. I was really excited about the chance to get my first byline, and so I started writing and I had something like 'sirens blaring through the Loop' and that sort of stuff. Mike came and looked over my shoulder and said, 'You know, this is a pretty dramatic story. You don't need to add anything to it. Write it straight.'

"He was right. That was his gift, using those crisp declarative sentences."

There were other papers besides the *Daily News* that wanted to print Royko's words, especially in Washington. In April 1975, Royko met with Ben Bradlee, who was trying to convince him to come to the *Washington Post*, which, after Watergate, rivaled the *New York Times* as America's premier newspaper. Although Royko scoffed at the notion that he needed to be in Washington to be recognized as the nation's preeminent columnist, he knew there was more than a grain of truth to it. He took a train to Washington to be wined and dined, talking with Bradlee about how he would cover the Capitol. On April 9, 1975, Bradlee wrote:

Dear Mike:

It was good—for me—to see you, and to express the keenest interest in what I can imagine as Royko hitting the Washington runway running.

If you stick to your game plan, writing irreverently about the people in this town who are falsely revered, or who falsely revere themselves, you will be working a new vein. No one's working it now.

As for your concern that there might be too many stars here already—some prick wrote something about "Ben Bradlee's All-Star Revue" the other day—there is dissatisfaction only that today's paper could have been better, that we aren't as good as we are going to be, and we aren't scared to say it. I know our stars would like you to join them any way you want—and I haven't said a word about you to any of them. I think you'd like to work with them.

I gather Bellows has heard of our lunch (one of our young historians called him to ask if it was true that he had hired Bill McIlwain...and Bellows told him, "Why don't you ask Bradlee why he had lunch with Royko." So be it. Let's stay in touch.

All the best, Ben

"Bellows" was James Bellows, one of the few rivals to Bradlee as the best editor in the country. Bellows had been hired to rescue the *Washington Star*, once the Capitol's preeminent paper, which had been surpassed in reputation and circulation by the *Post* under the leadership of Bradlee and publisher Katherine Graham. Bellows was also very interested in Royko and had also made his pitch, which Royko had apparently discussed with former *Daily News* sports writer David Israel, now working for the *Star*. Israel then spoke with Bellows, prompting this letter of April 24, 1975, two weeks after Bradlee's correspondence.

Talked to David Israel some after he talked with you, and I thought I'd better clear things up.

Look, I'm ecstatic over the idea of Royko in Washington, writing in the Washington Star. Today, next week, next month, September, or even in 1976. The sooner, the better.

All I was saying in the last conversation was that I'd hate to see you worked over (running) in the Post on the CDN (Chicago Daily News) service before we get you here in The Star on whatever syndicate arrangement is worked out.

Clear? I hope so.

David is doing very well. And improving all the time.

Yeh, I know, Washington ain't Chicago, but it is the seat of the world's power, etc., etc. And there are a hell of a lot of good crony bars in the South Capitol area.

Enough. I'm still counting on you sooner or later. We'll run you as you've never been run before and we'll do more than that in promotion ...

Take care. Jim

Royko never went. "A lot of things came up. My wife didn't really want to move and my kids didn't want to go. My wife's parents would have had to be moved and I didn't want to do that to them. So I considered all that said, ah, fuck it," Royko said in a 1979 interview for *Esquire* magazine.

But again, he was bothered by people who said he passed up the chance because he might not be able to nail Washington the way he had Chicago. "That's just crap. My political instincts are as good as anybody's. It would take a little time to do my homework in Washington because I'd get out and do it myself. But there are just as many crooks and hustlers there as here, and I'd love to write about them. I was the first guy to nail Agnew after he was nominated. I got onto Carter before most anybody else. I knew Nixon was a crook. So I don't buy it."

In 1974, Bob Duckman, Carol's brother, died suddenly. "My dad was devastated. He just sobbed," David Royko said. Bob Duckman had been divorced from Royko's niece, Barbara, by the time of his death at forty-four. Carol's mother, Mildred, was confined permanently to a wheelchair by muscular sclerosis, and Fred Duckman was suffering from diabetes.

Fred did everything for Mildred. Being retired, he had time to shop, cook, wash dishes, clean the house. In 1975, he bought a van and installed a wheelchair lift so he and Mildred could travel. But one day while he was doing yard work, Fred stepped on a thorn, which caused a foot infection that developed into gangrene. Both his legs were then amputated, and Carol found a nursing home for her parents.

"Mom went twice a week to visit them, and she made friends with everyone there. At Easter, she made baskets for all the residents. She got an idea from all the terminally ill patients. With her friend, Sue Schwartz, she started the I CARE greeting card business. Mom took the pictures and Sue wrote the verse. Mom edited it and actually rewrote a lot," David Royko said.

Royko hated the idea of Carol working, especially early in their marriage when her meager income as a receptionist was vital to them. But she continued to do various part-time jobs; for a while she modeled and painted. She attended Roosevelt University downtown to take art and interior decorating courses.

"Dad would drive her and drop her off and say, 'Have a nice day at school,' like she was a little kid. She hated that but he always said it," David Royko said. Carol later worked doing architectural renderings.

"And Dad was never home. It wasn't the column; it was the bar," David said.

There were regular fights between Mike and Carol, some of them about his late hours, some of them about the "young lovelies," and all of them about drinking.

Royko's sister, Eleanor, remembered when Carol decided she would not attend any more family holiday gatherings. "Mike and Carol had troubles. She started getting sick on Christmas Eves. Holidays are not always easy on a woman. Carol threatened Mike that if he was going to drink on Christmas Eve, she would not come to my house. So he told her, 'If you don't show up at my sister's, I won't show up at your mother's tomorrow.' I'd say, 'Mike, don't do this to your family.' Christmas Eve was hard on all of us because we knew Carol was unhappy."

For a while, Carol convinced Mike to try anabuse, a drug that would cause nausea when alcohol was consumed.

Lois Wille remembered the period. "He would tell everybody he wasn't drinking. He did that periodically, but he told me he was taking anabuse. Then a few days later he said he couldn't take it anymore and I asked why. He said he discovered that the salad dressing at Riccardo's was made with wine, and if he decided to have a salad he'd get sick. It was just an excuse. He never wanted to take it."

John Schackitano also remembered: "I was out with him a couple of times when he drank after taking anabuse. He got sicker than a dog. He had so much guilt about his drinking. He was a smart man. He knew what drinking was doing to his family, but he wouldn't quit. It would be like giving up control. But he tried many times. He'd quit for maybe a year, drinking hard, and there was a period in the late seventies when he told me he was going to a psychiatrist nearly every day. He had a lot of pain."

Despite Royko's liberated attitude toward women in the newsroom, or even in politics, and despite his unquestioned adoration of Carol, he was a chauvinist at home. He had inherited his father's work ethic, and he believed that his main role as a husband and father was to be the breadwinner.

"The guy did nothing at home," said Schackitano. "His backyard screen door was falling off for three years before somebody put some screws in it. I think Carol did. She did everything around the house. He never spent any time there. He wasn't a Little League kind of guy, never played ball or did homework with the kids. It couldn't happen because he was never there. Either he was at the paper or at the Goat or some other bar."

Royko was also demanding. His sister, Eleanor, said, "Carol would serve him chicken and he would say, 'Where's the paprika? Why isn't there paprika on the table. You know I like paprika with chicken.' But Carol was easy, she'd say just say, 'Relax, I'll get it.' Or if there wasn't any she'd go to the store."

Rob Royko remembered moments when she wasn't relaxed. "She made hot dogs, and he started bitching because there wasn't any relish. She threw the whole thing against the wall and walked out, saying, 'That mess better be cleaned up before I get back.' He cleaned it."

With Royko, as in his writings, there was always humor blended with anger. Rob Royko remembered when his mother was insisting on anabuse. "She made a mistake and gave him the dog's heartworm medicine, but she didn't realize it until after he had left for the office. So he's downtown working on his column and she called up crying and said, 'You won't believe this, I gave you dog medicine.' He had to go and get his stomach pumped."

Robbie also was creating anxiety for his parents. "I became independent at a young age. At nine, I got a job washing dishes in a Chinese restaurant. When I was in eighth grade I had a job working on the *Sun-Times* truck as a helper. One of the drivers was a bookie, and I became one of the highest paid helpers they ever had; every time I collected for the bookie, I got five or ten dollars. I found out later my old man knew I was collecting for a bookie and he didn't care. By then I was smoking, drinking. When we went to Europe, when I was nine, I had my first buzz. They put a few inches of beer in a mug for me, but when my old man wasn't looking I drained his, too.

"By fifteen, I was a pain in the ass, out until midnight all the time, drinking, smoking pot. Grounding me didn't work. When I was fifteen I stole his car and ass-ended somebody at Central and Elston. Like an idiot, I just took off, left the car there. Naturally they check the plate number and it comes up Mike Royko. Mom and Dad were in Wisconsin, and the police called them there. They threatened to send me to the Audy Home [juvenile delinquent facility]. They had some friend who was a counselor, and she suggested they stick me in Michael Reese Hospital. I was in a ward with the nuts, and they were smart enough to tell me I didn't belong. One of these places where they only gave you spoons to eat with and no matches allowed. I signed a five-day release saying they had to take me before a judge or let me out. I got out," Rob said.

Meanwhile, David Royko was enrolled in Roycemore, a private high school in Evanston, where "I spent the best four years of my life.

I think my Dad really got a kick out of the idea that he could afford to send his son to a school where the Lester Crown family and W. Clement Stone family sent their kids.

"I'm sure he was proud of whatever I was doing, but I know it disappointed him that I didn't enjoy sports. He would take me to baseball games, but I didn't enjoy it. Once he took me to a Chicago Bears game, and I kept asking when it was going to be over. He said, 'It's the third quarter.' I asked how many quarters were there and he said he would never take me to another football game. I said, 'Good.'

"I was a pretty big kid and I enjoyed sports, especially basketball and softball, but I didn't love it the way he did. I hated Little League baseball. But he accepted that. He never pushed us. He was not the kind of dad who was saying you had to do this or do that," David recalled.

Rob Royko said, "We did a lot of fishing. He taught me a knot I've never forgotten. But if you didn't do it right he'd yell at you. A day of fishing with him could be nerve-racking. He wasn't content with a bobber, worm, and hook. He had to have every single gadget, every type of lure, two tackle boxes, fish finders, depth finders, contour maps, temperature tables, everything."

Each of the two boys reacted differently to having a famous parent.

David said, "When I was in high school it started becoming a real pain in the ass being the son of Mike Royko. I really hated it. Before that I never felt it had any kind of an impact on how people related to me. It was a gradual thing. He got the column when I was four, and so I knew that people somehow knew who my dad was. But later I just hated being seen as simply an extension of somebody else. Robbie loved it. Robbie adored it. Robbie's the one who'd be sitting next to somebody on the train and say, "How's my dad's column today? You're reading it. That's my dad.""

By 1976, not as many people were reading the column because the *Daily News* circulation continued its tailspin. The *Daily News* had more than 500,000 readers when Royko began his column, in 1963, but by 1976 that number had plunged to barely more than 300,000. It undoubtedly would have been far less had not Royko appeared five days a week. When Royko took a rare day off, or was on vacation, circulation fell. The impact of Royko's presence was made obvious by front-page overlines heralding "Mike's Back" after every absence.

By the late 1970s, America's inner cities, particularly in the rust belt, were losing white populations to the suburbs, where afternoon papers arrived either too early to be of any value or too late to compete with the dinner-hour network television news. There was nothing the afternoon newspapers in America could do to halt dwindling readership, and papers were being shut down one after another in Boston, New York City, Philadelphia, Pittsburgh, Cleveland, and Detroit. The *Tribune* had given up on *Chicago Today* in 1974, putting a few hundred newsmen out of work and casting a brief pall over the various saloons where everyone—even rival *Daily News* staffers—mourned.

But Marshall Field V, then thirty-six, was not ready to surrender. His morning *Sun-Times* was thriving on the growth of a new black middle class, one that used public transportation and preferred the handy size of the *Sun-Times* over the bulky *Tribune*. The *Sun-Times* had surpassed the *Tribune* in city circulation; in total, the *Sun-Times* trailed due to the *Tribune*'s dominance in the suburbs. Only the *Daily News* had ever truly rivaled the *Tribune* for suburban readers, but in the decade from Larry Fanning to Daryle Feldmeir, there had been many fits of change. Fanning had made the *Daily News* serious and combative, but circulation continued to fall. Roy Fisher tried gimmicks such as games and features; circulation fell. Feldmeir attacked *Chicago Today* with old-fashioned splashes of stories on ax murderers and child rape, many of them in far-off locations reported by wire services.

The perilous situation at the *Daily News* was obviously on Royko's mind when he turned down the Washington offers. "My column is

needed here," he told *More* magazine in 1975. "If I left the Daily News
—if what they tell me is true—maybe it wouldn't sink the ship, but it
sure as hell would be listing pretty badly. Hell, yes, that enters into
my considerations. Do I want that on my conscience the rest of my
life—that I was the last nail in the coffin? Who the fuck wants that?
'Yeah, he was a really great newspaperman; he really helped put a
paper under.'"

In the fall of 1976, Field asked James Hoge, the editor of the *Sun-
Times,* to run both newspapers and save the *Daily News.* Hoge, a
handsome blond graduate of Yale, had been a startling success in
Chicago journalism, joining the *Sun-Times* in 1958. A decade later he
was the *Sun-Times*'s city editor and earned accolades for the paper's
coverage of the tumultuous 1960s. He was named editor of the *Sun-
Times* in 1968 at age thirty-three, and the "little paper," as Royko
always derided it, won five Pulitzer Prizes in the next eight years.

Although both newspapers shared the same building, there was
no love lost between the two news staffs, both of which were appre-
hensive about the new arrangement. *Sun-Times* staffers feared Hoge
might ignore them or, worse, use their resources to shore up the
Daily News. Staffers at the *Daily News* feared Hoge would run
roughshod over what were then the last remnants of a great newspaper literary tradition, one that was almost solely held together by
Royko and a few others. Royko had no reason to like Hoge. He told
More magazine: "Hoge and I have never been friends, let's say. Part of
this, I think, is because we come from different social backgrounds.
Hoge's had all the breaks. Good looks, rich parents; he came to the
Sun-Times right out of Yale. Me, I put in time at the City News Bureau
first, then had to bust my ass to get my job. But I'm very impressed
with Hoge. I'm impressed with his ability to grasp a concept and turn
it into an understandable, workable tool.... I don't think I could picture myself going on a canoe trip with the guy or anything, but that's
irrelevant."

Hoge did some smart things such as promoting Ray Coffey to run the Washington bureau following the death of the esteemed and honored Peter Lisagor; he put Lois Wille in charge of the editorial page; he purged the foreign staff, which had deteriorated from its prize-winning stature of the 1930s. *Daily News* staffers dubbed him "Attila the Hoge," and "Hoge the Butcher." He outlined a new strategy of ignoring mayhem and reporting lifestyle, a trend that other newspapers had already started and one that ultimately reduced the amount of space dedicated to actual news stories in American newspapers. His smartest move was to name Royko an associate editor.

"All day long there would be a string of people going in and out of his office," said Wille. "He used to complain he never got to work on his column until late at night. He was a job counselor, therapist, friend. I think he liked it. And he liked being able to sit in meetings and throw out ideas about what the paper should do."

Royko had often said that his ambition as a young reporter was to be a city editor, then perhaps an editor. But in later years, the experiences at the *Daily News* gave him a different outlook. "I would have made a bad editor. I could never fire anyone."

Despite his responsibilities, Royko's column was as good as ever, or better. In the spring of 1976 he revealed the "Royko Rule" for beating traffic tickets and created a bureaucratic nightmare for traffic court. Royko advised his readers that a little known Illinois law provided that all traffic offenses must be brought to trial within forty-five days of occurrence. Hundreds of motorists who had been assigned court dates longer than forty-five days from the time of their tickets won dismissal of all charges. Circuit Court Clerk Morgan Finley, a Daley protégé from the Eleventh Ward, asked that police normally assigned one day a month to court be assigned two days to clear up the backlog, but the number of cases to be called with so many policemen appearing to testify against violating drivers caused a mob scene in the courts. The law was eventually changed, but hundreds of drivers owed Royko for "fixing" their tickets.

In January 1977, Royko was in the paper, as usual, but as the subject of a story, not the author of a column. After one of his late nights of drinking, following a planning meeting with some members of the softball team, Royko found himself drunk and alone. None of his usual after-work friends remembered being with him the night of January 28, 1977, when he wandered shortly after midnight into one of the singles bars on Lincoln Avenue, an area he rarely visited.

He was drunk and strolled over to a table where six members of a theater company were sitting in a corner. One of them, Suzanne Quinlan, said Royko came over to the table, told them who he was, and offered to buy her a steak. She said no. Then Royko and the five men began insulting each other. The police report, which Royko kept in his files, said, "After a short period he became abusive and was asked to leave the table. He started yelling and picked up a ketchup bottle from the table, broke it and threatened those at the table. The manager, Dave Carmody, asked Royko to leave the tavern. At this time, the arresting officers, making a premise check, entered the tavern and were informed of what had transpired. The manager signed a complaint for disorderly conduct and the other five victims signed complaints for assault. Royko was transported to Area 18 where he was processed and then posted bond."

The "victims" claimed Royko threw ketchup at them, and Ms. Quinlan said it had stained her fur coat. Royko called Sam Sianis at the Goat, and Sianis paid the bond and drove Royko home. Royko did a number of things that foggy winter morning. He telephoned the City News Bureau shortly after 6 A.M. and talked with Paul Zimbrakos, the day editor who had worked with Royko at CNB twenty years earlier.

"He told me he had been in a bar fight and there might be a police report and he wanted me to kill it. I said, 'Mike, you know I can't do that.' Then he asked if I could leave his address out of the story. I told him I would see what I could do," said Zimbrakos, who would rise to become the top editor at CNB.

Royko then called Jim Hoge and resigned. It might have been the first time Royko resigned to Hoge, who, perhaps unaware that a Royko resignation was usually subject to immediate recall, immediately called Studs Terkel.

"Hoge says, 'You got to talk to Mike. He's threatening to quit over this ketchup thing.' I said, 'You're putting me on,' but I grab a taxi and head over to Mike's place on Sioux Avenue. Carol lets me in. She's got on a frock like she's been painting. Mike says to her, 'Get him some wine. Get out the best.' Then he starts talking about Carol's painting and I said, 'Hoge called and told me you're quitting.' He says, 'Yeah,' and I said, 'Don't you know you're a hero? What guy out there hasn't made a pass at a cute girl and gotten turned down? That makes you like everyone else. They love you.'

"He says, 'You're full of shit,' but he was going back to work anyway."

Royko said in several interviews, "I was serious about it. I was determined to quit and find a job someplace where it isn't such a crime to get in a fight. They made a big thing about the ketchup and it drove me up a wall. They made me out as some kind of weirdo throwing ketchup at a woman."

Royko also called one of Chicago's most flamboyant lawyers, Julius Lucius Echeles, to defend him. He called Lois Wille to tell her he was quitting.

"I didn't think he would do it," Wille recalled. "Carol knew how to soothe him after long drinking bouts. She could bring him out of his depressions. Dick Christiansen used to call her the last of the Christian martyrs."

The ketchup story was all over the Chicago papers and television stations. There were too many Royko rivals, and not a few enemies, for any editor or news director not to gloat over the opportunity to embarrass Chicago's top newsman. Royko was angry at his colleagues, embarrassed, and, as usual, felt guilty.

"Carol was all over his ass on that one," said John Sciakitano.

Carol went to stay with her parents for a few days, deciding whether she should leave him, Mike told friends. But her parents persuaded her that Mike really loved her and was a wonderful provider.

Royko would have appreciated that. Being a good provider was a badge of honor for first-generation ethnics. His son David remembers that when there were arguments about how much time the column took, Royko would declare, "The column is the family. Without the column, there is no family."

On the night he was accused of flinging ketchup on Ms. Quinlan's fur coat, Royko was in the first year of a $75,000-a-year, five-year contract. He was providing very well.

Echeles made a court appearance on Royko's behalf and read in court a note one of the "victims" had slipped in Royko's pocket. "It sure was fun bandying words with you. I'm writing this in catsup, heh, heh ... You can make up for this by a little publicity in your column."

"The woman's not the problem," Royko said during the case. "She's satisfied. I've written her a letter of apology. Christ, it's a collector's item. I said I was the worst piece of shit that ever lived. I groveled. We offered to buy her a new fur coat. She said she didn't need one, just pay for the cleaning on the old one. We did. It's the guys. They say, 'He behaved very badly. He was drunk.' Of course I was drunk. That's what I thought saloons were for." Royko paid a $1,000 fine for disorderly conduct. The misdemeanor assault charges were dropped.

Years later, Royko's eyes would narrow at any complimentary remark directed toward any of the half-dozen reporters who had been assigned to cover his court hearing. "He's an asshole. She's a slut," he would growl. "They were out to get me."

Throughout the 1970s, news appetites changed. Although Royko was the top columnist in Chicago in 1969, only CNB rookie Ellen Warren was sent to cover his traffic court case. Of course, the ketchup bottle incident was meatier than giving a traffic cop the finger. But the extensive coverage of Royko's 1977 ketchup affair was also indicative of how the news media were slowly becoming preoccupied with celebrities rather than news.

Over the course of the years, Royko had been involved in a variety of bar scuffles, which were de rigueur for Chicago's hard-drinking newsmen. Drinking and brawling were a part of the culture that included long hours, low pay, and obsession with finding an exclusive story. It wasn't until the 1980 corporatization that newsroom culture changed to prize sobriety and political correctness more than imagination and news beats.

Most of the *mano a mano* contests of the Chicago Press Club, in its various locations, or at Riccardo's, O'Rourkes, or the Goat were carried out in such a haze of alcohol that few injuries were ever reported. They simply became proudly applied telltale Band-Aids.

Terry Shaffer recalled one of those episodes. "Mike and I went into a saloon on Lincoln Avenue called the Vieux Carré and it had a real long bar and the only one there was a guy who worked for the *Sun-Times* and Mike didn't like him. So, what else, he goes all the way down the bar to sit next to this guy and starts in on him. Next thing I know they're standing up, and I figure if Mike hits him he might kill him so I grab Mike from behind and pin his arms and the other guy takes a swing and his fingernail clips Mike on the jaw. That's it. He's got this little cut. For days he went around telling everyone, 'I'm being pounded, this guy is killing me, and my best friend is holding my arms.'

During the Chicago blizzard in January 1979, Royko was at the Billy Goat when Dorothy Collin came in and tried to strike up a conversation. Collin, who had always been received pleasantly by Royko, had just started a new page-one feature column in the *Tribune*. Although it was not remotely like Royko's daily offering, she now qualified as a rival and he began to curse her viciously.

"Someone, I don't know who, told Mike he was out of line and suddenly chairs are falling over, and Larry Weintraub of the *Sun-Times*, who was Mike's friend, is stepping between people and he gets hit and knocked down. The only one who got hit was the guy trying to break it up," Collin said.

While overserved newspapermen were occasionally flailing at one another harmlessly, a colder, deadly violence had gripped Chicago. In May 1977 Royko covered the trial of Chicago hit man Harry Aleman:

> Harry Aleman is something special. People come to the courtroom and stare, fascinated. When there is a break, he struts coolly around the hallway and they stare some more.... What makes Aleman special, as accused killers go, he is said to be a genuine professional hit man for the Chicago Crime Syndicate. That's right. Just like one of the characters who kissed Marlon Brando's ring.
>
> Watching him, it's hard to see him as the police do—wearing a ski mask, pointing at someone's head with a shotgun. But that's what they claim he does for a living. And at the rate he's going, they say, he might fill up an entire cemetery on his own. They credit (I'm not sure that's the appropriate word) Aleman with as many as 22 killings.

Royko wrote seven consecutive columns on the trial. He wrote about the testimony of a woman whose brother Aleman was on trial for killing. The woman's husband had also died, and she was raising four children on her own. He wrote about Louis Almeida, a scummy

murderer who was the principal witness against Aleman because he
had helped him do the hit; he wrote about a witness who said Aleman offered her $10,000 if she lied about what she saw the night he
murdered her neighbor, William Logan; he wrote about the trial
judge, Frank J. Wilson, who apparently knew about the charges but
refused to let the witness testify on those subjects; he wrote a column
that took the form of a letter to Police Superintendent James
Rochford, chiding the police for not keeping tabs on Aleman. After
all, they had spent twenty-five years carrying out surveillance on
independent alderman Leon Despres, who always irritated Daley at
council meetings:

> Not once did your sleuths discover Despres shooting anybody, col-
> lecting juice loans, taking bets or hijacking trucks. To the contrary—
> Despres himself got shot in the leg one night. That seems to be the
> way most of your surveillance work turns out. You follow such dan-
> gerous characters as Studs Terkel, Len O'Connor, Despres and me.
> While some of us must provide fascinating sights—especially me—
> none has done anything to get into Sing Sing. That's why I mention
> Harry Aleman.

Besides ridiculing the police for their ignorance of Aleman's activity, Royko also poked fun at their spy list, which revealed that Royko
and others were watched by the department's infamous "red squad,"
the same bunch that had told the *Tribune* in 1968 that the hippies and
yippies were going to put LSD in the city's drinking water. Royko
loved the idea that he was under surveillance. Occasionally, he would
spot one of the undercover cops in a restaurant and confront him in a
friendly way to embarrass him.

Royko printed a huge section of the court transcription regarding
the witness who claimed she was offered a bribe by Aleman. The
transcript showed Judge Wilson repeatedly ignoring any efforts by

the state's attorney to have the witness testify. As usual, Royko simply let the judge's quotes show how truculent he was.

Royko's skepticism about Judge Wilson's handling of the notorious case erupted on June 2 when the judge found Aleman "not guilty." Royko called the case a fix, and his passion for the story contributed to the public outrage over the verdict. It was unusual for Royko to focus so much on a story that was being covered on the front pages, but his unerring sense of political chicanery and his courage in repeatedly attacking a known mob killer added to the public interest. It wasn't until 1992 that Royko found out that his suspicions had been correct: During a federal investigation called Operation Greylord, an informant testified he had paid Judge Wilson a $10,000 bribe to fix the case, and that his orders came from First Ward boss Pat Marcy. The hapless Wilson, when informed by the FBI of the stool pigeon's revelations, killed himself. Royko wrote:

This isn't about an alderman grabbing a wad for a zoning vote, a judge selling Traffic Court acquittals, or city bids being rigged. Harry Aleman was the Chicago mob's ultimate weapon, their version of The Bomb. Forget the nonsense about the mob having business sophistication. What their sales pitch boils down to is, "Do what we say or we'll kill you." That's why Aleman is important. He's good at killing people.

So if the government has the evidence, what we'll hear at Marcy's trial is how the mob's top political operative kept the mob's top killer out of prison by corrupting a respected judge. Think about it. Someone goes to college, then law school, gets credentials as a lawyer and takes an oath as a judge to uphold the law. A judge. A respected figure in our society.

Then one day that judge is somehow persuaded by a well-tailored old thug like Pat Marcy to sell his soul. That's not Hollywood, it's Chicago.

Royko's hatred for the mob sometimes had a soft spot, at least when their families were trying to bury them. The family of Sam Annerino, who was killed by a shotgun blast, was doing just that when a WMAQ-TV camera crew showed up to record the proceedings.

One might also say that Channel 5 was insensitive, boorish, ghoulish, intrusive and dumb. But of all the things that can be said, the family of Mr. Annerino said it best. They summed it all up by knocking the hell out of the TV reporter, and busting the camera. It was the best job of TV criticism I've seen in a long time and I hope the family is given an award.

In September 1977, the *Daily News* got a face lift. Hoge ordered a new design, an innovative modular make-up with deep black rules separating the modules and running vertically and horizontally across the page. In some ways the new design was the shape of things to come, but changing the look of a newspaper that had been around for eighty years was risky. Circulation fell again. Critics of Hoge blamed the new design, but readers may have liked it and still stopped reading the paper. Or it may have been one more excuse to cancel a subscription.

On February 3, 1978, Marshall Field V stood on a desk in the *Chicago Daily News* city room and announced that the 102-year-old newspaper would fold in a month. The Chicago journalism community descended into a frenzy of pity and glee, empathy and sadness, triumph and hollow echoes of praise. First was the anachronistic commentary about the loss of a great voice in the community. Editors from the *Sun-Times* and *Tribune* went through this maudlin charade as the rivals of the *New York Herald-Tribune* had done. There was little of it when *Chicago Today* folded. It had lost its heritage as one of Hearst's great muckrakers long before its demise. But the *Daily News* had a grand tradition, and the obituaries were obligatory if not sincere. In fact, Chicago no longer needed another voice. Suburban newspapers, especially weeklies, were growing in all directions. They had helped bury the *Daily News,* and their impact on the remaining newspapers would be severe in coming decades. The *Daily Herald,* which started in Arlington Heights as a triweekly, had gone daily with a circulation of about 25,000 in 1971 and was selling 65,000 daily by 1980. Before the end of another decade, the *Daily Herald* would reach the 100,000 mark, threatening *Tribune* dominance in the northwestern suburbs.

At Riccardo's and the Billy Goat, there was another kind of commentary, the gallows humor of a staff that was being tossed out of work, and the month-long speculation about how many and which of the *Daily News* staffers would be hired by the *Sun-Times.* The angst was not restricted to the *Daily News*; *Sun-Times* staffers were wondering which of them might be fired or demoted to make room for a *Daily News* stalwart. The same kind of misery visited the news community when the *Tribune* folded *Today* and took its best people to the *Tribune.* A decade later there was still mumbling in the *Tribune* newsroom about "*Today* people," who never really fit in at the *Tribune,* and mumbling among the "*Today* people" about the arrogance of a newspaper that didn't realize it had been strengthened by fresh talent.

After Field's announcement, the brutal selection of who would stay and who would go began. There were legal factors and talent considerations and economic considerations. If there were two critics with similar skills the higher paid writer was gone. The rumors floated that pension funds would pay for severance.

"It was vicious," said Rick Kogan, who after years of being promised a job was finally hired at the *Daily News* in August 1977, the year his father, Herman, retired from the *Sun-Times.* "Jim Hoge and Greg Favre [managing editor] would call people in the office and tell them they were staying or they were gone. Hoge called me and said he had arranged a job for me with the *Los Angeles Examiner,* and if I went there he would try and bring me back in six months. I said, 'Fuck you,' and left. The night after the paper folded there was a party at the old Sheraton, and some little guy went after Mike because there was this weird notion that he and Lois were deciding who got kept. This guy kept it up and Mike started chasing him and I don't know if he hit him or not but the next thing I saw was this guy passed out at Ann Landers's feet."

Twenty years later, there were still former *Daily News* employees who believed that Royko had kept them from being hired. Wille said, "That was ridiculous. We had nothing to do with it."

WBBM-TV ran a story saying that Royko had compiled a "hit list" of staffers he wanted excluded from the *Sun-Times*. Another station broadcast the report that *Daily News* staffers were drinking out of bottles hidden in brown paper bags while Field delivered his death knell. Among the *Daily News* staffers who weren't hired by the *Sun-Times* were Hanke Gratteau and Wade Nelson, two of Royko's legmen.

"That should have been the proof that Mike wasn't asked," Gratteau said.

The *Tribune*, rebuffed in its overtures to Royko, who it considered worth 12,000 papers a day, succeeded in hiring Ray Coffey and Dick Christiansen, although both had been offered jobs with the *Sun-Times*. It was the end of Coffey's long friendship with Royko, exemplifying the acrimony that was created by the death of the *Daily News*.

Coffey recalled that Hoge told him he would move to the *Sun-Times* and fill the Washington bureau-chief post that he had held with the *Daily News*. "I was upstairs in Hoge's office and he didn't say anything about Loy Miller who was the *Sun-Times* bureau chief. Meanwhile, I find out later, Miller was downstairs being fired. That was on a Friday. When I got back to Washington on Sunday I got a call from Loy Miller, and he was wounded that I was taking his job and he was fired. Max McCrohon of the *Tribune* [managing editor] had already called me about moving to the *Tribune* so I called Dick Trezevant at the *Sun-Times* and told him I was gone.

"After that, Royko spent three or four years pissing all over me, because I hadn't gone to the *Sun-Times*. He and I didn't speak for ten years. Finally, someone got us together at Riccardo's in 1986 when I was back in Chicago with the *Tribune*. I was mad at him, particularly since five years later he walks across the street to the *Tribune*. Our relationship soured in 1978, and it stayed sour," Coffey said.

Newspapers such as the *Los Angeles Times, Washington Post, Dallas Times-Herald,* and others sent talent scouts to Chicago to interview *Daily News* staffers who were out of a job. Gratteau, who had gone to work in the sports department in one of Hoge's innovative efforts to bring gender diversity to a previously all-male bastion, was offered a spot on the *Dallas Times-Herald* but took a public relations job to stay in Chicago. While individuals and their friends fretted over their futures, the major speculation in the larger journalism community centered on Royko, who was not a certainty to join the *Sun-Times*, which he disliked almost as much as the *Tribune*.

Ben Bradlee's oar was in the water first. On February 1, two days before Field's announcement, he scrawled a brief note: "Mike, If I can help, call. Ben."

The *Tribune* made an immediate inquiry; so did the *Miami Herald*.

Ralph Otwell, who held the title of editor of the *Sun-Times* but who reported to Hoge, wanted Royko to know that he would be loved at the "little paper."

Otwell, aware of the many *Sun-Times* staffers who were jealous of Royko's stature, wrote, "You'd be surprised to know how many of our people admire and respect what you do, and how you do it. In fact, that goes for all the staffers who are truly interested in the future of the *Sun-Times*...and therefore everyone who really counts. When all the silly hub-bub dies down, I'd like to buy you a drink and try to convey some of our feelings in person. Meanwhile, I had to let you know how eager we are to put the league's best pitcher in the *Sun-Times* lineup."

There was no question the *Sun-Times* wanted Royko, but it also had a dilemma. It already had two columnists, Bob Greene and Roger Simon. Greene, at thirty-one, was the rising star, second only to Royko in reader appeal.

Royko was going through another period of anxiety. He wrote in February to his literary agent, Sterling Lord:

> As you may have heard, the *Chicago Daily News* is ceasing publication in early March. This doesn't effect my column, since I've already been assured of a prominent place in the *Chicago Sun-Times*, the *Tribune* and have had offers from most major papers and syndicates. However, I'm not sure I want to keep my column going. I've done just about everything I can do with it, so this seems as good a time as any to consider other options.
>
> For a long time I've been mulling over a novel. I put very little on paper, because I've been too damn busy with my column. But I have it worked out in my mind and I think it would be good on both commercial and critical levels....
>
> So my question to you is this: What kind of advance can I get? I'm in no position financially to take this on without a sizable advance. But if I keep doing my column, I won't have time to do the book. So I'll need enough money to work on the book full time for six months or more.

Almost every time Royko felt frustrated, he wrote to Lord about a novel. He had written in October 1976:

> I have worked out the outline for a novel: a political-crime-thriller-suspense-type thing, set in Chicago. It's a good plot, interesting char-

Mike Royko with his mother, sisters, and brother in the early 1940s.
From left: Eleanor, Helen, Robert, Mike, and Dorothy.

The Royko family in the early 1950s. From left, sitting: Helen, Mike, and Mike Sr. Standing from left: Robert, Eleanor, and Dorothy.

Royko's mother, Helen, with the flapper look in the 1920s.

Airman Mike Royko
in 1953.

Carol Royko sent this posed studio photograph to
Royko at his Air Force base in Washington in 1954.

Carol and Mike out on
the town in the 1950s.

David (left), Carol, and
Robbie Royko in 1965.

Royko standing on an overpass near Grant Park with the Chicago skyline in the background in 1972.

Sam Sianis and Royko at the Billy Goat, celebrating the 1971 publication of *Boss.*

Royko at a fund-raising event in the 1960s where he met John Belushi (left), the nephew of a close friend.

Royko in 1972 with Marshall Field, owner and publisher of the *Chicago Sun-Times* and *Chicago Daily News.*

Royko in his office at the *Chicago Daily News.*

Royko with his closest friend in journalism, Lois Wille.

Royko in formal attire spoofing at a party for his legmen.
Left to right: Hanke Gratteau, Terry Shaffer, Ellen Warren,
Wade Nelson, Helene McEntee, Pat Wingert, John Fennell.

An unlikely reunion as Royko and Studs Terkel (center) meet with Mayor Richard M. Daley, whose father's administration was constantly criticized by Royko and Terkel.

Royko and Mayor Harold Washington at a 1986 party.

Royko playing softball, his main after-work passion at the *Daily News*.

Royko sampling entries at his Ribfest in Grant Park in the early 1980s.

The Roykos at home in Winnetka.
From left: Judy (holding Kate),
Mike, Sam, and Griz.

The great killer of the mighty bullfish,
removing a hook.

Mike holding Sam in 1987.

acters, with a topical flavor. If I write it, I'm going to have to take
four to six months off from my column. So I'd need a pretty good
sized advance to cover my loss of income for that period. It would
have to be in the neighborhood of $25,000 to $30,000. I don't think a
publisher would be taking a great risk, since *Boss* showed I can handle a book, and the Chicago-area sale alone would assure the publisher of getting his money back.

There isn't much doubt that Royko could "handle a book." But
was he simply bored at the time, or was he facing a career crisis such
as having to take his column to the "little paper"? In the end, the column always won.

On March 3, the *Daily News* published its last regular edition.
Royko wrote:

When I was a kid, the worst of all days was the last day of summer
vacation, and we were in the schoolyard playing softball, and the sun
was down and it was getting dark. But I didn't want it to get dark. I
didn't want the game to end. It was too good, too much fun. I
wanted it to stay light forever, so we could keep on playing forever,
so the game would go on and on. That's how I feel now. C'mon,
c'mon. Let's play one more inning. One more time at bat. One more
pitch. Just one? Stick around, guys. We can't break up this team. It's
too much fun.

But the sun always went down. And now it's almost dark again.

He took his column to the *Sun-Times* on March 5, 1978, the day
after he wrote his farewell to the *Daily News* in its final souvenir edition. He ticked off the usual reasons for failure of an afternoon newspaper—growth of expressways, flight to the suburbs, suburban
competition—and he recounted how the *Daily News* had alienated
readers by its courageous coverage of civil rights and its equally

courageous criticism of Mayor Daley and the botched Black Panther raid.

> Courage didn't kill the *Daily News*, but it undoubtedly shortened its life.
> And so did apathy. In the Chicago area, 1.6 million people will turn on Welcome Back, Kotter. About 2.1 million watch Charlie's Angels. Wonder Woman draws 939,000. There's a big market for mental cotton candy.
> But out of 7 million who live in the Daily News circulation area, only 315,000 of them thought one of the better papers in America was worth 15 inflationary cents.
> When a new dictator takes over a country, one of the first things he does is seize or close the newspapers.
> Apathy isn't as heavy-handed as a dictator. But it can get the same job done.

The next day, his first column in the *Sun-Times* was about sixteen-inch softball. His second column on March 6 was about the fight that Rick Kogan described at the *Daily News* farewell party in a ballroom at the Sheraton Hotel just north of Tribune Tower.

The column described a little guy calling a big fellow an obscene name and kicking him in the shins. The big guy popped the little guy, and most of the people at the party hissed, "The big bully." The rest of the column was a defense of big guys hitting obnoxious little jerks.

A week later, there was a real surprise, because in the move from the *Daily News* to the *Sun-Times*, Royko's picture had gotten bigger. It was now bigger than Bob Greene's picture. Royko's byline was bigger, too. Not many readers would have noticed, but rulers suddenly were the most popular tool in every Chicago newsroom.

Three weeks later, Bob Greene was at the *Tribune*. "I have never been a person to whine about that kind of thing—complain about the size of my picture or whether there's a box around my column … When it came time to make my statement, my statement was to leave," he told *Chicago Magazine*. He also said that he had begun negotiations with the *Tribune* shortly after Field announced the *Daily News* would fold. He told *Chicago Magazine*, "A lot of my friends were treated very shabbily."

Greene, who was still one of the most widely syndicated columnists in the country as the new century arrived, experienced a journalism nightmare the first week he appeared in the *Tribune*, which had hastily pasted his picture on all its trucks and newsstand boxes. He wrote a column quoting a woman who said that two of her friends had been murdered on a visit to California, and she had wanted Greene to write a column warning young people of the perils of travel. The *Sun-Times* assigned four reporters to check out the story and found no evidence of any murders. Greene had been hoaxed, and the *Sun-Times*, true to the vicious newspaper traditions of Chicago, made the most of its opportunity to gloat. Royko was not one of them. "He always liked Bob Greene," said Lois Wille, noting that Royko rarely liked and usually despised any rival columnist or anyone who might have the potential to be one. "Maybe he thought Greene going to the *Tribune* was deferring to him," Wille said.

Greene's defection was the first salvo in what media watchers were declaring the "great newspaper war" in Chicago. The *Tribune* and the *Sun-Times* were now alone, and the *Sun-Times*, anchored by Royko and beefed up by the automatic subscriptions it gave to *Daily News* subscribers, jumped 131,000 in circulation in the May 1978 reporting period. But just as 300,000 readers had virtually disappeared when *Chicago Today* had gone under, there was no guarantee that either the *Sun-Times* or the *Tribune* could keep any of them.

The *Southtown Economist*, owned by Bruce Sagan, was a biweekly that became a daily in late February when it was official that the *Daily News* was closing down. The *Daily Herald* started a commuter edition that was sold at the downtown train stations. In a crass bit of exploitation, WMAQ-TV took out a full-page advertisement in the *Chicago Tribune* declaring, "The News is Dead. Long Live the News!" It invited readers to replace the disappearing *Daily News* with Channel 5's nightly newscasts. Royko said, "It was one of the most disgusting things I've ever seen."

The *Sun-Times* now claimed about 700,000 daily readers compared with the *Tribune*'s 750,000, but the *Tribune* had the big edge on the highly profitable Sunday editions: 1.3 million to 800,000 for the *Sun-Times*. There was no question that the *Sun-Times* dominated city circulation and circulation among minorities. But the *Tribune* had its prestige and its wealth, its dominance in the suburbs, and was planning a modern printing plant on Goose Island in the Chicago River north of Grand Avenue, a critical step to its future success.

When the *Daily News* folded, Royko called Terry Shaffer, who had been one of 147 *Daily News* staffers that were let go. Shaffer recalled the phone conversation. "He said, 'Look, I know you're out of a job. They're only letting me take one person to the *Sun-Times*. I can pick my own legperson. I know it's a step back—you've done it—but it's a job. I want you to have it.' He didn't have to do that. It kept me in journalism. I worked for him for about a year," Shaffer said.

"It was a great time. Jane Byrne got elected and she gave Mike anything we wanted. I had all her numbers, the office, at home, in Palm Springs. I could get through to her anytime. I'd call and she'd answer the phone and say, 'What does Mike want?'"

Although Royko was at first uncomfortable at the *Sun-Times*, the column showed no ill effects. He took on the city's biggest real estate mogul, Arthur Rubloff, to save the Chicago Theater from being replaced by a parking lot. He argued that while there wasn't anything architecturally stunning about the Chicago Theater, it was the place where many Chicagoans went on their first downtown date, where they first necked in a balcony or first saw a live stage show. The

Chicago Theater, proudly renovated, still stands. He also tried to save another childhood memento, the Buffalo Ice Cream parlor on Irving Park Road. But the developers won and turned it into a gas station.

In the fall of 1978, Royko wrote quite frequently about his friends. Not all of them were pleased. He upset Tim Weigel by writing about how many hours in a lifetime television newscasters spend blow drying their hair. He also wrote about Studs Terkel's encounter with a mugger, which involved Studs trying to interview the mugger for his forthcoming book, *The American Dream*. Studs was delighted to provide a column, but he asked Royko to leave out his address, because his wife was worried that the mugger might want to reprise his act.

"He said, 'Okay, besides property values might go down if everyone knows you're living in the neighborhood.' He meant it as a joke, but he also meant it. He could hurt when he wanted to," Terkel said.

The winter of 1979, with the great blizzard, the ineptness of Mayor Michael Bilandic, and the zany campaign of Jane Byrne, was another of those Chicago stories that reporters and readers relished. And Royko was the ringleader in humiliating the city administration and touting Byrne's election.

But it only took Royko six weeks to realize that Byrne's victory was not going to bring an end to machine politics. The mayor quickly embraced her former enemies and indicated it would be business as usual at City Hall.

On April 4, 1979, Royko wrote prophetically:

That vision was nice while it lasted. But it didn't last too long. For me it began ending the day Eddie (The Sewers) Quigley planted a kiss on MS. Bossy's lips and she neither slapped him nor had him arrested as a public nuisance.... After that it was a month-long love fest between Ms. Bossy and every ward boss and alderman who would slobber on her shoes. By now, it is hard to remember just whom she was so mad at during the primary campaign....

She learned the political trade from her hero and teacher, the late Mayor Daley, and I'm sure he taught her that a political machine feeds on patronage and power and is a natural enemy of people who preach independence…. So if you haven't figured out what we bargained for only five weeks ago when the snow was still on the ground, it will be someone who hopes to be the female version of the late mayor. But if that disappoints you, don't let it. The victory in that memorable primary was yours, regardless of how Ms. Bossy turns out. It will remain yours because you let them know that you did it once, so if the time comes you can do it again.

And that's something for Mayor Bossy to keep in mind.

On August 14, 1979, while driving home from one of his many routine bar stops, Royko was arrested for drunken driving as he sped on Irving Park Road near Austin Boulevard. The *Sun-Times* and the *Tribune* each ran a three-paragraph story about the incident. The television stations managed to find a minute to slip in Royko's latest misadventure. He was embarrassed again, but his discomfort would soon be forgotten, as would the painful final days of the newspaper he loved and the difficult adjustment to a newspaper he never cared about.

chapter 12

Carol Duckman Royko died on her husband's forty-seventh birthday, and he never celebrated a birthday again.

On the morning of September 17, 1979, John Schackitano's telephone rang. "Mike called from his office about eleven in the morning and said, 'John, Carol had a stroke. They're taking her to Columbus Hospital. Meet me there.'"

Carol had been on vacation in Florida. When she returned, Carol went to visit a friend in suburban Schaumburg, where she collapsed and was rushed to a nearby clinic; she was then transferred by ambulance to Columbus Hospital on the North Side.

By the time Schackitano reached the hospital, Royko, David, and Robbie were there. "She was on a respirator, and the doctor was showing Mike graphs showing her vital signs were gone. Mike explained to David and Robbie that she was being kept alive by the respirator. I was dating a girl who lived just around the corner from the hospital, and Mike and I stayed at her place for a couple of nights and spent the day at the hospital. A couple of times he said, 'Shack, do you want to go in and see Carol? She looks just like she's sleeping.' I couldn't do it. I didn't want to see her that way."

Royko told the doctors to take Carol off life support on September 19—his birthday. One of the many Royko myths that have been held sacred by his circle of friends was that he waited until his birthday so the day would be seared into his memory.

"That wasn't it at all," David Royko said. "The doctors told us there wasn't any hope. I think they wanted us to wait a few days to assure ourselves of that. They finally said we had to make a decision and we did. That's all it was."

The burial service was private, and only the family attended.

Mike's sister, Dorothy Zeltmeier, remembered Carol telling her that she and Mike had made a pact when they were married: "Together, forever." Dorothy said that when Carol died she asked Royko if he wanted any words on the coffin. "He said, 'Give me five minutes.' And then he came back and told me, 'Together, forever.' It was their pact."

Royko received condolences from thousands of readers and from every public official, including the mayor and the governor. Ben Bradlee wrote, "Please accept my heart-felt sympathy at the death of your wife. It has been a season of sadness with the loss of too many friends, or friends of friends. Although I had never met her, she lives for me through your column this weekend, so graceful a statement of love and regret."

The column Bradley and thousands of others read ran on October 5. It was the first Royko had written in nearly three weeks since Carol was stricken. It was the shortest he ever wrote:

It helps very much to have friends, including so many whom I've never met.

Many of you have written me, offering words of comfort, saying you want to help, share the grief in the loss of my wife, Carol.

I can't even try to tell you how moved I've been and I wish I could take your hands and thank each one of you personally.

Others have called to ask when I'll be coming back to work. I don't know when. It's not the kind of job that should be done without full enthusiasm and energy. And I regret that I don't have much of either right now.

So I'm going to take a little more time off. There are practical matters I have to take care of. I want to spend time with my sons. And I can use some hours just to think and remember.

Some friends have told me that the less I look to the past the better. Maybe. But I just don't know how to close my mind's door on 25 years. That was our next anniversary. November.

Actually it was much longer than that. We met when she was 6 and I was 9. Same neighborhood street. Same grammar school. So, if you ever have a 9-year-old son who says he is in love, don't laugh at him. It can happen.

People who saw her picture in this paper have told me how beautiful she appeared to be. Yes, she was. As a young man I puffed up with pride when we went out somewhere and heads turned, as they always did.

But later, when heads were still turning, I took more pride in her inner beauty. If there was a shy person at a gathering, that's whom she'd be talking to, and soon that person would be bubbling. If people felt clumsy, homely and not worth much, she made them feel good about themselves. If someone was old and felt alone, she made them feel loved and needed. None of it was put on. That was the way she was.

I could go on, but it's too personal. And I'm afraid that it hurts. Simply put, she was the best person I ever knew. And while the phrase "his better half" is a cliché, with us it was a truth.

Anyway, I'll be back. And soon, I hope, because I miss you, too, my friends.

In the meantime, do her and me a favor. If there's someone you love but haven't said so in a while, say it now. Always, always, say it now.

One of the condolence notes Royko saved was from Rosalyn Carter: "Jimmy and I were saddened to learn of Carol's death. Please know you will be remembered in our prayers at this difficult time."

No one, not even Royko, could have realized how difficult the times would be.

"Several of his friends were seriously worried that he might commit suicide," said Hanke Gratteau. "Although I was no longer working at the newspaper, we had become good friends. I know this was the low point of his life. He began drinking very heavily."

His sister, Eleanor Cronin, said, "He drank more after Carol's death than he ever drank."

Eleanor and her husband, Eddie, made plans to move in to Royko's house to help him with the boys. David, then twenty, was attending Lake Forest College, and had moved back home because he had a job at WFMT in downtown Chicago and the commute was easier from the Sioux Avenue house. Robbie, then sixteen, was still in his free-spirit mode.

Royko spent the first few weeks after Carol's death with his brother, Bob, who was then living in Bartlett, a western suburb, with his wife, Geri, and their children, Steve, who was seven, and Amelia, five.

"He was devastated and he was drinking a lot, every night. He was sleeping in Stevie's room, and he would lie there sobbing and would rock Stevie in his arms," Bob Royko said. When Geri poured more water than scotch in the glass, Royko snarled, "Don't think I don't see you watering down my drinks." Geri, who came from a family of

Sicilian restaurant owners, replied that she was the boss in her house, and Royko did not argue.

Mike and Bob Royko were always close. "We talked every week or every other week all our lives," Bob said. Bob had quit DePaul to sell photocopy equipment in the 1960s, while Mike was working at the *Daily News*.

"One day I walked into a bank where I thought I was going to sell five machines and the purchasing managing told me he was no longer interested. He had just bought something called a Xerox and showed it to me. I said to myself, 'That's the machine for you, pal.' I applied for a job with Xerox but they turned me down. They thought I'd be a spy for the photocopy company I worked for. The people at Xerox didn't even realize that within a year they would put that company out of business.

"I was kind of depressed so on the way home I bought a new wine they were advertising on television, Thunderbird. I was living in my sister's building, so I asked Dorothy and her husband, Cliff, to try it. We all liked it, and I decided I was going to sell wines." Bob Royko was hired by the Gallo Wine Company. He was based in Madison and Oconomowoc, Wisconsin, from 1967.

"The first time I overnighted in Madison I asked someone to recommend a good restaurant, and I wound up at the place owned by my wife's uncle Joe. It was a great steak house, and they had all these gorgeous waitresses, one of which turned out to be my wife. I went in there a lot, and at some point Geri and I started dating and then she proposed and we got married. She loved Mike and the two of them got along great."

Bob and Geri Caravello were married in January 1971, with Mike as their best man. While they were on their honeymoon in Aspen, Mike sent Bob an early copy of *Boss*.

"He sent along a note in which he said, 'Hey, Kid, This is a genuine first edition. I figure all of them will be first editions. So you get the first of the first. I ask one favor of you: Do not let anybody else read this. It won't be published until March and I don't want anybody reading it free. Let them buy, buy, buy, so I will be rich, rich, rich. I'm also enclosing a copy of Breslin's book (*The Gang That Couldn't Shoot Straight*). It's about Sicilians. His wife is a Sicilian. I read it and it makes my blood run cold. I caution you, Bobby, to frisk her before retiring at night. Those people use stilettos.'"

While Royko was drinking up much of his grief, he was also gulping pills until Geri Royko discovered all the various medicine bottles and flushed the pills down a toilet.

After two weeks, he returned to his home and wrote his column to his readers. Then he started another binge. He showed up one morning about 6:30 A.M. at Schackitano's house. "I was living at 3750 N. Clarendon, right across the street from a Walgreen's, and he telephoned me. 'You got company?' he said. I said no. He said, 'Can I come over?' I asked, 'Where are you?' He said he was at the payphone in the Walgreen's. He came over and his face was the color of a red flag. I thought he was having a stroke. He wanted a drink. I said, 'How about if I make coffee?' He grabbed the vodka bottle and filled a half glass and started drinking. His blood pressure must have been as high as it could go without killing him. And he was smoking one cigarette after another. It reminded me of one our fishing trips to Bull Shoals where we shared a room. I'd wake up during the night, and I'd see the glow from the tip of his cigarette. In the morning this big ceramic ashtray would be filled with butts. He never could sleep, and he must have smoked a pack of cigarettes during the night. Then he'd get stiff sometimes and burn his fingers."

In the weeks following Carol's death, Royko stayed out later and longer.

"His drinking was steady," said Schackitano. "He'd start with a couple of beers, then go to vodka on the rocks and finally finish up with

Tullamore Dew or Bushmills. It was brutal. But he had an amazing constitution. He was a strong man. I never saw him throw up or complain of a hangover."

Bob Royko said, "Mike once told me he might not drink so much if he ever had hangovers. He didn't."

Royko's evenings at the Billy Goat lasted even longer. Perhaps he felt it was his home. Perhaps hanging around the taverns his parents ran during his youth had blurred the distinction between a social gathering place and a home. As a teenager, they were the same place. As an adult, Royko still treated them as the same. But the Goat was more like the Roykos' family tavern than his other watering holes.

"He was like my brother," said Sam Sianis, the nephew of the legendary "Billy Goat," who took over the bar when his uncle died. The feeling was reciprocated. Royko clearly felt a deep attachment to the Sianises. He eulogized Billy Goat in a 1970 column. He wrote that Billy had been driven home one night after the tavern closed at 3 A.M. and had gone up to his room, fallen over, and died:

> It was typical of Billy Goat that he would die during the only five hours of the day when his place wasn't open for business....He ran his tavern by stern rules: cash, no fighting and printers from the newspapers weren't supposed to sit on stools if they had ink on their pants....Billy became famous for his stunts with his pet goat which he kept when his place was located across from the Chicago Stadium. He liked to smuggle it into places where one normally wouldn't expect to see a goat, such as a World Series game or a hospital room....The whole city should take a few minutes off and go into a tavern and have a drink to the memory of Billy Goat.

William "Billy Goat" Sianis had been something of a celebrity before Royko made him a Chicago legend. When his goat was barred from entering Wrigley Field during the 1945 World Series, he declared that he had put a curse on the team and the Cubs would never return to the World Series until he removed it. David Condon of the *Tribune* and John Carmichael of the *Daily News* joked about it, but as the years and decades passed without a series, the Billy Goat curse was revived at least once or twice annually in some vain attempt to explain the Cubs' ineptitude.

Royko wrote about the comforts of the Goat often. So much so that he turned the tavern into a tourist spot. Over the years the printers with ink on their pants vanished, and many of the shot-and-beer patrons were replaced by Yuppies and women in furs getting off Gray Line tourist buses to see the place where grill men shouted "cheezborger, cheezborger," and where Mike Royko drank, sometimes signed autographs, and sometimes cursed at strangers. But the walls remained adorned with pictures and enlarged stories of the Chicago newsmen who passed by and often passed out. None was more prominent or prevalent than Royko. The Billy Goat had enough Royko memorabilia on the walls that it might have been Royko's own den.

After the Goat, Royko usually headed for the Acorn on Oak, where he would lean against the piano bar and listen to Buddy Charles play jazz. Rick Kogan was a latenight partner one evening when Royko was presented with two bowling balls for finishing second in a Bowling Writers of America contest. "I'm sure he never entered—they'd just spotted some column he had written on bowling," said Kogan, who was commandeered to carry the balls around all night. "At the Acorn, after a few cocktails, we got ten people to stand like human bowling pins by the bathroom and we had a bowling match. We didn't roll them very hard but the rule was if the ball touches you, fall over like you were pushed. Everybody got into it." Kogan said.

But even when the Acorn on Oak closed at 4 A.M., Royko was not always ready to call it a night.

Terry Shaffer was living in the western suburbs, which was a long dry ride from the North Side. "I found a place in Cicero where you could get a membership card and buzz the door after hours. If you weren't too drunk they would let you in. It was called Tiffany's, and it was a rough spot. Guys were in there packing guns. One guy was shot in the parking lot. One night Mike and I were drinking on Wells Street and the place closed, and I took Mike out to Cicero. He had on Levis, which was rare for him, and I told him they might not let him in. I buzzed the door and was let in, but when the owner saw Mike had Levis on he was going to throw us out. I introduced them and of course he knew who Mike was and he went in the back and came out with a raincoat and told Mike to put it on so no one noticed the Levis. He gave Mike a card, and after that he used to go out there quite a bit."

Schackitano said, "In those days a lot of different guys went out with him. There was no way I could keep up every night. No way one man could keep up with him night after night."

At home, Eleanor and Ed Cronin were trying to keep a household together. "I'd cook every night but Mike never ate much. He was always so critical. He'd say, 'The soup's not the way mama made it,' or 'I can't eat that.' The boys were used to eating out all the time," Eleanor said.

"When my mom passed away, I was only sixteen, but I was an adult, a drug-addicted adult, but an adult," Rob Royko said. "And here comes Aunt Eleanor. She walks into my room; the walls are painted black, the carpet's bright red, and the lights are yellow. Two window panes are busted. I used to break them when I'd lose my keys. There's a chain with a combination lock around the window so my friends could come in when I wasn't home. I had my own telephone. Aunt Eleanor thought I was dealing drugs because of all the people who were in my room. I kept beer in the refrigerator, and when I'd grab one she'd say, 'What are you doing, taking your father's beer?' I'd tell her, 'That's his beer on that side. This is mine.'"

David Royko remembered coming home and finding his father and Robbie and Robbie's friends there. "Dad was drunk and Robbie and his friends were stoned and Dad was regaling them. Robbie spent some time with him, but Dad and I didn't see each other much, didn't talk much. But I'm sure having kids was one reason he didn't jump off a bridge after Mom died. He lived in terror at the thought of being without her. All those nights out—I never remember him being home before I went to bed—I'm sure all that stuff that went on outside the house didn't mean anything to what he felt about her. There was never in his mind a question of his devotion to her. He was always in fear of losing her."

Eddie Cronin sometimes took Mike's place as the disciplinarian. "One night about midnight I went up to Robbie's room. He must have had fifteen guys in there smoking pot. Shack Jr. was one of them. I said, 'Okay, everybody out.' They wouldn't go so I looked at Shack Jr., he was the biggest, and said, 'You're going first if I have to throw you down the stairs.' Robbie said, 'He'll do it.' They all left.

"About once a week Robbie would break a window and I'd have to replace the pane. Mike saw me doing it one time and didn't say anything. He knew," Cronin said.

After Carol's death, Mike slept on the sofa.

"He never slept in my mother's bed again after she died," David Royko said.

"And he stopped listening to classical music after she died. I think it hurt too much. There were a lot of things that changed after she died, and music was the big one."

"Classical music was central to his life. I remember once we were in the car, I was about thirteen, and the Beatles were my obsession. I asked him, 'Don't you think the Beatles made a lot of great music?' He said, 'If you took all the Beatles albums, how many hours of music would that be?' I said they made about twelve albums of a half hour each so that'd be six to eight hours of music. He said, 'That's less than all the Beethoven symphonies and that's not all he wrote.'

"He took a great deal of pride in being able to differentiate performances of Beethoven symphonies or whatever," David continued. When (Georg) Solti took over the Chicago Symphony he loved it. It's not something he wrote about much, but he was really into it, and we went to the Symphony a few times. The music was his scene, but the scene itself wasn't his scene."

Royko had told his brother, Bob, that he was attending Central YMCA when he heard his first opera, *La Bohème*, and was enthralled, but no one is certain where or why he became an aficionado of classical music.

"My mom always thought it came from his desire to improve himself," David Royko said. "I can imagine him reading George Bernard Shaw or other great writers who mentioned Mozart or Beethoven or Wagner, and he decided he had to know what they talking about. But I think he was driven only if he really loved something or that it fit the image he wanted. Of course, if he liked it then he would become obsessed with being the most knowledgeable."

Royko loved blues the way Buddy Charles played them. He loved swing era music, Louis Armstrong and Earl Hines and Erroll Garner. He hated bebop, or modern jazz and performers like Miles Davis, Charlie Parker, and most of the works of Duke Ellington.

"He thought it was an arrogant form of music, that musicians stopped giving a shit about the audience and only played for themselves," David said, which would have been consistent with Royko's universal disdain for pomposity in all fields.

On November 22, after being out of the newspaper for nearly two months following Carol's death, Royko returned with a poignant story of a man closing his summer cottage for the final time.

The two of them first started spending weekends at the small, quiet Wisconsin lake almost twenty-five years ago. He worked odd hours, so sometimes they wouldn't get there until after midnight on a Friday.

But if the mosquitoes weren't out, they'd go to the empty public beach for a moonlight swim, then sit with their backs against a tree and drink wine and talk about their future.

He wrote that the relatives had sold that first cottage but the man had more money than he ever dreamed of, and one day they bought another cottage on the same lake:

> They hadn't known summers could be that good. In the mornings, he'd go fishing before it was light. She'd sleep until the birds woke her. Then, he'd make breakfast and they'd eat omelets on the wooden deck in the shade of the trees….
>
> This past weekend, he closed the place down for the winter. He went alone. He worked quickly, trying not to let himself think that this particular chair had been her favorite chair, that the hammock had been her Christmas gift to him, that the lovely house on the lake had been his gift to her.
>
> It was the last time he would ever see that lovely place. Next spring there will be a For Sale sign in front and an impersonal real estate man will show people through.
>
> Maybe a couple who love to quietly watch sunsets together will like it. He hopes so.

His next column was a blast at the television networks' coverage of the Iranian hostages, and the day after that he ripped Ronald Reagan. He was back.

While Royko was on his sabbatical, his legperson, Patty Wingert, who had been with him since Shaffer's brief rerun ended in the fall of 1978, had been assigned to the city desk. Hanke Gratteau had been working in public relations, hoping to get a call from the *Sun-Times*, which maintained a rehire list of those who had been let go from both papers when the *Daily News* folded. Royko needed another leg-man. Gratteau wanted back in the newsroom.

"Mike also needed help," said Gratteau. "He wanted, maybe for the first time, someone to lean on, and he wanted to be near his friends. There was socializing but there was also caretaking. It got to the point where we almost had a schedule of who would be with him at night. None of us could take it every night. I remember one night, about one in the morning, I had an apartment at Sandburg Village and Julius the doorman buzzed me and said, 'I have Mr. Royko down here.' I got on the phone and said, 'Look, you can come up here and drink all you want but I can't go out. I just can't.'

"Then there were the Royko Rules," she added. "No matter how drunk he got or how late he was out, he was always in the next day for the column and no matter how close we got or that I was hung over because I was with him, you had to be there in the morning. A hangover was no excuse. The column came first."

Even Royko was having trouble following the Royko Rules. He tried again to quit drinking by entering the alcohol rehabilitation program at St. Francis Hospital. He called a few close friends, Lois Wille and Shack, to let them know he was there.

"He lasted about four days," Schackitano said. "Mike couldn't take a program like that where you had to let someone else be in control. He always had to have control."

Wade Nelson became a regular member of nightly tours. Nelson was not hired by the *Sun-Times* when the *Daily News* folded, and he moved to Washington and married Ellen Warren, who was covering the White House for Knight-Ridder newspapers. He was hired by the *Baltimore Sun* and then accepted an offer from Illinois Secretary of State Alan Dixon to work in his 1980 campaign for the U.S. Senate seat that opened with the surprise announcement by Adlai Stevenson III that he would not seek another term.

"I moved back to Chicago on January first, 1980," Nelson said. "I hadn't seen Mike for a while and hadn't talked to him since Carol had died. I called him and we got together for a drink and then it became a regular thing. He had absolutely nothing to do beyond his column. He wasn't spending much time at his house. He was either at the office or out with his pals which was a changing cast. But Tony Campbell and Hanke and me were always around. We'd meet five nights at week after work at the Goat's or Riccardo's.

"Sometimes we'd have a couple of drinks and go home. Sometimes, particularly Friday, might be an all-night session. Some would drop off, but others would stay. We'd start at Ric's, go to dinner sometimes, then maybe O'Rourkes, which closed at two, then to the Acorn on Oak to hear Buddy Charles until four and then down to the Oak Tree on Rush and Oak for breakfast.

"If he was still in a mood to drink, we'd go out to Cicero and we'd come out of the Tiffany at seven in the morning in bright sunlight and look at the people going to work," Nelson said.

"Sometimes we'd go back to his house and go have some lunch. Then he'd rent Fred Astaire and Ginger Rogers movies, and we'd sit there with these incredible hangovers and figure out whether it was time to start drinking again. The fact that he was cranking out five columns a week of incredible quality not only while he was in personal pain but while he was drinking heavily showed what an amazing constitution he had. He was a world-class writer and a world-class drinker," Nelson said.

One of the items Royko had saved in his many boxes of memorabilia was a hand-scrawled four-page list of rules for "Lonely Guys."

It was written at the bar of the Mayfair Regent Hotel one night in the winter of 1980 when Nelson, Gratteau, and Campbell, the *Sun-Times* financial editor, were with Royko.

"Tony didn't have a girl friend. He and Hanke had just broken up, and Hanke didn't have anybody. Ellen was in Washington and I was living with my mother, and Mike considered himself a lonely guy, and, for whatever reason we came up with a list of characteristics for guys who were lonely," Nelson recalled.

When Mike was drinking he always put on an accent and mimicked a guy called Chisel who used to come in his dad's saloon on paydays and shout, "Cham-pag-nee for all vomans. I no chisel."

"We all copied it and Hanke is shouting, 'Cham-pag-nee for all vomans,' and we drew up this list and we put a lot of names on it who couldn't be members of the Lonely Guys, mostly whoever Mike was mad at," Nelson said.

The group effort, scribbled in several different handwritings, was distinctly Roykoesque, as some of the rules indicate.

- The ultimate condition for a Lonely Guy is to be the only member.
- We are courteous to models, starlets, stewardesses and other weak creatures.
- Lonely Guys have meetings every night—or more often in emergencies.
- Any Lonely Guy who pisses in his pants during a meeting is suspended until 5 P.M. the next day.
- The "Temples of Loneliness" where the club will attempt to arrange discounts are: Billy Goat, Riccardo's, O'Rourke's, Acorn on Oak, Oak Tree, Tiffany, Stop and Drink, Cricket's.

The Lonely Guys Club was created and disbanded on the same night, which was a good thing for Royko, or he would have had to resign because he was involved with Pamela Warrick.

Pam Warrick grew up in Indiana, and her first job in 1969 was as the second reporter in the Goshen bureau of the *Elkhart Truth*. "I really got an introduction to the newspaper business," she recalled. "The bureau chief kept a bottle of Kessler's in his drawer and chain-smoked. One of my assignments was to check the fire department every morning so I'd be there at dawn having breakfast. One day the alarm rang and the firemen took off and it turned out to be my office. The bureau chief had fallen asleep with a burning cigarette."

Her next stop was at the Chicago suburban chain, Pioneer Press, where she met Ellen Warren and eventually met many people who worked at the *Daily News*. From there, Warrick was hired by *Newsday* in Long Island, New York, where she spent five years working on a high-powered investigative team that won several major journalism awards.

"I also got to spend winters in Albany covering the legislature. It was very exciting. I remember when New York City was going broke, I interviewed Mayor Abe Beame who was sitting in the hallway of the capitol with hat in hand, literally," Warrick said. By 1977 she wanted to return to the Midwest because of the ailing health of her father, who lived in Wisconsin.

"I was the last person hired at the *Daily News*," Warrick said. It was November 1977, only three months before Marshall Field would announce the closing of the newspaper. "Bobby Shriver (the son of Kennedy clan members Sargent and Eunice Shriver) was hired in the morning, and I was hired in the afternoon. When the *News* folded everyone was anxious to keep their jobs. We were all trying to figure out how to convince Jim Hoge to take us on at the *Sun-Times*. I baked a big tray of cookies and delivered them warm to Hoge. He said, 'Thanks, this is great. See what Bobby Shriver brought me.' It was a box of Cuban cigars. I thought I was screwed, but Bobby went to law school and I went to the *Sun-Times*."

Warrick began working with Pam Zekman, for twenty years Chicago's top investigative reporter at the *Sun-Times* and later WBBM-TV. Zekman had spearheaded a fascinating series in 1978 when the *Sun-Times* bought a bar, named it the Mirage, and waited for the various city officials to ask for kickbacks to keep the place open. It was the kind of story Royko loved and the entire city gobbled up each installment. The *Sun-Times* was convinced it had a Pulitzer Prize in the bag, but the rules of engagement for journalism were changing. The *Sun-Times* reporters worked undercover, never revealing they were reporters. The Pulitzer Prize board rejected the jury's selection of the Mirage series because it felt the newspaper engaged in deception. Royko was outspokenly critical of the Pulitzer board and often blamed the fact he never received a second Pulitzer on his rude remarks. Royko had been a Pulitzer finalist again in 1978, but the board that year selected *New York Times* writer William Safire as the commentary winner. In fact, the commentary prize has come to represent something of a lifetime achievement award because no one has ever won more than one.

That did not deter Royko from flaring up in anger whenever anyone won, or worse, when anyone suggested one of his peers had actually written something worth reading.

Warrick said, "I made the mistake once, just once, of suggesting that something Russell Baker wrote was humorous. Mike got hurt."

Warrick and Zekman worked on several series including one on abortion profiteers. "That's when I first met Mike. He would come up to Pam and me and tell us he thought the stuff was terrific. He seemed to be very encouraging professionally and I was never intimidated by him," Warwick said.

"I used to stop at his office if I saw he wasn't on the phone or writing. I went in once and he invited me to sit down, then he pulled a six-pack of beer out of a file drawer, broke it in half and gave me three and kept three. Then he talked about journalism in Chicago for about an hour. I think that was on a weekend when not many people were around. That was the beginning of our friendship," Warrick said.

"I remember years later I stuck my head in a couple of times at the *Tribune*, and he still had a six-pack in a file drawer," she said.

"He also had a drawer with all kinds of clothes sticking out. I think it started with softball stuff; then it became any kind of clothes. Sometimes he'd reach in blindly and pull out a tie and say, 'I haven't seen that in twenty years.'

"And he always had his sign that was a quotation from Samuel Johnson that said, ONLY A BLOCKHEAD WOULD WRITE FOR ANY REASON OTHER THAN MONEY. I don't think he really thought that way."

Through the spring of 1980, Royko and Warrick became very close. "We were engaged. A lot of the time we just stayed at home. We kept separate places but were living at either his place or my place. I lived on Fullerton Parkway in a brownstone across the street from (Governor) Jim Thompson. I had a landlady who lived in the basement. One day she came to get the rent and asked, 'Was that Mike Royko I saw on the patio last night?'

"Sometimes he would lay out on the patio. He was pretty much an insomniac and sometimes he would fall asleep. He also had this thing that sometimes he would stop breathing in his sleep. It really scared me.

"I was deeply in love, and I thought he was, too. But he was still torn. He still wore Carol's wedding ring on a chain around his neck. He had pictures of her all over the house. They were on the inside of the kitchen cabinets. He talked on and off about Carol. In his column he had created a positive portrait of a marriage, but it wasn't perfect. He'd say things, 'It was my fault, I caused it, I drove her off.'"

Royko went into one of his penitential moods, trying anabuse again to stop his nightly forays. "At the beginning of our relationship he was taking anabuse, not because I asked him to, but he tried a lot of different things. When he was drinking, I was drinking and that made for a very volatile relationship. We would both say the next day that we were sorry about everything that happened last night, let's try again today," Warrick said. "He was insecure about so many things. He always worried that today was the day the whole thing would fall down, that he wasn't worthy of all the praise and everything he got. He was insecure about his lack of college, especially when he looked at Jim Hoge, and he used to get pissed off that he looked older than he was."

Royko was still going through a period of adjustment at the *Sun-Times*. Seventy *Daily News* employees had moved with him, and he thought their presence made the *Sun-Times* a much better paper, but he still was not comfortable there. Although he and Hoge would come to have a solid relationship and a deep appreciation of one another, Royko still felt wary of him.

"He had this thing that Hoge had been born to the manor, that Mike was clearly brighter, more clever, but he never could be Jim Hoge," Warrick recalled. "He was almost dismissive of the people at the *Sun-Times*. He really thought that it wasn't a paper that appreciated a great writer. He was always quitting. When I was dating him, he would spend a lot time with a pad of paper, figuring out what his assets were, if he could afford to just quit, to stop writing his column. He could have afforded it, but he never could quit the column."

Of course, for whatever reason, Royko often resigned.

Warrick said, "One night he fell asleep on my sofa, and suddenly he woke up and grabbed the phone and called Ralph Otwell at home, wakes him up and quits. I was in shock. I thought, 'My God, this is terrible. The greatest writer in Chicago has just quit and it happened in my house. I was crying and begging him to call Ralph back. He said, 'I'm not taking any more of their shit.' The next morning he went in and wrote his column."

Royko had been granted access to the executive lunchroom when he was made associate editor of the *Daily News,* and he had enjoyed going there and holding court for some of his close friends. Many of the friends that he used to go out to lunch with in his earlier days were gone, either retired or out of the newspaper business. But he quickly commandeered the *Sun-Times* softball team and continued trying to relive his youth, or perhaps more accurately, reinvent it.

His sister, Eleanor, said, "Mike was never competitive as a boy. He was a loner, a bookworm. He didn't become competitive in sports until he was a man."

On April 20, 1980, he played one of his great tricks on Chicago. He became a White Sox fan. He claimed he could no longer put up with a team whose players pouted about their salaries. He went to Comiskey Park on opening day and with legendary promoter Bill Veeck, the Sox owner, he went to the pitcher's mound. Of course, the moment became a column.

> It was an uplifting spiritual experience. Veeck bought me a couple of beers; Harry Caray welcomed me and bellowed "Holy Cow" in my left ear, and I went home and learned the words to "Na, Na, Na Na."

It made news all over Chicago. But a few months later, the Sox were sold to an Ohio shopping-mall magnate, and Royko reneged.

In August, Royko made one of his dreaded trips to the Democratic National Convention in New York City. He hated political conventions, hated competing against people who might possibly know as much about the politics and players as he did. Besides, he had to drive, and Pam Warrick, who was assigned to cover the convention, wasn't able to join him until a few days later.

Robert Novak, the Washington-based *Chicago Sun-Times* syndicated columnist, remembers Royko's arrival at the newspaper's work space at the Statler Hotel across from Madison Square Garden where the convention would be held.

"He came in and all he could say was 'I hate New York, I hate conventions, I hate Democrats. I'm not writing one goddamn word.' He stalked out," Novak said.

He undoubtedly stalked to one of several bars in the hotel, then to the liquor store, then locked himself in his room and refused to answer the telephone. The *Sun-Times* put out an SOS to Hanke Gratteau, who flew to New York and talked Mike out of his room. At least that was the story that circulated in the *Sun-Times* newsroom. "What really happened," Gratteau recalled, "was that I really wanted to go to the convention, and I hadn't been assigned. Mike knew if he threw a tantrum they would send me to New York to get him back on track and that's what happened."

One of the things that Pam Warrick and Royko argued about a lot was politics. She was mad that he had reduced one of her favorite presidential candidates, Jerry Brown, to a pithy phrase. Royko finally agreed with her:

I don't think many people noticed, but somebody finally made a highly intelligent speech to this convention....The speech was made by Jerry Brown, governor of California, who is sometimes referred to as "Governor Moonbeam." I have to admit that I gave him that unhappy label. I'm sorry I did it because the more I see of Brown, the more I'm convinced that he has been the only Democrat in this year's politics who understands what the country will be up against in the future.

Having placated his conscience (or his sweetheart), Royko went off to have fun:

I told the cabbie to take me to Umberto's. And it turned out to be another reason I envy New York. Umberto's is on Mulberry St, which is the heart of the old "Little Italy" section. Block after block you have one great Italian restaurant after another. Umberto's may not be the best of them. But it's the only joint on Mulberry St. that can boast that Crazy Joe Gallo was ventilated while eating its magnificent linguine with clam sauce.

His column on famous mobsters and the New York restaurants where they were gunned down was much better than anything else he, or most others, wrote about the convention, which was foreordained to renominate President Carter despite a few late, unrealistic gasps of hope from supporters of Edward Kennedy.

Warrick had gone through a rental agency to find a place on Long Island for a vacation after the convention.

"We drove out to West Hampton to this lovely place with the bay on one side and the ocean on the other. When we got there we weren't alone. The owner had rented us the summer house, but he was shacked up in the pool house with some teenage bimbo. And the house is full of pictures of guys like Carlo Gambino. Mike started yelling at me, 'What the hell did you rent?' Then the phone rings and it's Jimmy Breslin who knows the guy who owned the place.

"Breslin recognized Mike's voice and said, 'Royko, what the hell you doing there?' Mike said, 'Breslin, what the fuck are you calling me for?' We had a great time, just walking on the beach, laying around the pool. We drove back to Chicago in his Chevy Blazer and it was a great trip. We listened to tapes the whole way, blues, lots of blues, but he also had musical comedies, *The King and I*, which he liked but would never admit," Warrick said.

But the romance was shaky.

Warrick received an invitation to apply for a journalism fellowship at Stanford University for the 1980–81 year. "I was in my early thirties, and my idea was that I'll take this fellowship and have a last fling at freedom before I become an instant stepmom and caretaker. By now I knew the situation with Robbie and David and what life would be like."

In September, Warrick drove out to Stanford where she and Royko continued their relationship through a long series of letters and occasional telephone calls. "He would pick up the phone in that gruff voice and when I said it was me his voice would get two octaves higher. On his letters he would draw little hearts with arrows in them. I don't know how many people saw that side of him."

In January, during the break between semesters, Warrick joined Royko and his sons on what had became an annual Florida fishing vacation. Robbie brought along the girl he was dating, but David's girlfriend at the time couldn't make the trip.

All of them remembered one particular night when Royko and Pam were both drinking and she asked him, "Are you drunk?" He replied, "You are looking at a man who is lighting a cigarette in one hand, smoking another one in the other hand and you have to ask if I'm drunk."

David, who liked Pam, was upset over the trip and flew home. "Dad would always give you twenty dollars to buy a Coke or a loaf of bread and say keep the change. He did this so much after a few days I had enough money so I just left and caught a plane."

David Royko said, "I found it amazing how easily Pam could accept Dad. Sometimes he would drive me nuts, but she'd just tell him to fuck off or laugh. After a lifetime of seeing my mother trying to just sort of survive, those long stretches when Dad was not a whole lot of fun to live with, it was amazing to see how relaxed Pam was and how much fun she had. With Pam it was like a big sister thing. Of all the women he dated after my mom died I got to know Pam best, and she was very nice. Our lives didn't overlap much even though we were living together, but she was very sweet and I just couldn't see her trying to cope with Dad."

Warrick went back to Stanford and finished the fellowship in time to return for David's graduation from Lake Forest College, where Royko was given an honorary degree. "He really loved that day," David Royko said, "and he really got a kick out of it when they were handing out all the diplomas and we were marching across the stage. When the dean called out my name he turned and gave my diploma to Dad."

Warrick had struck up a relationship with another fellowship student and was beginning to have doubts about the future of her relationship with Royko, which she voiced to him often during that summer.

"We would spent a lot of time up at the Wisconsin cottage—he had a shag rug and round bed which he thought was the height of cool—and we would talk and cry. While I was at Stanford, I thought, can I really do this? I really didn't want to stay in Chicago. I've always moved around. I loved New York. I loved California. I said to Mike, 'You can do your column anywhere, why not move? He said he had to be in Chicago, that he couldn't do his column anywhere else. I knew I was going to lose that argument.

"Mike said he had really severe headaches most of his life. He said the only thing that made them feel better was to go out into the Chicago wind that he called the Hawk. 'I need the Hawk,' he would always say."

"But drinking was the main problem. When he drank, I drank. I couldn't be his policeman. I don't know if it was the age difference, but I couldn't be that strong. The other was that I felt he needed someone who would be subservient. He never said that, but he always talked about how his mother let his father run things. He saw that as nurturing.

"When you got close to him there was a tendency to want to save him but I never felt I could tell him to clean up his act. Is a cleaned up Mike Royko the same Mike Royko? Maybe not. I couldn't do it. I found a quote by one of Hemingway's sons that always made me think of Mike, 'He drank. It was not the core of who he was.'"

Warrick moved to Washington, D.C., in early 1982. Their affair was over. She later moved to California and worked part-time for the *Los Angeles Times* while raising two children and in 1999 joined the staff of *People* magazine in Los Angeles.

When their romance ended, Royko went for consolation to Lois Wille.

"He came to my office and was crying, I mean real tears. He cried rather easily. He told me, 'I've loved two women in my life, and now I've lost both of them.'"

chapter 13

On any given day, Mike Royko could find someone to hate.

To his friends and family it seemed incongruous that someone who could cry watching Fred Astaire dance could spew so much venom at almost anyone he perceived as an enemy.

Throughout his career Royko went out of his way to demean and attack almost anyone who tried to do what he did in print, although few of them even approached his excellence. He was relentless in trying to destroy insignificant, virtually anonymous writers, who he believed had stolen his thoughts or phrases. Some of that response was pride of authorship, for he slaved over every sentence and searched long for the right word, and he was furious that someone in some small town would steal his work.

Nothing could anger Royko more than someone stealing his thoughts and his words.

In 1974, Robert Vare wrote a book called *Buckeye* about Ohio State football and its then legendary coach, Woody Hayes. Someone brought it to Royko's attention. Horrified, Royko devoted a column to what he called the "the neatest literary ripoff since Clifford Irving's caper." Royko printed excerpt after excerpt of Vare's opening chapter on the daily routine of Hayes, which was remarkably similar in form and language to the opening chapter of *Boss*.

He concluded by noting, "The blurb on Vare's book says that he is now working on a novel. Whose, I wonder." The *New York Times* noted that after Royko's column the Vare book took "a harsh pounding" and its credibility weakened.

There were many newsmen Royko dismissed as frauds, scorning them in the same manner as the politicians and insensitive bureaucrats he savaged. And then there were the people he just didn't like, a great deal of them radio and television broadcasters whose celebrity matched his own. He believed they displayed little talent or value.

Part of Royko's orneriness was for the good of the column. Sensational crime and high tragedy have always been the grist of newspapers, and Royko knew it better than anyone. The column needed conflict. It also needed new protagonists to change the pace. He had to give the politicians a day off. He needed new enemies, and when there weren't any handy, he created them. Sometimes it was a place—New York and California were favorites. Sometimes it was a group—rednecks or gun nuts or joggers. Many times it was a celebrity—Muhammad Ali, Frank Sinatra, Anita Bryant, Howard Cosell. But once Royko created an enemy, they stayed enemies.

Royko's disdain for television usually took the form of sneering at Walter Jacobson, the WBBM-TV co-anchor, who was the subject of more Royko columns than almost any of the other celebrities he chose to attack. Royko often called Jacobson "Walter Babytalk." He was the standard answer in Royko's annual Cub quiz to the question, "Which Chicago television personality used to be a Cub batboy?"

And the answer would be something like, "Walter Jacobson, who hated the job because the players threw their underwear at him."

After five years of Cub quizzes, Jacobson wrote his own tongue-in-cheek rebuttal in a letter to the *Daily News*:

> Your Mike Royko has gone just about far enough. His gratuitous slap at me for being a Cub batboy is demeaning to every law-abiding citizen who, in the process of growing up, got himself a summer job. I worked hard from nine in the morning until six at night for $2.50 and a ham sandwich....Furthermore people in glass houses shouldn't throw underwear. The truth of the matter is that Mike Royko was a batboy himself. Everyone knows how he played in sandlots and alleys—throwing things, always throwing things; he's still throwing things.

Royko once explained why he didn't like Jacobson. The column began:

> Walter Jacobson made a shocking and scurrilous personal attack on me. Jacobson flatly stated on the air that and he and I are "friends."
>
> I suspect Jacobson smeared me because of a grudge that goes back many years when were both young reporters covering the courthouse. Then, as now, it was his dream to someday come up with a "scoop," which is an exclusive news story. However, he did not know how to get one. He used to run up to strangers on the street and cry: "Gotta scoop? Gotta scoop?"
>
> He came to the pressroom early in the morning before any of us had arrived and crawled into my desk drawer, hoping to overhear an exclusive story.... Most people don't realize this but Jacobson is quite tiny—barely three feet high.

Jacobson once opined he might like to leave television broadcasting and "work with his hands." Royko jumped all over the remark, dedicating a column to finding something to free Jacobson from the "burden of being a $150,000-a-year TV anchorman."

"I hope he doesn't really feel that way," Royko wrote. "Next to the Muppets, he is my favorite television creature."

Jacobson also once told an interviewer that he was underpaid at $140,000 a year, since two other anchormen at his station earned $200,000.

Royko, who at the time was making about $60,000, wrote, "I have no idea why Walter makes only $140,000, and how he scrapes by on that amount. With the rising price of hair spray these days, I can see why he is upset."

The people on Royko's enemies list ranged from anyone he perceived as a rival, which ranged from the august pundits of the *Washington Post* and *New York Times* to anyone who wrote any kind of a column in any Chicago newspaper. It also included reporters he believed fabricated news stories and almost anyone who didn't work at the *Daily News*. He also despised, almost unrealistically, anyone who stole a line or a paragraph from him.

While many of Royko's peeves and hatreds were personal, much of his ire stemmed from what he perceived as journalists betraying their public: cloying at the public figures who provided them tidbits, reporting rumors without checking facts, failing to allow targets of criticism the opportunity to respond, and in the case of television news, the superficiality of its performers. And he didn't care how high or low his targets sat on the journalism ladder. He attacked Dan Rather for riding to and from political convention arenas in limousines while CBS was laying off reporters. He went after an obscure columnist in El Paso, Texas, who he swore stole three paragraphs from his Iranian hostage piece.

He never forgave the *Sun-Times*'s Tom Fitzpatrick for winning a

Pulitzer Prize for his 1969 coverage of the radical Days of Rage window smashing spree on Chicago's North Side. In a 1990 *Chicago Magazine* piece, he said of Fitzpatrick, "He's the dirtiest little man I've ever known and I ran him out of town twice. If he comes back, I'll run him out again."

Royko had written that the Days of Rage was an overblown media event and that the Students for a Democratic Society had merely run down Clark Street, smashed a bunch of windows, and retreated rapidly when police lines blocked their path:

> The police had pulled a sneaky trick. Instead of flailing anything that moved, as they did during the Democratic National Convention, they acted coolly, professionally, made quick arrests, kept the head-butting to a minimum. The SDS found itself scattered all over the Near North Side. And it is harder to be brave, and charge police lines, in groups of four or five. The rear rank is closer to the front.

Fitzpatrick said, "I admired him, but I never liked him and he never liked me."

When *Boss* was published, Hal Scharlatt of E. P. Dutton sent a possible list of people to be invited to a book party. Royko replied, "There are two names I'd like removed from the list, both radio jerks. 1. Jack Eigen, a nasty little SOB. I hate his guts. 2. Wally Phillips, with whom I have feuded. I hate his guts."

Another of his ongoing feuds was with John Madigan, a former city editor of the *Chicago American*, who was both political editor and media critic of WBBM radio. Ellen Warren recalled, "[Royko] taught me to hate John Madigan with a passion."

Madigan incurred Royko's wrath by pointing out in his media commentary that the *Daily News* had buried the story about the ketchup incident in 1977 and that the *Sun-Times* did not write a story on Royko's drunken-driving arrest in 1979. Moreover, Royko thought

Madigan treated the City Hall bosses far too kindly and pointed it out in 1977 when Mayor Michael Bilandic proclaimed December 7 as John Madigan Day.

"A bunch of politicians are even holding a banquet for him," Royko wrote. "I hope Tom Keane is paroled in time to make it....If ever anyone in the news business has earned that type of recognition from City Hall, it is John Madigan. It shows what can happen if a fellow keeps his nose to the whatchamacallit. No, I don't mean the grindstone."

Another fellow high on the list of enemies was George Will, the Washington columnist who was First Lady Nancy Reagan's dinner companion:

> George Will is more than a columnist. He is the chief egghead of conservatism. He not only writes and pontificates, he has helped craft some of President Reagan's speeches and has helped coach him for debates. He doesn't stand on the sidelines and watch. He bulls right into the middle of the cricket game.

Royko once wrote that Will, a notorious Chicago Cubs fan, had called him and informed Royko he planned to write a column about the Cubs season opener. He wondered if Royko could send him some of the columns he had written about the Cubs. He quoted Will as saying, "I'd like to take a look at them. Why don't you bundle some copies of them together and mail them to me?" Royko said his secretary refused because she thought Will would use it for himself.

Royko didn't dislike David Broder, the nation's leading political writer, but he did take him to task for Broder's "brazen misuse" of the word "clout." Broder used "clout" to mean power. "Clout is used to circumvent the law, not enforce it," Royko wrote. That brought William Saffire of the *New York Times* into the fray, who sided with Broder.

Years earlier, Royko had picked a fight with *Vogue* magazine for its misuse of "clout." *Vogue* wrote that President Johnson and Ho Chi Minh had clout. Royko was furious. "Clout means influence," he wrote, and then typed out an example that became a stereotype for Chicago politics throughout the country: "Somebody beefed that I was kinky and I almost got viced, but I saw my Chinaman and he clouted for me at the hall." Translation: someone complained about a patronage employee and he almost got fired but he went to his political sponsor who took care of the problem with the mayor.

Another of Royko's feuds was with priest, columnist, and author Andrew Greeley, who wrote of Royko, "He is crude enough to fit the stereotype of the Chicago Slav...." That attack happened, according to Royko, after he was in a television studio where Greeley was appearing on a news show. "Well, at least there are two of us on the show," Greeley said. Royko asked, "Two of what?" Greeley replied, "Two Irishmen." Royko asked, "How many Poles are there?" According to Royko, Greeley said, "None. They don't know to talk."

The feud continued after Royko made an appearance on the *Today* show in 1976 and was, as usual, highly critical of the Daley machine. In his *Chicago Tribune* column, Greeley labeled Royko's criticism anti-Daley venom. Royko wrote Greeley, "I have warned you in the past about this kind of irresponsibility, you thin-lipped, constipated, quivery twerp."

David Halvorsen of the *Chicago Tribune* wondered if the whole feud was a sham. It might have been except for Greeley's most grievous sin. Years earlier, he had written several negative reviews of *Boss*. Any minor item in any publication that Royko perceived as criticism could set him off.

Lois Wille said, "He would see some remark in *Chicago Magazine* or the *Reader*, and he'd call and ask if I saw it and then he'd say, 'I'm going to get that guy.' I'd say, 'Remember, the eagle does not hunt flies,' but he would go on about how someone was out to destroy him. Nobody could touch him, and he knew that, but sometimes he just couldn't ignore all the little cheap shots. He couldn't brush it off. It got to him."

As usual with Royko, nothing was consistent. In 1968, Dorothy Stork, a former fifteen-year air force veteran who had junked her military career after a successful free-lancing period brought her a job offer from *Chicago Today*, began writing four columns a week. It was *Chicago Today*'s latest effort to challenge Royko's supremacy.

Royko should have detested Dorothy Stork.

"I was so nervous about doing the column I came down with the hives," Stork recalled. "I was walking down Michigan Avenue, and I ran into Royko who I wasn't sure knew me. He said, 'How're you doing?' and then hauled me off to someplace I had never been called Billy Goat's. The place stuck to the soles of my shoes.

"He gave me great advice," Stork said. "He told me, 'You have got to remember you are doing what you are doing because you love to do it and you are good at it and doing it makes a difference in your life. And one more thing, never go beyond noon without an idea.' "

Stork also sought advice from Jimmy Breslin, who told her, "Ask for more money. Take one word out of every sentence you write. Only write about money, murder, and cats. Don't drink booze until you finish."

Stork particularly took to heart the last bit. "You remember in those days booze was so much of the culture. We'd go across to the Goat and get martinis in paper cups and bring them back to the paper while we worked. Royko could get nasty when he drank too much— I remember him slinging plates—but he never turned on me. I'd call and ask for advice or we'd meet at Riccardo's and I'd complain that the column was ruining my life."

Royko replied, "The column is your life."

About the only sportswriter Royko ever admired, except for his old pal Ray Sons, was the *New York Times*'s columnist Ira Berkow, who had grown up in Chicago. In 1991, while many New York scribes were writing memorabilia about the famous Bobby Thomson home run that captured the pennant for the New York Giants forty years earlier, Berkow wrote about his childhood reminiscence of a home run hit by Phil Cavaretta to win a game for the Chicago Cubs.

The game itself was pretty much meaningless, but Royko wrote to Berkow and congratulated him on his writing and explained that he also remembered exactly where he was when Cavaretta had hit the game winner. He said he was on a beach in Indiana with a cousin and a couple of girls, listening to the game on a portable radio. The day was memorable for him because it ended with a romantic tryst. Royko sometimes described it to friends as "losing my virginity."

He also wrote Berkow, "I once had the pleasure of meeting Cavaretta. I was 15 and a student at Montefiore....Cavaretta came to Montefiore to hand out Christmas gifts. Most of the kids were kind of wild and didn't know who he was. They were almost mugging him for gifts until he yelled: 'Siddown, ya' little fuckers, it's Christmas...'

"You write great stuff. If there's anyone doing sports who is even close, I haven't read him," Royko wrote Berkow.

In later years at the *Tribune*, he also showed his warmer side to Mary Schmich, who had started writing a column in the Metro section. Like so many others, Schmich had been wary of strolling into the Royko den.

"One day, one of his assistants came out and said Mike wanted to talk to me. I went back and he talked about writing. He was very encouraging and flattering. He asked me if I had ever read *And Quiet Flows the Don*. He said it was the greatest novel ever written. A few days later, his assistant dropped off a copy at my desk."

It was not surprising that Royko loved *And Quiet Flows the Don*, a novel by Nobel Prize–winner Mikhail Sholokov, which portrayed the resilient, indomitable spirit of the Russian people before and after the revolution. Royko's ethnic pride was swelled by the story of his forebears, whose toughness he admired and emulated.

Despite the mutual admiration he shared with Ben Bradlee, the *Washington Post* did not escape Royko's scorn. The scorn was piled

on when the *Post* was forced in 1981 to return a Pulitzer Prize it had
won for a story about an eight-year-old boy who was being turned
into a heroin addict by his mother's boyfriend. The story was a
fake.

Royko's response was to bash the *Post*'s editors for not demanding
the name of the child even when the District of Columbia police
chief asked for it:

> I'll tell you what I would have done if I had been the editor and a
> young reporter came to me with that same story. I would have said
> something like this:
>
> "I want the name of the kid now. I want the name of the mother.
> I want the name of guy giving the kid heroin. We're going to call the
> cops right now and we're going to have that sonofabitch put in jail,
> and we're going to save that kid's life. After we do that, then we'll
> have a story."

In newsrooms all over America there were plenty of editors and
reporters who would have done the same thing as the *Washington Post*
did. The sense of empowerment that the news media felt in the 1970s
had distorted the responsibilities of newspapers and elevated individual achievement beyond the simple task of supplying the public with
what it had a right to know. There were many newsroom seminars on
public responsibilities, and many young journalists were surprised to
learn they were first of all citizens who had a responsibility to notify
police of a crime just like any other citizen, not to save the facts until
a story was nominated for a Pulitzer.

Coverage of the *Post*'s fake story raised questions about the credibility of all American journalists, but Royko demurred:

> Only the Washington Post is tainted by this affair. After all, when
> Bradlee, Woodward, Bernstein and the Post became national idols

because of Watergate, they didn't say that the rest of us were also national idols. Now that they're bums, they can keep that distinction for themselves, too....

What would the Post have done if it had discovered that a congressman knew an 8-year-old child was being murdered, but had given the killer his word that he wouldn't reveal his identity? I know what the Post would have done. It would have demanded the congressman's scalp.

Royko wasn't really looking into a crystal ball. After the *Washington Post* broke the Watergate story, media critics were proclaiming a new era in journalism, an end to the *Front Page* era of Hildy Johnson. Royko said the new era lasted about nine weeks. It ended when the *Post* got wind that an Ohio congressman, Wayne Hays, was keeping a woman on his payroll who couldn't even type but was more accomplished after hours.

Hildy would have loved it. The Post's dedicated reporter—with the backing of their fearless editor—used or considered using just about every sneaky trick in the book. They tiptoed about, following Hays when he and Ms. Dim Wit went out on the town. In collaboration with Ms. Dim Wit, they listened on the telephone to her conversations with Hays. And, according to one published report, they even considered, but rejected, planting a recording device somewhere in the vicinity of Ms. Dim Wit's mattress.

It is worth noting that when Royko blasted the *Post* in 1981 he had long since ended his dalliance with the idea of moving to Washington and taking his column national. But in 1976, when he criticized the *Post* for the Wayne Hays story, he was on very friendly terms with Bradlee and was enjoying, if never seriously contemplating, the continual flow of offers to join the *Post*. Those offers were just as constantly leaked to various publications, and if they did nothing else, they certainly served as the ego fix Royko always needed.

Another of the great conflicts for Royko also took place in 1976, when he discovered that one of the entertainers he admired most was being "clouted" in Chicago.

On May 4, 1976, Royko wrote that Frank Sinatra had a twenty-four-hour police guard outside his penthouse suite at the Ambassador West Hotel. He filled his column with the usual inane explanations from various police officials, wondering if someone would threaten Sinatra because he didn't like the way he parted his hairpiece, and concluded, "Frankly, I'm surprised that Sinatra, who has such a tough reputation, would need somebody standing outside at all hours. He's an absolute terror when it comes to punching out elderly drunks or telling off female reporters."

Sinatra sent a letter that Royko reproduced in full the next day. Sinatra denied requesting police protection and called Royko "a pimp." Then Sinatra wrote:

> Lastly, certainly not the least, if you are gambling man:
>
> (a) You prove, without a doubt, that I have ever punched an elderly drunk or elderly anybody, you can pick up $100,000.
>
> (b) I will allow you to pull my 'hairpiece.' If it moves, I will give you another $100,000; if it does not, I punch you in the mouth. How about it?

In the column, Royko said he had to admit it was a thrill to receive a hand-signed, copyrighted letter from Sinatra "even if he did call me a pimp":

> For thirty years, I've considered him the master of pop singers. Why, in 1953, I played his great record of "Birth of the Blues," so often that a Korean house boy learned every word. And he probably taught the

song to his children. So if Sinatra has a fan club in the Korean village of Yong Dong Po, it's because of me. I mention this only to show how deeply it pained me to be critical of him. The pain may have been brought on by French fries at lunch, but I prefer to think of it as sentiment.

The feud made news all over the world. Friends sent Royko clippings from British and Italian newspapers. Royko enlisted an old pal to keep a good thing going.

His next column consisted of a visit from Ben Bentley, a popular sports promoter who suggested Royko take Sinatra up on the offer to have a fight:

> "I can see it. It would be the greatest promotion I was ever involved in. I could sell 25,000 seats in the north end of Soldier Field and the cheapest would go for a half a yard."
>
> "How much is half a yard?" I asked.
>
> "Half a yard, in the language of the rubes, would be $50."
>
> "When we had the weigh-ins," he said, "you and Sinatra could argue and threaten each other. Then, when we've picked out the gloves, you could argue about that.... Then there'd be the most exciting moment when you step into the ring. That's the most exciting moment in sports. And I'd introduce you: In this corner, weighing 185 pounds, in the white truss....
>
> "In the white truss? What do you mean by that?"
>
> "Did I say that? I meant trunks...."

Royko got three more columns out of the Sinatra affair, raffling off the original Sinatra letter, copyright and all, for charity. He got a lot more columns and laughs out of Bentley.

Bentley was one of the few real characters in Chicago. He had gotten involved in the fight game after World War II. As ring announcer, promoter, and friend he had a myriad of stories that he delivered in Damon Runyonesque language, which matched his appearance; basset-hound eyes and a few strands of slicked-back hair.

"Sugar Ray Robinson," Bentley would say, "never tipped me after a fight. He always gave me a tie. He says to me, 'If I give you a C-note you'll spend it and won't remember where you got it. If I give you a tie, you'll think of me every time you wear it.' Smart man, right? How come he ends up on the boulevard of broken dreams?"

Ray Coffey remembered when Bentley would stroll the *Daily News* city room, hustling stories on some upcoming fight. "Royko would see him and jump up on a desk, shouting, 'Oh no, it's a mob hit man. It looks like Tony Accardo.' "

Coffey also remembered when Bentley was doing promotion for the Chicago Bulls in their first years during the mid-1960s. "You couldn't give Bulls tickets away then, and Benny was always looking for ways to promote the team. One time Royko called him and puts on an upper-class accent.

"'Mr. Bentley, I'm having a very formal party at my Lake Forest home for some very important people, and I would like to have several of your Chicago Bulls for the evening,' Royko said. Bentley got all excited. 'Sure, sure, how many? When, where?'

"Royko said, 'Oh, about four of them, around six o'clock.'

"Bentley asked if he wanted them to wear their uniform jackets or anything.

"Royko said, 'Oh no, we have dinner jackets for all the waiters here.' Bentley figures out he's been had and hangs up the phone."

Bentley appeared in a Royko column when Royko's friend, Bob Billings, and his wife, Loren, opened an art gallery that featured a casket:

Just then, Ben Bentley, chewing on a foot-long cigar, walked in. Bentley, who bears a startling resemblance to Tony Accardo, is a former prizefight promoter and has been described as a Damon Runyon character. Bentley thinks Damon Runyon is a West Side intersection.

"What are they looking at?"

"A casket," I said.

He looked sick.

His wife whispered: "He hates caskets. Anything to do with death. I have to drag him to wakes and funerals."

"Why do they want to look at a casket?" Bentley asked, nervously chewing his cigar.

"There's a man in it," I said.

"A stiff?" he asked.

"No, he's alive," I said.

"Oh, gosh," Bentley said. "That's awful."

Then the lid popped open and a man, in dark formal clothes, sat bolt upright.

Bentley cried, "Let's go," and fled down the stairs. If art can create a feeling of emotion, Bentley had clearly been moved. His wife, rushing behind him, said: "Really, when a cemetery lot salesman phones the house, he just hangs up on him."

Bentley was also a prop in a column about Chicago's changing culture. By the late 1960s and early 1970s, the downtown and the Rush Street night-life area were becoming desolate. The suburban residents who once tarried in the Loop or Near North Side for an after-work cocktail were racing directly to the trains and expressways. The civil rights fights of the 1960s frightened whites. Jazz lovers who used to wander to South Side black nightclubs wouldn't go near those places anymore. And the suburbs, once a culinary wasteland, were now welcoming restaurants that hoped to rival some of the more traditional downtown dinner spots.

One of those traditional downtown restaurants was Fritzel's at Lake at State. Fritzel's had been the place to be seen in Chicago during the 1950s. It was the home of the Loop's wheeler-dealers, the Vegas tan, the phone in the booth, the blonde on the arm of a baggy-eyed man. Then Fritzel's announced it was opening a place in far-off Arlington Heights, where, in Royko's mind, no one ever visited unless the race track was open. He decided to go there and took Bentley along:

> He loves the original Fritzel's. "It's the best place in town," he says, "I know it is, because if it isn't, what would I be doing there?"
>
> Bentley goes in there at least twice a week, and he is always treated with respect. If you look like Bentley, you are treated with respect anywhere you go in Chicago. He once went into a strange restaurant just to use the phone, and a man ordered three vending machines, a year's supply of napkins and asked Bentley to be godfather to his oldest child.
>
> On the drive out, he had said: "The greatest thing about Fritzel's is the way they greet you. Even your grandmother don't greet you the way they do."
>
> But inside, a young man hardly glanced up from his reservation book as he said: "Two?"
>
> "A mathematical genius," Bentley said.
>
> "It's too dark," Bentley said, stumbling across a varicose-afflicted leg. "Lookit, nobody in the joint even turns to see who came in. If there was a celebrity here, nobody would even see him. Boy, I feel like I'm on the road selling something."

Royko had nailed it. The days of Fritzel's and glad-handers and people gawking at celebrities were over. They did that on television. The night-life era of the post–World War II generation that had spawned the Winchells and Earl Wilson's and Kup in Chicago was over. Everyone went to restaurants and all the restaurants would be the same.

Bentley was in one of the various orbits of people that circled Royko. He was not a legman, or a softball buddy, or a close friend, or a late-night companion. But he was more than a piece of stagecraft. He was somewhat like Slats Grobnik, except he was real.

There were other orbits that Royko reached out to for his columns. For example, there were thugs that he loved, none more than the Panczko brothers: Pops, Butch, and Peanuts. Pops Panczko got up every morning, washed his face, brushed his teeth, and went out to steal something. That had been his routine since the depression. Some days he wound up with a box of onions, other days an armful of watches. There was a hint of admiration in the way Royko wrote about Chicago's version of the "gang that couldn't shoot straight." Pops was in many ways just a Chicago guy dealing with the cards that fate had given him. He put on no airs.

The Panczko's were arrested more than 250 times, and convicted only 12 times. Each time Pops faced an arresting officer filling out his rap sheet he would be asked his occupation.

"I'm a teef," Pops would reply.

The youngest Panczko brother, Peanuts, was more ambitious and decided to be a jewel burglar in Miami, which brought him to the attention of the feds, who put him away for a long stretch. Pops and his younger brother, Butch, were content to deal with the Chicago police, who were either not as skilled or not as interested as the feds:

> For Pops, it might be a department store safe on a good day; a head of cabbage from a produce truck on a bad day. Then a jewelry store. Then maybe a box of nylons from a salesman's trunk. For Butch, a cement mixer on a good day, a bag of S&H green stamps on a bad day.
>
> They were culturally deprived, socially disadvantaged, and not too smart to begin with. They couldn't afford to sit around and ponder their problems. Every day it was the same old grind; get up, get out and look for something to steal.

Pops once beat a jewelry-store robbery charge by explaining to a jury that he found the store's door mysteriously open at a time when nature called, and said that he had been taught not to relieve himself in the street. When Butch was caught with a cement mixer, he was charged with having no license to drive construction vehicles.

Royko was so taken with the colorful Panczko brothers that he told Hal Scharlatt of E. P. Dutton that he wanted to write the biography of Pops Panczko. Sharlatt loved the idea, and Royko contacted Pops's lawyer to discuss the deal. The meeting turned into a column:

> They set up a meeting with Panczko, who arrived carrying 10 leather attaché cases under his arm. "Here, you can always use these," he said, handing them to his lawyer.
>
> "Where did you get them?" the lawyer asked.
>
> "They fell out of a delivery truck downstairs," Panczko explained.
>
> The writer outlined the proposal. Panczko would simply talk about his life, how he began as a teen-aged thief around Humboldt Park and worked his way up to burglarizing the best apartments on Lake Shore Drive. The book would be a cinch best seller. There probably would be a movie. Maybe a TV series. Panczko could become rich....They made a date to begin taping the following week in the lawyer's office.
>
> Panczko showed up but he said the deal was off.
>
> "My sister don't want me to do it," said Panczko, a bachelor.
>
> "But why?" asked the writer.
>
> "She said it would embarrass the family."
>
> The writer pointed out that the family is already well known.
>
> "My sister says people throw newspapers away. But a book goes into libraries. She says that would be embarrassing."

In the spring of 1981, Royko reached into another of his orbits for friends who could join him in a business venture. The nature of the newspaper business is that it creates instant experts. Reporters who spend a great deal of time covering a field or an individual eventually come to believe that they could do that job as well or better than their subject, whether the job is managing a baseball team or governing a state or running a corporation or defending a murder suspect. That explains why many newsmen leave the business and become part of a political campaign or accept a stepping-stone position in a major corporation. It also explains why so many of them in the 1990s spent more time on television talk shows than covering their respective beats. Most of them, however, usually wind up as little more than coat holders or spokespersons. Royko, being smarter and more imaginative that most newspaper people, thought he could do almost anything, and he enjoyed discussions that raised such possibilities. He liked to imagine someone buying a newspaper and making him the publisher, or someone financing a golf course and letting him design it. Even his self-promotional flirtation with a Republican party offer to run for mayor had a hint of Walter Mitty.

So when Royko's buddy, Charles O. Finley, approached him with a baseball business proposition, Royko leaped right in. Charles O. Finley was a frequent habitué of the Goat. Finley, a successful insurance salesman, had barged into major league baseball when he bought the Kansas City Athletics in 1966. He named the team's mascot donkey Charlie O., and hustled off the franchise to Oakland where he built the best team of the 1970s. Finley was both a maverick and a visionary. He could also be insufferable. He and Royko became friends, sometimes to the exclusion of other friends.

Terry Shaffer was sitting at the Goat's bar one night when Finley, accompanied by two young women, was holding court at a table. "Mike came in and said we should go back and sit with Charlie, so we did. Finley was just dumping shit on these two women, and I told him he was an asshole. One of the girls reached over and kissed me on the cheek, but Charlie got mad and Mike took his side. Finally, Sam comes over and says, 'Terry, I put you in my office and lock door. You sleep there. Don't turn on lights after we close. Cops see lights, maybe shoot you. Call you wife. Tell her you sleep here.' I wasn't in Mike's inner circle that night."

By 1981, Finley was out of baseball. He was ahead of his time because he saw what soaring salaries could do to the game, and he tried to break up his championship team by selling players, such as pitcher Jim "Catfish" Hunter and slugger Reggie Jackson, to reduce his payroll and increase his profit. Baseball commissioner Bowie Kuhn forbid the sale and a federal judge later handed down the landmark decision that made Hunter a free agent and opened the door to the multimillion-dollar athlete salaries of the 1990s. Finley didn't get to sell his players and ultimately sold his franchise before he went broke. But Finley wanted back in baseball, and in early September 1981 he heard that a certain team was up for sale.

William Wrigley had inherited the Chicago Cubs after the death of his father, P. K. Wrigley, the man who planted ivy on the outfield walls and presided over the most unsuccessful sports franchise of the forties, fifties, sixties, and seventies. Eventually, estate taxes made the sale of the Cubs necessary. Finley and Royko got the bright idea that they should buy the ailing team. And they should buy it with Marshall Field's money. Royko began pestering Field about the scheme. He argued that Field could move the Cubs from WGN-TV, owned by the Chicago *Tribune*, to Channel Thirty-two, which Field owned. It was a brilliant strategy, foreshadowing the synergy strategies of the 1990s that melded content companies with communications and entertainment enterprises. Field's advisers told him it was a bad idea.

Royko went on a fishing trip with Field. "I kept bugging him. I told him he could change the name of Wrigley Field to Field Field, or Marshall Field, whatever," Royko often reminisced. The plan was for

Field to purchase 51 percent of the team, which Finley thought would cost about $21 million. Finley would buy 5 percent and Royko would buy one-half of 1 percent. Finley would find buyers for the remaining shares. "I would have been on the board of directors," Royko would muse.

"But then, the tarpon started running and Marshall, who was a great fisherman, had to be in Florida when the tarpon were running. Charlie Finley was rounding up the buyers and kept asking me when were we going to do the deal and I said when Marshall gets back from chasing tarpon. Then one day someone shouts at me, 'Did you hear about the Cubs?' I said, 'What?' 'The *Tribune* bought them for $21 million.'"

Royko recovered from his heartbreak to write a column demanding to know if the *Tribune* Company planned to install lights at Wrigley Field; whether the *Tribune*, which had editorially been preaching for a new superstadium, would now offer to move their baseball team from Wrigley Field; and if the Cubs would shortly be taken off free TV and moved to cable, which the *Tribune* Company was investing heavily in at the time.

The *Tribune* did install lights at Wrigley Field and no longer supported a stadium that would be used by all Chicago sports teams, and although it was not until the mid–1990s, it did move some of the Cubs games to its cable station.

If Royko was somewhat stunned by the sale of the Cubs to the dreaded *Tribune*, his readers and most of Chicago were equally stunned by his announcement in November 1981 that he had become a high-rise man.

Throughout his career, Royko had carefully cultivated his image as Chicago's ultimate blue-collar persona, the guy who could spit between his teeth, bowl, drink beer, pitch pennies, play sixteen-inch softball, suffer with the Cubs, wear long underwear, put linoleum in the parlor, and define the recipe for the Chicago hot dog (no ketchup). That image was about to be shattered.

But it made for a great column:

I was born Bungalow Man. Or Bungalow Baby, to be more precise. Later...I became Basement Flat Child. Still later, I became Flat Above a Tavern Youth. For a while, I was Barracks Man. Then, in early manhood, I became Attic Flat Man. Then Two-Flat Man. Most recently, I was Bungalow Man again....

And I had an extensive understanding of such mutants as Suburb Man and such lesser creatures as Downstate Man.

High-Rise Man was a different matter. I could study him only from afar, getting a fleeting glimpse as he jogged past or whizzed by on his 10-speed bike....So to study High-Rise Man, I set out to join the Lakefront Tribe and become High-Rise Man.

Royko then listed twenty-six items he would need as High-Rise Man. A picture accompanied the column, which showed Royko with two of his leg people, Hanke Gratteau and Helen McEntee. One of the required items he listed was High-Rise Dolly.

Then he added a second High-Rise Dolly. "High-Rise Man always carries a spare."

The column was a great read, but for Royko it was not all that humorous. He truly was facing a life change.

David Royko recalls the move clearly. He had spent most of July and August in Europe after his graduation, and when he returned he learned that the house on Sioux Avenue had been sold and his father had bought a huge condominium at 3300 Lake Shore Drive.

"I wasn't surprised. He had to get the hell out of that house. There were too many memories of Mom for him. The only reason it took two years is because it just took that long for him to be able to get up the energy to do something like move. He couldn't live there. He never—maybe once or twice—slept in the bedroom after Mom died."

David was starting graduate school and planned to get an apartment downtown, but Royko persuaded him to move into the apartment. David's cousin, Debbie, also moved into the condo. "She was going through a rough period, wild times, and Dad wanted to help her out, and she stayed in the servant's quarters which had its own bathroom. The place was huge. And Robbie was there. It was a wild time.

"There were four separate entities living there. Sometimes Dad would find things in the cupboard and mix them up to make dinner. It was usually horrible, and he would equate it to living in the depression when people had to make do with whatever they had, but it wasn't really about the depression. It was about living as a single guy. He had his shelf in the refrigerator and I had mine. It wasn't like you couldn't eat his food or anything but we lived more like roommates."

"Some mornings I'd get up and go into the kitchen and see women I'd never seen before in some form of dress and I'd say, 'Oh, Hi,' and get some coffee and leave. It felt kind of foolish since some of them were my age, and then when I came back they'd be gone."

Although he wasn't spending much time at the condo, Royko bought a grand piano and resumed the lessons he had started taking while living on Sioux Avenue. But he mastered only one song.

David recalled, "You'd be woken up at six or seven in the morning by him playing 'Ain't Misbehavin'.' He'd play it sometimes twenty or thirty times in a row. Then he'd sing it in different voices. I think he knew some other songs but every morning, or whenever he sat down at the piano, it was always 'Ain't Misbehavin'.'"

John Schackitano had the same memories. "We'd be out drinking and then he'd rent some videotapes and we had to go to the condo to watch his favorite movies and he'd recite all the dialogue. He had memorized the dialogue to all these movies and you had to listen. Then he'd go to the piano and play 'Ain't Misbehavin'.'"

Rob Royko lived off and on at the condo and was going through a high-school experience not unlike his father's. "I got kicked out of Roycemore for hitting some kid who broke my guitar. Then I wound up going to Central YMCA where he had gone and he would drive me downtown. Those were some wonderful mornings listening to him talk about things, and sometimes I'd read his column and it was about what we had talked about. But sometimes he would really rag me, and I remember one morning I got out of the car and I was practically crying. Then I went into the school and there's his picture on the wall."

David was attending the Illinois School of Professional Psychology, from which he received his doctorate in clinical psychology in 1989. He continued to play in rock bands, as did Rob, who kept getting into various legal scrapes.

"I got stopped one night by the police, and when they saw my license and asked if I was Mike Royko's son, the guy gave me a pass. I thought, 'This is neat.' But when dad was really ripping Jane Byrne I got stopped by a cop who was about to let me go. Then he said, 'Are you related to Mike Royko?' and then he gave me five tickets; cracked windshield, running a yellow, no headlight, he made stuff up and said the police department hated my dad's guts."

Royko was exasperated with his youngest son. He often refused to post bail, sending Pam Warrick or Phil Krone in his stead.

In 1982, Rob married Kip Everett, a Northwest Side neighbor. Royko was against the marriage but went to the wedding reception, where one of the over-served guests tried to pick a fight. Bob Royko was also a wedding guest and remembered when the incident began. "The guy started yelling at Mike and wanted to fight, but my brother said, 'Wait a minute. You know my name but I don't know your name. You have me at a disadvantage.' Then he took out a piece of paper and said, 'I want to know your name.' The guy gets second thoughts—he's already seeing his name in a Royko column— and he took off across the floor like a scalded rat.'"

It was not always easy being Mike Royko. And it was not easy being his son. For one thing, Royko and Rob's new wife did not get along.

"Dad always hated Kippie. For one thing, she never read his column. He thought she didn't even know he was a columnist. That got to him," Rob said.

There was irony in Royko's disappointment over his son's marriage, since romance was the subject of his most popular column in 1981.

It was a sentimental, touching love poem to the year's most famous couple, Charles, the Prince of Wales, and Lady Diana.

> Nobody is really sure what love is. Shrinks mess around with trying to define it, and just make it sound more complicated than it is. Poets, as neurotic as they are, do a much better job. I'm not sure what it is myself, except that is leaves you breathless, makes everything else seem unimportant, and can cause you ecstasy and misery and drive you crazy. And also drive you happy. I hope despite your cool, English manners, that this is what you feel. I hope both of you feel crazy and happy.

The public service bureau of the *Sun-Times* was swamped with requests for reprints of the column. Most of the requests, which were received for weeks, were handled by the office's newest employee, Judy Arndt.

chapter 14

Judy Arndt grew up in Rock Island, Illinois, where she became a passionate champion tennis player. She was the second daughter of Sam and Dorothy Arndt, both lawyers, both reform Democrats and loyal supporters of Adlai Stevenson II. The Arndts had supported Stevenson in his 1948 gubernatorial triumph and worked for him in his two unsuccessful presidential campaigns against Dwight Eisenhower. Sam Arndt ran for mayor of Rock Island twice and missed winning once by four votes.

Judy Arndt was born in 1946, five years after her sister, Connie. Although she was aware of the family's active involvement in politics, she put her early energies into tennis. She won her first tournament at age eleven and kept winning after that. In 1964, the slender blonde enrolled at Trinity University in San Antonio, enticed by the fact that a recent Trinity graduate, Chuck McKinley, had won Wimbledon, the world's most prestigious tennis tournament.

After graduation, Judy decided to move to Washington, D.C.,

where her older sister was working as a legislative aide for Senator Paul Douglas of Illinois, the great liberal who was considered the conscience of the U.S. Senate. Arndt worked in the office of California congressman Charles Wilson and later in the research department of Common Cause, the public interest lobbying group. Her sister married another Douglas aide, William Singer, and moved to Chicago, where Singer became a major force in the independent lakefront movement opposing the Daley regime. Singer eventually won an aldermanic spot and ran against Daley for mayor in 1975. Like everyone else, he lost.

In 1972, Arndt decided to join her sister in Chicago. She was hired at the Lake Shore Club as a teaching tennis professional. She stayed there until the blizzard of 1979 that caused the collapse of the roof of the tennis club. Tired of cold weather and out of a job, Arndt moved to California. She returned eighteenth months later and taught tennis again before being hired at the *Sun-Times* in the customer relations department. Although she spent her first days on the new job sending out reprints of the Prince Charles–Lady Di column, she did not meet Royko until the following summer.

Among the many things Royko believed he had mastered in life was the ability to barbecue ribs. In the summer of 1982 he wrote a column boasting that his ribs were the world's finest and that his sauce dated to a seventeenth-century Polish recipe. He got what he wanted. Scores of calls and letters, many from African Americans, challenged the idea that a honky knew how to cook ribs. Royko immediately challenged back, announcing his first rib cook-off at Grant Park, home of his glorious softball exploits.

Hanke Gratteau had finished her second tour of duty with Royko when she heard about his latest promotion. "I called and asked if he was out of his mind, inviting all of Chicago to Grant Park to cook ribs. 'Have you talked to the Park District? Have you done this? Have you done that?' Royko said, 'Oh, we can do this.'"

And he did. Again, Royko had caught the spirit of the times. One of the more successful things Jane Byrne had done in her fledgling mayoralty was to stage a huge food and entertainment festival at Grant Park called ChicagoFest. This weeklong public picnic attracted thousands of Chicagoans, and, surprisingly, many of them came in from the suburbs. For the first time in nearly twenty years, the city was enjoying a rebirth. The dread of going "downtown," which had been sparked by the years of civil rights and war protests, had become a fading memory. Perhaps the fact that a maverick woman was running the city, albeit in the same style as her predecessors, contributed to the aura of friendliness. For whatever reason, Chicago was changing and Royko saw a rib cook-off as the perfect way to draw both blacks and whites together in a party mood.

Interracial relations, however, were not at their best. Earlier in the summer, Byrne had triggered an outcry in the black community when she dumped two African-American school-board members and replaced them with whites. The Reverend Jesse Jackson protested and announced a boycott of the 1982 ChicagoFest, which blacks followed. Jackson also began a massive voter-registration drive aimed at electing a black mayor in the February 1983 election.

There was no boycott of Royko's Ribfest in September. Thousands attended his show, black and white. Rib cookers of all kinds appeared. Some contestants set up small tables with china and crystal. Others used paper plates and plastic forks and knives. Some used hibachis, some used kettles, and others brought huge cookers that were mounted on trailers. Royko enlisted Gratteau, Kogan, and other pals to serve as judges. David and Rob brought their band and played all afternoon. Beer flowed freely. Royko signed autographs, sampled ribs, and was the center of attention. The winner was a black man, Charlie Robinson, who was awarded an Illinois license plate—"RIB 1"—and went on to launch a successful string of rib restaurants.

Judy Arndt helped promote the event. "The first time I met Mike was at a meeting to discuss Ribfest. I remember wondering if he was serious when he said he was worried somebody might show up and put cocaine in the sauce. I thought it was tongue in cheek, but who knew?"

Arndt remembered the first Ribfest for a couple of reasons. "I remember seeing Mike autograph some young woman's midriff, and I was just appalled that such a serious writer would do something like that." It was also the weekend she and the man she had been seeing for four years ended their relationship.

When Arndt first moved to Chicago she lived in the same building as Rick Soll, a writer who worked at the *Tribune* and later the *Sun-Times,* where he dated and eventually married investigative reporter Pam Zekman. A few weeks after Ribfest, Arndt was visiting Soll in the *Sun-Times* newsroom when he told her, "I know somebody who wants to date you." Arndt said, "Yeah, who?" Soll replied, "Mike Royko."

Arndt recalled, "I knew this was all Rick's idea, and it was a while before he finally worked out a blind date. It would be Rick and Pam, Tony Campbell and Lynn Drozier and me and Mike. It was October 30th at Un Grand Cafe in the Belmont-Stratford Hotel. As it turned out, Rick had to be out of town on assignment. Pam and I met and had a drink in the lobby. Finally, Mike and Tony and Lynn showed up about forty-five minutes late and we went in to dinner where Mike talked nonstop for forty-five minutes. It turned out he was nervous. After dinner, Mike and Tony suggested we all go next door to another bar but Pam and Lynn were kicking Tony under the table and said, 'Why not let Mike and Judy spend some time alone?' We went and had a drink and then, as usual with him, ended the evening at the Acorn on Oak listening to Buddy Charles.

"We drove in my yellow Volkswagen Rabbitt—his car had been stolen—and about 3 A.M. I dropped him off at 3300 Lake Shore Drive. He got out, walked toward his door, and then turned around and with that silly grin, jumped in the air, and clicked his heels."

Royko had been wrong years earlier during his furious letter-writing courtship of Carol. He had quoted W. Somerset Maugham. "He says that some people fall in love many times but the fortunate, or unfortunate, fall in love only once. That's us." Mike Royko was falling in love again.

"The next day he called and we went to see a movie, *My Favorite Year*, with Peter O'Toole, and then we stopped in the Goat. It was a low-key night. Neither of us drank. We both had hangovers," Arndt said.

"My impression was that he was a such a nice, thoughtful man."

Arndt saw Royko in the office, but they did not immediately go out again. He was busy writing about the possible effect of the black registration drive on the 1982 fall election. Jackson's drive for the 1983 mayoral election meant that many new black voters were also eligible for the November 1982 gubernatorial election in which Governor James Thompson was seeking a third term. Thompson had been a heavy favorite against former Senator Adlai Stevenson III, but the black turnout was so overwhelming for the Democratic Stevenson that the final result was not decided for days, and Thompson was eventually declared the winner by less than 10,000 votes.

The day after the election, Royko called Arndt and made a date for lunch. Later in November, they went together to the *Sun-Times* awards dinner.

"After that, we were together almost every night," Arndt said.

In January, Royko took his usual Florida trip with another *Sun-Times* colleague, James Warren. While he was there, he sent Arndt a telegram:

"Please hang up your telephone and call me immediately. Love, Mike."

She called and he asked her to fly to Florida to be with him. She did.

They returned at the end of January to find Chicago divided by a raucous three-way mayoral contest.

Spurred by voter registration in the city's black wards on the South and West Sides, black leaders put the pressure on a pleasant veteran legislator, Harold Washington, to run against Byrne. Washington had grown up in the machine and had always been a loyal City Hall vote during his years as a state legislator in Springfield and as a congressman in Washington. His congressional record was marked mostly by absence, and he was rarely seen at city hall. He had held a job in the city's corporation counsel office, but those who worked there remembered he used to get his check by mail.

Washington was personable, bright, and had attracted little attention from the news media during his political career. But when he announced his challenge for the mayoralty, an old story that Washington had been given an eighteen-month prison sentence for failing to file income tax returns was promptly resurrected. Since Washington was always on one kind of public payroll or another, taxes had been deducted from his paycheck, but Washington simply never filled out a tax return.

Royko wrote: "It's not just that he didn't file those returns....It's that he's never provided what I consider a reasonable explanation. If he had said something like, 'Look, I was going through a mid-life crisis...my private life was in turmoil...I split with a girl friend and fell apart.' Anything would have done. But 'I forgot' as an excuse? That's pretty thin since his birthday is April 15."

Washington exploded. He demanded to see publisher Jim Hoge. He held news conferences. He charged, "If this campaign becomes racially polarized, I think we can look back to Nov. 11, the *Chicago Sun-Times*, page 2, and point a finger at Mr. Royko."

"That scared Mike," Lois Wille said. "I think he was surprised at the reaction."

Royko rarely got rebuttals from major political figures. Perhaps they had figured out that Royko's follow-up columns could be even more embarrassing than the original. In this case, Royko also thought his credentials in supporting antimachine candidates, his years of imploring the black community to sever their ties to the machine, and his previous, rather kind remarks about Washington made such racist charges unwarranted.

"I thought, 'My God in heaven, what a reaction!'" Royko recalled in the Tower Production interview for the 150th anniversary of the *Tribune*. "Hoge had seen him [Washington] that night and said, 'Do you really think that was a racist column?' Harold said, 'Mike had to write what he had to write and I got to say what I got to say. He knows that.' I felt a little bad about that. I would rather he said that I should explain why I didn't pay taxes one year. I'd rather have him do that than say I'm going to cause the city to be burned down."

Royko didn't think much of Washington's chances. In late January 1983, a month before the primary, Royko wrote that even if Washington got the bulk of the black vote, he would need to take about 20 percent of the white vote from Byrne and the third candidate, State's Attorney Richard M. Daley. Royko doubted he could pull it off.

He concluded: "It would be wonderful if Chicagoans put their prejudices aside and simply vote for the candidate who appeared to be the most intelligent, thoughtful and forthright and who presented the best programs. If that ever happened, Washington probably could start planning his [Washington's] victory party."

It happened. Washington won the three-way battle and now faced what was supposed to be the usual token opposition of the Republicans, this time, a former state legislator named Bernard Epton.

It wasn't going to be the usual. It was going to be ugly and racist, and Royko tried to take the edge off all those things with the very first line of his column the day after the primary: "So I told Uncle Chester—'don't worry, Harold Washington doesn't want to marry your sister.'"

In the spring, Royko and Arndt got engaged. "He was talking about marriage and I moved in with him at Lake Shore Drive. After awhile, I saw that there were a lot of things that had to be worked out and I moved back to my house, which fortunately I had kept. But we still were together all the time."

In April 1983 the Democratic party announced that next year's national convention would be held in San Francisco. Royko did not hide his displeasure:

> Now, everybody in America knows what California is. It's the world's largest loony bin. Whether you're in San Francisco or Los Angeles, it doesn't matter. Scientific surveys have shown that 97.3 percent of all Californians are deranged. And the others are kind of strange, too.

Royko conceded that Chicago politics were also kind of strange but at least Chicagoans didn't cast votes for someone named "Sister Boom Boom." Sister Boom Boom, a man who liked to dress up as a nun, had received 23,124 votes running for San Francisco's board of supervisors. He was not to Royko's taste.

The column prompted a flurry of letters and calls from all over California. The *Sun-Times* provided editorial page space for San Francisco mayor (and future U.S. senator) Dianne Feinstein to respond. She stated, quite kindly, that Royko didn't know anything about her city.

That fall, one of Chicago's legends, George Halas, founder, owner, and for more than forty years the coach of the Chicago Bears, died. Halas was more responsible than anyone else for the creation of the National Football League. He also was the only other person besides Pops Panczko that Royko ever offered to write a book about. Halas, like Pops, said no. At least Royko got a good column out of him. He wrote:

> He was in many ways, a classic Chicagoan. Like most us, he was an ethnic. He came out of a working-class family on the West Side and hustled and scrapped and worked his way to success. He might have heard of the eight-hour day but he didn't get close enough to it to pick up any bad habits. Like this city, he could be tough, even almost

brutal at times. He could be shrewd and conniving, pushy and loud, arrogant and overbearing. But he could also be generous, compassionate and direct. If he had something to say, he said it to your face, nose to nose, eyeball to eyeball.

In October 1983, Royko celebrated his twentieth year as a columnist:

It's not a record by any means, but it's a considerable period of time. If you like numbers, it means that since 1963, I've written about 240 columns year. That comes to 4,800 columns. The average length of a column is 1,000 words. So that comes to 4.8 million words. The average book is about 100,000 words long. That means I've written the equivalent of 48 books. That's more books than Hemingway wrote. But Hemingway's readers called him, "Papa." Mine call me "s– – – -head." Life is not always fair.

But the newspaper business in which Royko had flourished during the past twenty years was in the early 1980s in a critical transition from private ownership to corporate maze. When most American newspapers were held privately, owners did not have to account for their yearly earnings. As more and more papers were sold off to public companies such as Gannett, Knight Ridder, and a handful of other chains, Wall Street began to appraise newspapers the same way they looked at soap companies and car builders. Newspapers were considered a "mature industry," which in financial jargon meant they offered none of the massive growth the analysts saw in technological companies. Newspapers were also considered swollen with employees and wedded to old ideologies that did not translate to the bottom line. Papers such as the *Des Moines Register* and *Louisville Courier Journal*, which were once statewide newspapers of enormous influence, were losing money by trucking their newspapers to the far corners of their domains, and their advertisers would not pay for such

circulation. New owners would eliminate that unprofitable circulation. Innovations in computer technology would bring about vast
reductions in the production of newspapers, eliminating pressroom
jobs, and consolidating editorial positions. Newsprint prices were rising, and although newspapers had weathered the threat of radio in
the 1920s and television in 1950s, their grasp on American advertising
was being threatened by a combination of new technologies.

The cable industry, which began meekly as a service to rural
Americans, was encroaching on the big-city markets with the promise of forty or fifty channels. Two of them, ESPN, which started in
1979, and CNN, which came along a year later, would dramatically
alter the delivery of news in the coming decades and diminish the
impact of the newspaper tossed on the porch. At the same time, satel-
lites were flying into space, where they could pick up signals from one
part of the world and beam them instantly into American homes.
The slow days of the 1960s, when film had to be flown across the continent or from Southeast Asia, were gone.

All these factors seemed to indicate that only large corporate
operations holding interests in television, cable, and other new technological advances would survive selling newspapers. Entrepreneurs,
such as Al Neuharth at Gannett, became the new prophets for newspaper fiscal success. The biggest newspaper entrepreneur in the
world was a bold Australian raider named Rupert Murdoch.

Murdoch had been successful in Australia and London in a mixed
fashion. He used cheesecake and lurid stories in some of his newspapers and maintained high journalism standards in others, such as the
stately *Times* of London.

Murdoch's American invasion began in Boston when he purchased
the *Boston Herald* and in New York where he bought the troubled *New
York Post*. In no time at all, he managed to embarrass the newspapers
with such famous, tacky headlines as "Headless Body Found in Top-

less Bar." In newsrooms across America, holier than thou reporters thanked God they weren't working for Rupert Murdoch.

In November 1983, Marshall Field announced that he was going to accept an offer to sell the *Sun-Times* to Rupert Murdoch.

Lois Wille recalled, "Not long after the November announcement by Marshall, Jim Hoge asked several of us to make no commitments to leave—to sign nothing. He was mounting his effort to buy the paper. Naturally we agreed; we would have done anything for Jim, and to help him get the *Sun-Times*. A delicious rumor was circulating that Marshall pushed the sale to Murdoch from the beginning because he was intensely jealous of the adulation swirling around Hoge as publisher, which Marshall obviously (and rightly so) had not a whiff of when he was publisher; and he knew that if Murdoch owned the paper, Jim would have to leave."

The deadline for the sale was December 15, 1983, and Hoge enlisted Royko to help convince his old fishing buddy, Field, not to let the paper go to Murdoch. Royko called Field who, predictably, was fishing in Florida. Hoge had made a bid of $63 million for the newspaper. At the last minute, JMB Realty, one of Chicago's high-flying development companies of the early 1980s, put together a $100-million offer which would have been partly financed by selling a half interest in the Field Syndicate to the *Boston Globe*. But Field's lawyers advised him that Murdoch could sue for breach of contract if he wasn't allowed to complete the purchase, and on December 20, 1983, Murdoch bought the paper for $90 million.

Chicago, in general, and the *Sun-Times*, in particular, were in shock. There was hardly any room at the bar in the Goat and Riccardo's, where newsmen shook their heads, murmuring, "Never...Won't work for that guy...Great paper ruined." Many of the *Sun-Times* employees had gone through the turmoil of being out of work only five years earlier, when the *Daily News* was shut down.

Most people wanted to stay in the newspaper business, but there was a general belief that anyone who worked on a Murdoch newspaper would destroy the future possibility of being hired for a prestigious foreign or national assignment by such papers as the *New York Times* or *Washington Post*, which, in the early 1980s, were still the ultimate goal of many younger newspaper people.

There was a provision in the *Sun-Times*'s contract with the News-paper Guild that provided a fifteen-day window for employees to resign and receive full severance if the newspaper were sold. Many people plotted to take advantage of it. Not all of them were at the *Sun-Times*.

The editor of the *Chicago Tribune*, James D. Squires, smelled blood in the water. Squires, a Tennessean who had been the *Tribune*'s Washington bureau chief and, later, editor of the *Tribune* Company's paper in Orlando, Florida, had been named editor of the *Tribune* in July 1981 at age thirty-eight. He quickly became an able competitor for the flashy Hoge. Squires had been a Nieman Fellow and was regarded as one of the best young editors in the country. Among his priorities was moving women into higher-ranking positions at the *Tribune*, which was still largely ruled by white males.

Even before Murdoch's proposal to the Fields (Marshall's brother Ted was an equal owner of the *Sun-Times*), Squires had his eye on Lois Wille, the editor of the *Sun-Times*'s editorial page.

"Jim Squires called me the day Marshall Field told the assembled staff that he and his brother Ted (Field) had decided to sell to Murdoch. This was in November 1983. In fact, he made the call while I was listening to Marshall. I phoned him when I got back to my desk, and he said something extremely reassuring: 'I don't know what you'll want to do, or how it would work out for you there, but if you want to leave, just put on your coat and walk across the street.'"

Wille, like everyone else, was conflicted. She had spent her whole career battling the *Tribune,* and the transition from the *Daily News* to the *Sun-Times* had worked well, thanks to Hoge, whom she admired. But shortly after Murdoch's people began to take charge of the *Sun-Times*, she received a telephone call from Washington columnist Bob Novak. "He told me that the Murdoch people were worrying about the Marxist woman who ran the editorial page."

Wille had already made up her mind, but when she heard herself referred to as a "Marxist woman," her decision was final. She joined the *Tribune* as deputy editorial page editor in January.

The big question was whether Royko, who had a contract paying him $230,000 a year, would honor it.

On the day the sale to Murdoch was made final, Royko held a press conference and answered the question. "No self-respecting fish would want to be wrapped in Murdoch's publications. He puts out trash. I'm very disappointed in Marshall Field and his brother. They could have sold it locally. Marshall did it out of a sense of cowardice. Field now ranks with the name of Capone in Chicago history. He's betrayed the people who have given their adult lives in working for that man."

His remarks were printed everywhere. He took an immediate leave of absence. His telephone messages that week came from everywhere: *Time* magazine, Variety, NBC, CBS, ABC, the *Los Angeles Times*, all the local stations, his publisher, his agent. If there was ever any doubt that Royko was the journalist that other journalists looked to for a reinforcement of the values with which they hoped to earn public trust, it was the national response to his damning of Murdoch. Now the only question was what he would do.

On the morning of January 11, 1984, while television crews lingered on Wabash Avenue in front of the *Sun-Times*, Royko walked into the office of Robert Page, who had moved from the *Boston Herald* to become publisher of the *Sun-Times*, and handed him his resignation. Page tried to get Royko to reconsider, but Royko closed the door, marched out the back door of the *Sun-Times* and, with his lawyer, Leonard Rubin, walked across Michigan Avenue and into Tribune Tower, the citadel of all that he hated about journalism for all his adult life.

"A lot of people in Chicago couldn't stand the *Tribune*," Royko said in the Tower Productions interview. "It was against Roosevelt, against unions, against everything Democrats believed in. Journalistically, it was the last of the big powerful papers to catch up with the twentieth century. Reading the *Tribune* in the 1950s and 1960s you would have had a hard time understanding that this country was going through a major change—the civil rights movement. The *Tribune* didn't cover it at all. The *Tribune*'s idea of covering a riot in the Sixties was to count the number of broken windows and ask a police commander to comment. The *Tribune*, for a long time, never had a columnist. Colonel McCormick didn't want any other opinions in the newspaper but his.

"It was so politically involved. If you ran for office as a Republican you had to do whatever the *Tribune* said. People used to say during the 1950s and 1960s that its political editor, George Tagge, should register as a lobbyist."

During one of Royko's early forays, he went to Springfield for the Illinois State Fair in 1964, and he and his wife bumped into Tagge. "He started mouthing off," Royko recalled, "and my wife was trying to excuse us, hoping he would go away because she knew I could get easily provoked, but I thought he insulted her and I took a punch at him. Luckily, I didn't hit him. He was considerably older than me. But the next day, I was in the press gallery and all these people I didn't know kept coming up to shake my hand. They were afraid of the *Tribune*."

The *Tribune*, which McCormick had made the conscience of the Republican party in the 1930s and 1940s, was still acting that way in the 1960s. In 1968, when most newspapers were trying to avoid any perceptions of impartiality, the *Tribune*'s editor, Don Maxwell, got a hotel suite during the Republican National Convention in Miami and interviewed candidates he considered worthy of joining Richard Nixon on the GOP ticket. He forwarded his recommendations to Nixon.

When Royko walked into Tribune Tower on January 11, 1984, he must have thought about the time he turned down the *Tribune* a dozen years earlier by telling then-editor Clayton Kirkpatrick, "The thought of walking into that building makes my neck hurt."

I was managing editor of the *Tribune* at the time. Royko, his lawyer, Squires, and I met in Squires's office with a contract that was exactly the same as Royko had with the *Sun-Times*; a salary of $230,000 a year, a $20,000 car, and an interest-free, $100,000 loan for five years. Royko joked that the *Tribune* could have had him a lot cheaper back in 1959. Squires took us all to lunch and by the time we returned, a temporary office had been set up for Royko. He sat down and wrote his first *Tribune* column, with the headline, "New address sits fine, thank you."

That same day, the *Sun-Times* reprinted an old Royko column, and for three days Royko's columns ran in both the *Tribune* and the *Sun-Times*. The day Royko appeared in the *Tribune*, the *Sun-Times* filed suit. Publisher Page accused Squires of "malicious interference in our business."

On his second day at the *Tribune*, Royko wrote:

Around here, if somebody walks into the boss' office and says something like, "You're kind of a disreputable character and I don't want to work for you, so I quit and here is my resignation," the boss would understand....But apparently it doesn't work that way in the Alien's native land.

On Thursday, Judge Anthony Scotillo of Circuit Court dismissed the *Sun-Times* suit, agreeing with Royko that his contract with the *Sun-Times* was unenforceably broad and that Royko was covered by a Chicago Newspaper Guild contract permitting members to resign within fifteen days after the newspaper was sold.

At the time, the *Tribune's* daily circulation was 750,000, compared to 640,000 at the *Sun-Times*. Within weeks, *Tribune* circulation creeped up about 8,000, although Royko liked to boast that he brought 20,000 readers with him. Still, anything that gave any newspaper a boost of 8,000 readers was remarkable in the 1980s. No other single columnist in America could provide that kind of circulation jump.

Royko's arrival at the *Tribune* almost paralleled, as did his career, a major change in American newspapers. When Royko began his column in 1963, newspapers were entering their most distinguished period. They had, for the most part, tossed aside the lapdog position toward government that had been fostered, understandably, by national unity during World War II and had hung over during the copacetic 1950s.

By 1963, newspapers were dealing openly with the civil rights question. Southern editors, such as Eugene Patterson in Atlanta and many others, were openly critical of the South's odorous segregation policies and brutal attacks on blacks. The Vietnam War created a news corps that was openly skeptical of the U.S. military involvements abroad. While Lyndon Johnson was declaring that U.S. warplanes were bombing only military installations in North Vietnam, the *New York Times's* Harrison Salisbury was reporting from Haiphong that civilians and hospitals were being blasted from the skies. Walter Cronkite, probably America's most credible news voice, toured South Vietnam and announced to millions of viewers that America could not win that war, a remark which White House insiders said drove Lyndon Johnson from the presidency.

Then came the national disillusionment of Watergate and the role the press, notably the *Washington Post*, played in exposing the hypocrisy and deceit in the nation's highest office, revelations that forever changed the nature of politics and American citizens' trust in government.

The 1970s was also a time when the "boys on the bus"— political writers who covered the presidential campaigns—became the new stars of newsrooms. The new presidential primary system opened the process for dozens of candidates in 1976 and 1980, and turned what had been a week or two of political suspense at national conventions into a seasonal contest, not unlike baseball or football. The primaries offered dramatic spectator appeal from New Hampshire in February to California in June. And as the well-known faces of old politics disappeared, the new political reporters were charged with informing a nation about Jimmy Carter and George Bush and Bob Dole, not to mention fleeting fascinations with Morris Udall, Birch Bayh, John Connally, Fred Harris, and Governor Jerry Brown.

By the time Royko joined the *Tribune* in 1984 a new era was underway. Newspapers, and most likely their readers, were getting bored with government. Television and the expanding cable systems were providing "up close and personal" glimpses of all kinds of politicians the boys on the bus hadn't found. Entertainers and athletes, once remote, aloof, and idolized, were talking all day and all night, inviting cameras into their locker rooms and homes. In the 1980s, newspapers would devote more space to Donald Trump and other instant celebrities, often completely missing such major stories as the savings-and-loan boondoogle that cost taxpayers $500 billion.

When Royko moved to the *Tribune* his column moved from the Field Syndicate, where he had stayed out of loyalty, to the *Tribune* syndicate, where it would eventually run in more than 700 newspapers and where he was guaranteed a minimum of $50,000 a year for syndicate rights, a figure that ultimately went much higher.

He was given a glass-front office on the fourth floor of the Tribune Tower, which was the *Tribune* newsroom. The office was on the east wall with a view of Lake Michigan. He had a short walk down a row of similar glass fronts to the back door, which took him four flights down to the VIP parking lot, which was just across lower Michigan Avenue from the Billy Goat. But that was the only easy thing about the move.

Lois Wille said, "When he came to *Tribune* it was always irritating that he thought everyone resented him, hated him. When he was there a couple of weeks, he told me, 'You're not going to like it here. They hate us.' I found just the opposite. I thought everybody was great and I told him that. 'You're too naive to see it,' he said. But he made no effort to get to know anybody for a long time. I told him, 'It might be that when you ride up in the elevator in the morning and somebody says 'Hi,' you go, 'Ruff.'

"He had developed a very gruff, suspicious exterior that he didn't have when I first knew him. I don't know if *Boss* was the turning point, but it developed with fame and celebrity. I remember when a group of us went out to dinner, people might come and start arguments and he hated that. But if no one noticed him, he hated that, too."

In his WBEZ interview, Royko said he hadn't found the transition to the *Tribune* that difficult. "I had more trouble going to the *Sun-Times*. It was the same family but there was not a great deal of affection between the two staffs. When Marshall Field moved us into the same building, the *Daily News* didn't have enough space. We were resented by the *Sun-Times* people, and I used to run *Sun-Times* people out of our newsroom. I didn't like the *Sun-Times* or their people. It worked out better than I thought because when I left it was one of the best papers in the country.

"But the *Tribune* had changed. When Kirkpatrick became editor he made most of the changes. He was a genius at changing the *Tribune*. And Jim Squires was a wild-eyed populist, and the paper was certainly not the old right-wing GOP *Tribune* we used to beat up on. In some ways, I never had it so good. At the *Daily News*, I was the guy everyone came to with problems. I became an associate editor, and it was a lot of work to listen to people fret about their career. The *Tribune* is so big I can just do my job and leave."

But that was not what Royko had always done. He had been

involved with everything at the *Daily News*, and under Hoge he was involved to some degree with the happenings at the *Sun-Times*. At the *Tribune*, he was rarely asked his opinion and never volunteered any ideas about what the paper should or should not be doing.

One thing he settled before he signed his contract was that he could bring along a legman. Helene McEntee had been working for him at the *Sun-Times* when Murdoch arrived. Hanke Gratteau was back in the newsroom. But Gratteau was one of the many *Sun-Times* people who wanted to leave after Murdoch bought the paper. Except for the *Tribune*, she had nowhere to go.

"When Rupert bought the *Sun-Times*, Mike called and said, 'I'm going to the *Trib*. Would you work for me again?' I told him sure, as long they want me for myself. I wanted to know that I wasn't just going along because of Royko," Gratteau said. "At the *Sun-Times* he kept to himself more than he did at the *Daily News*, where he owned the place. At the *Sun-Times*, he was back by the editorial page with friends like Lois, John Fischetti, John Teets, Don Coe. But you couldn't get to his office without going by his assistant's desk.

"But the *Tribune*? That was behind enemy lines. A lot of people had built their careers over mutual dislikes at Riccardo's and the Goat's. The city room was huge, and when he would walk through with his head down, not looking at anyone, people thought he was stuck up, that he thought he was better than they were. He was just shy."

In March 1984, a few days before the primary election, Royko wrote a column asking readers to lie to pollsters. It received a typical Royko reaction. Political scientists, pollsters, and television stations argued that Royko was distorting the election, when, in fact, exit polling had caused television networks to announce winners in places where the polls hadn't closed, a factor that played a role in the presidential election controversy of 2000.

The *Washington Post* scolded Royko: "If everyone starts lying to the pollsters, the potential is there for a fiasco of proportions more memorable even than the coast-to-coast discomfiture of network pundits that Mr. Royko envisions."

The *Post* editorial exemplified, if nothing else, how much impact its editors believed Royko had nationally.

"Everyone was calling," Gratteau remembered. "People loved it. ABC wanted him on, Larry King wanted him. But he wouldn't talk to anybody. His standard reply always was, 'You want to know what I think. Buy the paper.'"

Royko had promised Gratteau she could end her third stint with him at the end of summer if she handled all the details of Ribfest, which *Tribune* promotions people were hungrily waiting to take away from the *Sun-Times*.

As it had been at the *Sun-Times*, the rib cook-out was a huge success, drawing more and more people. But it had changed. "The *Tribune* charged a $25 entry fee which went to the *Tribune* charities, and, suddenly, instead of a family gathering we had guys driving around in golf carts and security people with two-way radios. After two years, Mike gave it up. It was too corporate," Gratteau said.

In June, Royko concluded a typical evening by arriving at his condo entrance about 3 A.M. He was greeted by a mugger who stuck a gun in his mouth and robbed him of ninety-five dollars. David Royko said, "He was really upset. He came in and woke me and I said, 'What, it's 3 o'clock in the morning.' And he said, 'You know what happened to me. I just got robbed, a guy stuck a gun in my mouth.'

"I'm wondering did this really happen, is this one of the things where he's plowed and exaggerating. I went to sleep, and a few days later I thought, 'Jesus, I wasn't very compassionate.' He really had a traumatic experience."

Royko called Judy Arndt in the morning and she immediately went to see him. "I remember he was so upset, he kept saying, 'If that would have been my dad, he'd have chased that guy and run over him with his car. That's what I should have done.'"

Naturally, he wrote a column about it:

An hour after I was robbed, I was depressed because I realized I wasn't my father's son.

It happened to the old man many years ago. He was a milkman. One morning, before dawn, a guy with a knife started to climb into his truck. The old man kicked him in the face. The guy got up and ran. The old man slammed his truck into gear, drove on the sidewalk, floored the gas pedal, and—bump, bump—the world had one less stick-up man.

I could have done the same. My car was at the curb, only a few feet away. I could see them running down my street. In a few seconds, I could have caught them. I'm sure the insurance would have covered any damage to my bumper.

In July, Royko took the train to San Francisco for the Democratic National Convention. He was in a bad mood. He did not like conventions. His relationship with Arndt was uncertain. He wanted to get married. She wasn't sure. She had decided not to go with him to San Francisco.

The 1984 Democratic convention used up all its news before it opened. Walter Mondale, who had been Jimmy Carter's vice president, was the shoo-in nominee but was conceded little chance against the popular incumbent, Ronald Reagan, who had survived an assassination attempt. Mondale, looking for some kind of impetus, surprised everyone by announcing the week before the convention that Geraldine Ferrarro, a congresswoman from New York City, would become the first woman vice presidential candidate on a major party ticket.

That left the thousands of reporters, Royko included, with little to do except enjoy the pleasures of San Francisco and write about the hookers protests and the various gay organizations and the celebrity parties. It gave Royko time to brood.

It was Royko's first convention for the *Tribune*. When he arrived at the *Tribune* newsroom in the headquarters hotel, the San Francisco Hilton, he learned that the *Tribune* had only two of the prized floor passes that allowed reporters to go on the convention floor without waiting in a long queue. The dozen or so *Tribune* reporters covering the convention would have to share the passes, but Royko was told he could have one any time as long as he reserved it in advance.

"I won't need one," Royko said. "I can get to anyone I want on the telephone, and there isn't going to be any news inside the convention anyway."

On Monday evening, Royko had just finished his column when Jimmy Breslin arrived and suggested the two of them take a stroll on the convention floor. Breslin had his personal floor pass. Royko went to ask the *Tribune*'s Washington bureau office manager, Jean McGuiness, for a floor pass and she told him they were both being used.

Royko flew into a rage, possibly for Breslin's benefit, cursing McGuiness, cursing the *Tribune*, cursing the world. "I'm not writing a goddamn thing. I'm going to Pebble Beach and play golf the rest of the week. I quit." The *Tribune* newsroom, which was being used by all the Tribune Company newspapers, the *New York Daily News*, and the Orlando and Fort Lauderdale papers, fell silent. The score of reporters thought they had witnessed the end of Royko's brief career at the *Tribune*.

The next morning, Royko walked into the newsroom and continued his tirade to the *Tribune*'s editor, Jim Squires, who shouted back at him, defending McGuiness. Squires gave as good as he got. Royko stormed out.

Later that evening, Royko walked into the hotel bar and apologized for his behavior. He had already filed tomorrow's column. Whether he was momentarily embarrassed by the fact that Breslin one-upped him by having his own floor pass, or whether he was depressed by his separation from Arndt, Royko's outburst ended with a flurry of fine columns, particularly praising the keynote speech that night of Mario Cuomo, whose presidential aspirations Royko would push faithfully in future years.

And Royko's anger obviously was softened by a telephone call to Arndt. "He told me about what was going on and all the goofy people and about hanging out at the Washington Grill and then he casually said, 'I met Warren Beatty,' and I said, 'Warren Beatty!' So we patched things up, and he asked me to fly out at the end of the convention and I said yes."

Royko and Arndt drove down to Carmel for a week.

"It was just a magical time. We stayed at the Highlands Inn and walked and talked and ate. There was a beautiful nature preserve there. I was ready to move there. I rode back on the train with him. That was some trip. This little sleeping compartment and he was smoking three packs of Pall Malls a day."

Through the haze, they decided to get married. Royko wanted to elope immediately but Arndt thought they should wait until the following spring.

It turned out there was enough excitement for Royko in the fall of 1984 without a wedding. For nearly forty years Royko had been waiting for the Cubs to win something. He had devoted more than 100 columns to the haplessness of the team he fell in love with as a six-year-old. He had never forgiven the city of New York for 1969, the year the Cubbies were all but certain to win a pennant until the dreaded Mets overtook them in the final week.

In his third *Tribune* column he tried to explain his relationship to the Cubs as an employee of the newspaper that owned the team. He wrote, "I used to be against lights in Wrigley Field. Now I'm on the Trib payroll. Therefore, good company man that I am, I'm no longer against lights in Wrigley Field. So I'm a fink."

On August 6, Royko threw out the first ball at a Cubs game:

Inside the ball park, a vendor said: "You're going to be out there?"

Right.

"You ready?'"

Arm's a little sore...It's a rotator cuff condition. Lot of pitchers have it....

Then I was on the mound. The same mound where Don Cardwell threw a no-hitter in his first start as a Cub. Where Sad Sam Jones walked the bases loaded in the ninth, then struck out the side to keep his no-hitter. Where Fergie had worked them at the knees, and Lee Smith now threw smoke.

Jody crouched and put down one finger. I thought about shaking him off but he's a hard guy....So I just reared back and let go with the fast ball.

With 37,142 people watching, the ball went about two-thirds of the way to the plate, hit the ground and bounced weakly to Jody.

He walked out to the mound and handed me the ball. "Nice going," he said although he didn't sound too sincere.

"Uh, it's the rotator cuff," I told him.

When I got to my seats, the friends and relatives all looked away in silence. Finally a relative said: "Pathetic."

Another said: "God, I was embarrassed for you. I mean, you couldn't even get it to him on the fly."

"You really choked," someone else said.

It was the rotator-cuff problem, I explained.

Royko wrote a column announcing that his annual Cub quiz would not appear because he feared jinxing the team. In September, the Cubs ended a thirty-year drought by winning the eastern division and were set to open the National League playoffs on October 2.

Royko was offered seats in the *Tribune* luxury boxes by the company brass, but he turned them down. He wanted to wander the grandstands. He went to the first game with an entourage of fellow long-suffering fans, including his brother, Wade Nelson, Lois Wille, Dorothy Collin, and others. Collin remembers, "We were walking toward our seats and everyone recognized Mike. They were all cheering, 'Yeah, Mike, way to go,' as though he somehow was personally responsible for finally getting the Cubs to a playoff."

Royko was a product of a time when sports and the men who played them were not all commercial deities, when understanding and passion for the game were restricted to those who had played and watched and knew the nuances of the hit-and-run, the screen pass, or the switching man-to-man defense. It was a time when the Dodgers were still in Brooklyn, the Braves in Boston, and there were no teams in places like Charlotte and San Jose and Vancouver. He was like the rest of Hot Stove America in the depression and World War II eras, fascinated with the exploits of men called Giants and Pirates.

Royko was able to convey for his generation the importance of being a fan, and being a fan of the most inept sports franchise of the century was something that only Chicagoans understood until Royko spread the mantra of Cub heartbreak all over America. Royko was in a great way responsible for the national popularity of a team that was the exact opposite of the Yankees, Cowboys, and Canadiens, dynasties built on success. The Cubs were the lovable losers. Their owners made them losers; Royko made them lovable.

He was almost hard-pressed to deal with their victories in the first two playoff games. It was more natural after they lost three in a row to the San Diego Padres, dashing their hopes for their first National League title since 1945:

> Then it was over. And if losing wasn't bad enough, we were beaten
> by a bunch of wimps from a beach-bum city. People who were actu-
> ally silly enough to make a wave in the stands. And the Cubs had set
> another typical Cub record: the first National League team to win
> the first two games of the playoffs and then lose the next three. Why

did it have to be us? Why do we always get stuck with such embarrassing records?

Royko picked himself up in time to do his selection of a team to root for in the World Series, an annual dilemma, for the Cubs were never a possibility. Royko always rooted for any team that played against New York or Los Angeles. When those two cities weren't involved he rooted for teams from cities like Chicago. In 1984 it was a natural that he root for Detroit:

> We're both cities of shot and beer drinkers rather than wine sippers. We're still cities where you're more likely to find a guy with an eagle tattooed on his arm than a flower tattooed on his hip. And we're more likely to go to a tavern to pick a fight than to our back yard to pick an orange. We're cities, where football fans cheer louder for a toothless, snarling middle linebacker than for a delicate-natured quarterback. As cities go, Chicago and Detroit are sort of like cousins. Just the way San Diego and, say, Disneyland, are cousins.

Of all his baseball columns, the one that has been most reprinted in anthologies and collections is about the day Jackie Robinson played his first game at Wrigley Field. Royko wrote the column when Robinson died in 1972. It reflects Royko's undying empathy with the struggles of black Americans. But it also paints a landscape of the way baseball parks looked and sounded in the uncomplicated days after World War II. It is nostalgic but optimistic. It begins with a crowd of white men standing outside a bar, wondering what it will do to baseball:

> I hung around and listened because baseball was about the most important thing in the world, and if anything was going to ruin it, I was worried....I had never seen anything like it. Not just the size,

although it was a new record, more than 47,000. But this was twenty-five years ago, and in 1947 few blacks were seen in the Loop, much less up on the white North Side at a Cub game....

They didn't wear baseball game clothes. They had on church clothes and funeral clothes—suits, white shirts, ties, gleaming shoes, straw hats. I've never seen so many straw hats.

Royko wrote that he never forgot two things during the game. The first was that a Cub runner tried to spike Robinson, who was playing first base. The runner was Royko's hero, but Royko thought his behavior was rude. The second was that Royko said he caught a foul ball off Robinson's bat and that he sold it to a black man who offered him ten dollars.

I handed it to him and he paid me with ten $1 bills. When I left the ballpark, with that much money in my pocket, I was sure that Jackie Robinson wasn't bad for the game. Since then, I've regretted a few times that I didn't keep the ball. Or that I hadn't given it to him free. I didn't know, then, how hard he probably had to work for that ten dollars. But Tuesday, I was glad I had sold it to him. And if that man is still around, and has that baseball, I'm sure he thinks it was worth every cent.

It was Studs Terkel's favorite Royko column, although Terkel thought the bit about catching a ball off Robinson's bat was fiction. "Mike's piece was not about Jackie. It was about Jackie's people who were in the stands that day. Who cares whether he caught the ball or not. He made that day and that time so real."

chapter 15

In January 1985, Royko began preparing for his wedding. He had joined Ridgemoor County Club on the Northwest Side a few years earlier, and he wrote to club president Stanley Banas.

Dear Stanley:

I'm writing to you to request permission of the Ridgemoor Board of Directors to use the club house for my wedding reception and dinner on Sunday, April 21, 1985. Actually, I would have preferred using the half-way house and serving hot dogs, which would have saved me a couple of dollars, but my prospective bride, Judy Arndt, said she did not think it was spacious or grand enough, since we anticipate having 150 to 200 guests.

I told her that most of my relatives were married in far less grand surroundings than the half-way house. As I put it: What the heck, it has separate toilets for the men and women. What more can you ask? But, being a modern woman, she was adamant, so the shindig will have to be in the main building.

David Royko recalled, "I remember when Judy moved into the condo, but I didn't think much of it one way or the other. Dad never discussed her with me. Sex and the male-female relationship was not something we ever discussed. I heard from someone else that they were engaged, then I heard that Judy broke it off, and Dad seemed kind of depressed at that time. I was wary of anybody who wanted to marry Dad, if they were marrying him for his money?... the way he was living, who would want that?

"And then when I heard they were getting married—it wasn't from Dad—I really didn't give it a second thought. Dad seemed to be trying to get it together, clean up his act so to speak. I still really didn't know Judy, but I thought maybe this is what he needs. I didn't have any sense that he shouldn't get married again, my mother's memory or anything like that," David Royko said.

Rob Royko said, "I almost told him not to marry Judy. Like he would have listened to me. I thought he had enough money and enough women around to have a different one every night. Why get married and ruin that? But I thought Judy was cool."

The marriage ceremony took place Sunday, April 21, in the Lake Shore Drive condominium. The reception was held at the Ridgemoor Country Club, where all of Royko's family, legmen, and friends attended.

When Mike Royko married for the second time, his lifestyle changed. He still made his nightly stop at the Billy Goat, but the stops got shorter. He enjoyed going home. Judy was still working at the *Sun-Times,* and she would meet Royko at the Goat after work and they would go out to dinner or go home.

Royko's late-night carousing became a rare event. Except for his close relationship with Shack, few of his former late-night partners saw much of him. One of the people he became closest to in the 1980s was an Irish immigrant who had become a successful contractor in Chicago.

Dan Hurley was born in Ireland in 1930 and immigrated to Boston in 1948. He was drafted in 1950, spent two years in the army, and then moved to Chicago, where he worked as a plasterer, a trade he had learned in Ireland. In 1963, he opened his own contracting business, and a friend advised him he should take up golf as a good way to make business contacts. Hurley joined Ridgemoor Country Club, where he took lessons and hit balls, eventually becoming one of the club's better players. Hurley, whose touch of brogue never disappeared entirely, fit in well and was elected to various committees.

"The first time I met Mike was when the membership committee interviewed him. He came in and was rather quiet, almost shy. The committee asked the usual questions and he gave short answers. He wasn't like some guys who came in and acted like we were lucky they wanted to join Ridgemoor."

Ridgemoor was filled with politicians such as the chief judge of the Circuit Court, Harry Comerford, U.S. Representatve Dan Rostenkowski, Alderman Terry Gabinski, former county assessor Thomas Tully, Police Superintendent James Rochford and his top assistant, Jack Killackey, and a variety of state legislators and judges. But it was also the home of people like Hurley.

"There was no old money at Ridgemoor," Hurley said. "These were guys who had worked in factories in World War II, or been in the service, and they learned about business, became tool and die makers or plumbers and started their own companies."

Unlike many of the North Shore private clubs, ethnic names were a plus at Ridgemoor. Royko was voted in unanimously despite the fact that he made some members, especially the politicians, nervous.

Royko became a regular visitor at the driving range, but his work schedule restricted him from playing as much as he would have liked. He always tried to play on Fridays since he had no column due for the weekend. Hurley tried to fit his rounds in between his own erratic schedule, and the two of them kept meeting on the driving range.

"We finally started playing, and from the beginning we were a regular twosome. We always played Sunday at eight o'clock and on Saturdays about eleven. Then, he'd call from the office sometimes and say, 'Can you get out?' and we'd meet. Mike was an accomplished golfer. He had streaks of being a great short putter. One day I told him if my life depended on someone making a five-putt I would pick him because he was the best pressure short putter I had ever seen. He gave me that sly grin and said, 'Pressure? You dope, pressure is when it's 6:45 and the deadline is 7 P.M. and I haven't got a goddamn thought in my head. That's pressure.'

Royko also invited Jim Warren, Rick Kogan, and Tim Weigel to play Ridgemoor, but most of the time he played with Hurley. The pair always teamed up for competitions. "Mike loved competition. He loved to bet, always had a Nassau going. He wanted to win, but he didn't care that much if he lost. He just loved the competition."

Royko's humor was often on display. "One Sunday, we were playing the twelfth hole and his ball landed by some geese. He walked toward the ball and one of the geese charged; maybe he was close to a nest or something. He jumps back in the cart and backs it up into a tree. The next week we were playing and I was driving the cart and we got near some geese and he said, 'Stop, that's the one that attacked me last week.'

"I said, 'Mike, how the hell do you know it's that one?' He gives me that grin and says, 'I never forget a face.'"

Like most clubs, Ridgemoor had a Thanksgiving Day event called the Turkey Trot, a modified game that was mostly one last chance for the members to have another celebration before winter closed the course. "Mike was always trying to win this thing," Hurley said, "although we never did. One year, he came out with a thermos full of boiling water and twelve golf balls in it. 'If they're heated, they'll go farther than anybody else's balls,' he said. So we get to the first hole and he has this little fish net you use for fish tanks, and he pulls one of these balls out of the water—it's still bubbling in there, and I tee it up. All of a sudden the ball starts to make noises and all the paint begins to peel off. Then he decided the balls were too hot for the cold weather, so he tells one of these poor kids who were freezing that he has to keep the ball in his mouth to keep it warm between shots."

Royko was rarely funny on the course. Tim Weigel was one of his regular playing partners. "He would get very mad at his own game. He'd also get mad at my game. If it were just him versus me he would not get mad at my game. He'd be very happy that I was screwing up. However, if we were partners, he was brutal. You really had to be on your best game because he could be incredibly demanding," Weigel said.

John Schackitano had never played much golf, but when Royko joined Ridgemoor he wanted Shack to begin playing so he could have a close friend on the course. "He said I was the worst student he ever had. He said I was unteachable. Of course, with him, it became an obsession. I was always whacking it out of bounds. But there was one time when I was blocked from the hole about 220 yards away behind some trees. I got out a four-wood and he said, 'You can't make that shot.' But I hit it anyway and it went over the trees and landed on the green and rolled about eight feet from the hole. I made the putt for a birdie. You think he was happy? No, he told everyone in the clubhouse after that, 'This guy is so dumb he didn't know he couldn't make that shot.' Everyone laughed. I made the shot, but he became the focal point of the story."

It was a golf game in 1990 that ended Royko's long and close friendship with the enigmatic Bob Billings. In August, Billings had arranged to play with Royko at the best public golf course operated by the Cook County Forest Preserve District. It was called Forest Preserve National, and its green fees were ridiculously low, but there was a catch. There were no reservations taken. Golfers would begin lining up at 3 A.M. to book tee times for later in the day.

A few weeks after they played, the *Tribune* ran a front-page story detailing how the Cook County commissioners used the course as a private club, allowing special friends to play without going through the laborious waiting period. Royko may have felt guilty, or worried that someone might accuse him of being a special guest of a politician, or simply following up a good frontpage story.

On September 5, 1991, he wrote:

A few weeks ago, I strolled onto the grounds of the Forest Preserve National Golf Course and teed it up.

Unlike most of the golfers who play there, I hadn't arrived well before dawn and snoozed in my car in order to sign up for a precious starting time.

Not at all. My starting time was the product of clout. Not mine, but a friend who had a friend who knew a county official who knew a Cook County commissioner.

The county commissioners run the forest preserve system. And each of the 17 county commissioners, under a law they themselves passed, can dole out one starting time a day.

Did I feel guilt? Yes, a pang or two. It did seem slightly unfair that I should be arriving at 11:30 A.M., leisurely hitting a few practice shots, and teeing off at noon, while some poor stiff who was there before dawn was told, sorry, all filled up, try again some other day.

On the other hand, as a Chicagoan I felt a certain pride. In what other city do politicians dispense tee times as a form of patronage? The Machine might be dead, but some of the nuts and bolts are still scattered about.

John Sciakitano remembered what happened. "The guy who had set up the golf date, who was an old friend of Billings, got fired. Even though Mike never mentioned his name in the column, someone at the county figured out who had clouted him and Billings on the course. Billings was really pissed at Mike. They had a fight and they stopped talking. Every once in a while, Mike would ask me, 'You see Billings?' 'What's Billings up to?' He knew I was still close with Bob. One day Mike told me to try and patch things up between them. I went to Billings and told him, 'You know, Mike says he thinks it would be terrible if the next time one of you guys saw each other was at the other guy's funeral.'

"Billings looked at me and said, 'What makes him think I'd go to his funeral?' They never spoke again," Schackitano said.

One of Royko's very first columns in 1964 was about the perennial winter anxieties of golfers. The second column he wrote about golf was also in 1964.

"Can Negroes and whites play out of the same sand traps, putt the same greens and curse the same cruel fate that caused them to dub a shot? The answer appears to be 'no' on many of the courses in Cook County that seemingly are open to the public—the white public."

Royko explained that many public-fee courses had a gimmick, asking black golfers if they had a membership card, a question never put to a white golfer. Royko wrote that in his final days in the air force, while based at O'Hare, he and several others tried to persuade the owner of a course near the base to give a season rate to servicemen. The owner created a new racist cliché: "I don't mind they're being in the service but I don't want them playing golf with me."

Royko also debunked the idea that golf was healthy:

Not only is golf not good for you, it's bad for your health. . . . Golf, as it is played by most cart-riding wheezers, is not only less exercise than walking the dog or knitting, but it can destroy a healthy person. Ninety-nine per cent of all golfers suffer from stress. The bad golfers, the great majority, are filled with stress because they are bad. And somebody behind them is yelling that they should play faster. And every time they hit a ball into a pond they have lost $1.50. What other game fines you $1.50 for doing something that has already plunged you into depression?

The few good players suffer from stress because they are playing with or behind bad players. And they know that to remain good, they must practice for long hours at a time. This makes their wives unhappy, and they retaliate by making their husbands unhappy.

That's why golfers rush from the 18th hole to the nearest bar. Their nerves are frazzled. So they drink to calm their nerves. And their noses get red and their hands shake....

It happens that I am that rare creature, an outstanding golfer. Or, at least, I could be. It all depends on whether my new sticks with the shafts made of compressed moon fragments are as good as they say.

Mike and Judy took turns at each other's sport. He took up tennis and she began to take golf lessons. They often played a game of tennis after work if Royko finished his column early. Royko also set up mixed doubles games with friends.

Dan Hurley said, "Mike loved playing doubles with Judy as his partner. He just stood there while she beat up the other team."

Royko's routine at the *Tribune* rarely included going out for lunch. He often had a turkey sandwich at his desk. He spent most of his evenings at home, talking to Judy about his day and watching his favorite movies.

"Mike talked about everything, always," Judy said. He had this ability to talk about the good and bad for forty-five minutes every night. He needed to unload. It was his way of getting rid of the stress. A bad conversation on the phone, someone didn't treat him right, whatever it was. When he was in the office, he'd call and say, 'How do you think this sounds? What do you think about this idea?' And there were other times when he didn't want to talk about his column. The timing was always his, but if you hit him when he was ready to talk, it was great."

The *Tribune* may not have been fun, but it was comfortable. Royko moved from the east end of the newsroom to an office in the section where the editorial writers had private lairs. He was only a few doors away from Lois Wille. Royko's office was impressive with high ceilings and tall, vertical medieval windows overlooking Michigan Avenue, but it had none of the neat, clean-desk look that was the approved *Tribune* style. Royko quickly filled the room with boxes of clippings, assorted plaques and pictures, a sofa, and a huge ashtray.

Royko's presence added journalistic prestige to the newspaper. In 1971, when he had turned down Clayton Kirkpatrick's offer, the *Tribune* was still clouded by its McCormick heritage. *Time* magazine, in a piece about the Chicago newspaper wars, noted that none of the Chicago papers were in the same class with the *New York Times* or *Washington Post* and that they did not even approach the next tier of quality represented by the *Los Angeles Times*.

That began to change in 1973 when the *Tribune* pulled off a logistic coup by printing the first of the Nixon tapes in a special supplement the day after they were released. In 1974, Kirkpatrick convinced the *Tribune* bosses to do what seemed unthinkable for the Republican-oriented newspaper: call for the resignation of Richard Nixon.

During the 1970s, the *Tribune* recruited a wealth of talented reporters: Jon Margolis, Bill Neikirk, and James O'Shea in Washington, and foreign correspondents such as R. C. Longworth. In the early 1980s, under Squires, the *Tribune* expanded its foreign coverage. It opened bureaus in Rome, Nairobi, Mexico City, Tokyo, Hong Kong, and Warsaw to supplement its traditional outposts of London, Moscow, and Jerusalem. Following the lead of the *New York Times*, the *Tribune* expanded its national coverage with bureaus in Atlanta, Dallas, and Denver to go with New York and Los Angeles.

In 1979, the *Tribune* opened a state-of-the-art printing plant at Grand Avenue and the north branch of the Chicago River. It would soon allow for color reproduction, making the *Tribune* one of the first big metropolitan newspapers to frequently use color. It also gave the *Tribune* a huge advantage over the *Sun-Times*, which would use outdated presses for the next twenty years.

Charles Brumback, publisher of the Tribune Company's newspaper in Orlando, was named president of the *Chicago Tribune* in 1981, and within six months he brought in Squires, who had been editor of the *Orlando Sentinel* since 1977. Brumback was a no-nonsense numbers-cruncher determined to hike *Tribune* profit margins. In a few short years the newspaper jumped from 6 percent annual profits to double digits; impressive but still far below some media companies such as Gannett, where some newspapers earned 30 percent, and far below Orlando, which earned 25 percent. Squires, who had worked closely with Brumback in Florida, pried loose some of the new profits for the *Tribune*'s expansion and was able to maintain the precious news space that publishers everywhere were sacrificing to more advertising dollars. But the bulk of reinvestment was headed toward new technologies, which eventually cut jobs and reduced payroll.

By 1985, *Time* magazine was calling the *Tribune* one of America's ten-best newpapers, ranking it with the *Los Angeles Times* and behind the *New York Times, Wall Street Journal,* and *Washington Post.* No less an authority than Al Neuharth, chairman of Gannett, called the *Tribune* the best newspaper in the nation.

In 1985, printers at the *Tribune* went on strike. The printers' jobs were about to be eliminated by desktop computers operated by copy editors. The *Tribune* had guaranteed the printers lifetime employment, but they would be forced to accept whatever jobs they were offered, mostly maintenance and janitorial, and they would not receive any further annual salary increases.

The strike did not cripple the paper. Technology made it possible for a small band of management personnel to actually print the newspapers. The only problem was getting the paper delivered, which depended on truck drivers who were members of the Teamsters Union. *Tribune* executives had realized that drivers were critical to any labor problem and had worked out a contract years earlier in which the Teamsters would not honor picket lines set up by another of the other production unions. The *Tribune* hired replacement workers at a lower wage, and the printers lost their jobs en masse. The strike had little effect on the newspaper other than to eventually swell its profits. The picket lines made it uncomfortable for most *Tribune* newsmen who had worked in the composing room with the printers. It was certainly uncomfortable for Royko, a longtime member of the Chicago Newspaper Guild, but he did believe that the union leadership was leading the printers astray.

In the fall of 1985, High-Rise Man became Bungalow Man again. The Royko's moved to Sauganash, an affluent Northwest Side neighborhood. Royko hired a few friends as contractors to do some remodeling.

Although his work address had also changed, many of Royko's old friends still found their way to the *Tribune*. Phil Krone, Studs Terkel, and Ben Bentley would all drop in to gossip, seek advice, or suggest that Royko write some column praising their latest projects or benefactors. Royko was particularly happy when Terkel was awarded a Pulitzer Prize for *The Good War*, a gathering of oral history from ordinary Americans about World War II. He wrote a column poking fun at the lofty professors who were critical about the prize giving because they did not consider Terkel's work scholarly:

Studs should have immersed himself in the private papers of some prominent but dead person. That's where a real historian goes for history—into those musty old letters and files. And, of course, he should have read books written by other historians and lifted material so he could weave it into his own book.

He made another mistake. His book is easy to read. But a real work of history should be almost impossible to read. It should be loaded with footnotes, and footnotes to footnotes, and an enormous bibliography and an even longer index.

If Studs were a real historian he would have picked up a book like "The Good War" and lifted parts of it for use in his own book. And that would have made it legitimate history. You know, I think that with a different start in life, Pops Panczko, the famous burglar, would have made a pretty good professor.

Royko did not dwell much on the happenings at City Hall. Harold Washington was not the machine, and he was black, and Royko probably didn't want any more accusations that he might be responsible for a race riot. Beyond that, there was only the usual incompetence and petty thievery. When Washington's cronies were caught with their hands in the cookie jar, he blasted them, and he even came down on the side of the old machine stalwarts in the city council who had a 29–21 majority and were blocking most of what Washington proposed. While Washington and the liberals tried to paint this as more of the same old racial arrogance, Royko stood up for the old democratic principle of majority rule.

Some observers contended that Royko's column began to change at the *Tribune* because he was concerned about his syndicate clients and tried to write about events of national significance. Royko did write less and less about the bureaucratic idiots who were causing heartbreak and mayhem, but he did it for the simple reason that the insensitive louts who were the subjects of his tirades in the sixties and seventies had mostly disappeared. Even when city hall or the governor's office tormented some poor soul, a call from Royko's legman would no longer receive some flippant remark. Public relations consultants scurried around to explain and apologize and sometimes even remedy the problem.

"He would really get mad when you made so many phone calls that the complaint he got in the morning—which seemed like it would be a terrific column—would have been taken care of by the afternoon," said Hanke Gratteau.

"And he didn't get all the telephone calls he used to get at the *Daily News* and *Sun-Times*. Most *Tribune* readers lived in the suburbs. They weren't Chicago cops or precinct captains and they didn't know who was screwing who like his readers at the other papers," Gratteau added. "And Mike began to watch CNN just the way he used to read all the wires and other newspapers. In a sense, he became a part of the audience, or at least, he became more aware of the what the audience was watching and his columns tended to have more national interest."

But he was always on the hunt for official stupidity and he found it in October 1985 when a year-old-child was taken to Cook County Hospital with a concussion, burn marks, and an arm broken in three places. The child was beaten by the mother's boyfriend, but a Cook County judge decided the infant was not in "urgent or immediate danger" and returned her to her mother.

After a month during which Royko wrote several powerful columns, the judge's ruling was reversed:

> I suppose I should take satisfaction out of the fact that Lashaunda, the battered baby I've written about, is finally going to be placed in a foster home where it's unlikely she'll have any more broken legs, concussions, burns, internal injures, and other miseries....It also can be considered good news that Judge Davis, who initially botched this case by refusing to take the baby out of a dangerous environment, is being whisked away from juvenile court to another assignment,

where he can ponder such things as contracts or parking tickets. Things that don't bleed and cry out in pain.

In December, Royko made another of his doomed efforts to quit smoking, beginning at a clinic that promised an 85 percent success rate.

I've been smoking 35 years and the longest I ever went without a cigarette was two and one half hours. It was a long movie.

I didn't want to go, but my wife told me that if I smoked myself to death, she would punish me by going directly from the funeral to the Bahamas and spend all my money on skinny gigolos, rum drinks and wild living. Who wants to croak with that as a last thought?

So I went through the clinic's program...and tossed my last pack of smokes outside the car window.

By Sunday afternoon, when it was time for me to go to the office and write my Monday column, my nerve endings were humming louder than the Mormon Tabernacle Choir. And I knew that if I left the house, I'd go straight to the gas station and buy a pack of smokes. So I called in sick and spent the day swallowing pills and chewing on more carrot sticks. By Monday morning, I cranked up enough will power to come down here to do my job.

Now, if this program works and a month from now I'm still chomping cinnamon sticks instead of Pall Malls, I'm going to write about the program in detail. If as spineless a person as me can quit smoking, anybody can. But if it fails, these will be the last words I'll write on the subject.

Except, of course, to change my will.

Two days later Royko again wrote about his battle, fretting that he would get fat. The following week he pledged not to become one of those ex-smokers who become a proselytizer for good living. At the end of the month he wrote about a woman who asked him to stop writing boring columns about quitting smoking: "So I did what I often do when I am the recipient of unsolicited professional advice from a stranger in a public place. I reached into my pocket, took out a quarter and handed it to her....That's a refund. It's the price of the newspaper in which I had a column that bored you. Goodbye."

The mid-1980s was a happy period for Royko. He was finally settling into middle age. He still drank and smoked but he watched his diet and went through a series of abstinences from booze and tobacco. He was with a woman he loved, and the *Tribune* wasn't going to fold. The *Tribune* was not the friendly newsroom of his youth, but neither was it a place where daily conversation focused on the terminal illness of the *Daily News* or the second-place status of the *Sun-Times*. He had escaped from the clutches of Rupert Murdoch. He was no longer compelled to indulge in softball or put himself on display at Ribfests. All the efforts he had put into softball, fishing, and barhopping were saved for golf, and his every spare moment was spent at the driving range or on the course. He talked often of retirement, finally signaling that he at last realized he had achieved more than even he believed possible. His appetite for fame and praise was never sated, but he had won every journalism award imaginable and he often feigned being bored when told of his latest. He was now an avuncular boss to his legmen, not the aloof taskmaster. He loved regaling them and other visitors to his office with stories of the Chicagoans he had made famous in his column, and his discourses were as entertaining as his writing. He was making more money than he had dreamed he would make, and he was back in a house instead of a high-rise neighborhood where he worried about being run over by joggers in gold chains.

He wrote a hauntingly romantic column that blended both his current contentment with the best times of his past. Slats Grobnik told the story:

"Awright. It was a long time ago, maybe thirty years. I was in the lot and it was the night before Christmas Eve, about a half hour before I was going to close up. I hadn't seen a customer in two hours. I had maybe a couple dozen trees left, and most of 'em weren't much to look at....So I'm standing by the kerosene heater when this young couple comes in and starts looking at the trees.

"He's a skinny young guy with a big Adam's apple and a small chin. Not much to look at. She's kind of pretty, but they're both wearing clothes that look like they came out of the bottom bin at the Salvation Army store. It's cold as a witch's toes, but neither of them have got on gloves or heavy shoes. So it's easy to see that they're having hard times with the paychecks.

"Well, they start lifting the trees up and looking at 'em and walking around 'em, the way people do. They finally find one that was pretty decent. Not a great tree. But it wasn't bad. And they ask me the price.

"It was about $8 or $9. They don't say anything. They just put it down. They keep looking. They must have looked at every tree in the lot. Like I said, there weren't many that were any good. But every time I gave the price on a decent one, they just shook their heads. Finally, they thank me and walk away. But when they get out on the sidewalk she says something and they stand there talking for awhile. Then he shrugs and they come back.

"I figure they're going to take one of the good trees after all. But they go over to this one tree that had to be the most pathetic tree we had. It was a Scotch pine that was OK on one side, but the other side was missing about half the branches. They ask me how much that one was. I told them that they'd have a hard time making it look good, no matter how much tinsel they put on it. But they could have it for a couple of bucks.

"Then they picked up another one that was damned near as pathetic. Same thing—full on one side, but scraggly on the other.

They asked how much for that one. I told them that it was a deuce, too. So then she whispers something to him and he asks me if I'll take $3 for the two of them.

"Well, what am I going to do? Nobody's going to buy those trees anyway, so I told them they had a deal. But I tell them, what do you want with two trees? Spend a few dollars more and get yourself a nice tree. She just smiled and said they wanted to try something. So they gave me $3 and he carried one of them and she took the other.

"The next night, I happen to be walking past their building. I look down at the window and I can see a tree. I couldn't see it all, but what there was looked good. The lights are on, so I figure, what the heck. I knock on the door. They open it and I tell them I noticed the tree and I was just curious. "They let me in. And I almost fell over. There in this tiny parlor was the most beautiful tree I ever saw. It was so thick it was almost like a bush. You couldn't see the trunk.

"They told me how they did it. They took the two trees and worked the trunks close together so they touched where the branches were thin. Then they tied the trunks together with wire. But when the branches overlapped and came together, it formed a tree so thick you couldn't see the wire. It was like a tiny forest of its own.

"So that's the secret. You take two trees that aren't perfect, that have flaws, that might even be homely, that maybe nobody else would want.

"But if you put them together just right, you can come up with something really beautiful.

"Like two people, I guess."

Royko was feeling so good that he often wrote and reminisced about some of his favorite pranks. He wrote about the time in Mississippi, during his air force training, when he and some friends decided to get even with a Kentucky lad who was always teasing the city guys about their fears of snakes, lizards, spiders, and other things that crawl. "So when Smiley came in one evening and pulled back the blanket on his bunk to turn in, his screams rattled the windows. We could hear him bouncing off the walls in a frantic effort to escape. That was understandable. We had placed a rubber rattlesnake about three feet long, under his blankets."

Once, he stopped at a sporting goods store and bought a long, realistic plastic worm, which he dropped on Hanke Gratteau's desk.

"People who work in offices don't expect long brown worms to drop from the sky so her reaction was predictable. After her hysteria subsided, she described me in terms that ladies of my grandmother's generation seldom used."

One of his best practical jokes involved Gratteau as the straight man. In the early 1980s, the *Sun-Times* editorial board was interviewing judicial candidates. One day, Royko noticed a judge going in to be interviewed by Don Coe, an editorial board member. Gratteau remembered, "At the time, Mike used to play in a monthly poker game with some of the *Sun-Times* business executives. He always sent me to the bank in the Wrigley Building to get cash for the game. On this one day, he saw the judge going into Coe's office and starts to grin. He takes out a couple of hundred dollars and puts it in an envelope, seals it and tells me to wait until the judge leaves and then walk in and hand Coe the envelope and tell him, 'The judge left this for you.'

"It took about a minute before Coe came flying out of his office with the cash spilling out of the envelope, shouting, 'Where did he go? Which way did he go?' Mike finally stopped laughing and went and got his cash back."

Once he mailed a phony letter to erstwhile legman Terry Shaffer:

Dear Mr. Shaffer,
    Perhaps you have heard of me and my nationwide campaign in

the cause of temperance. Each year for the past fourteen years, I have made a tour of southern Florida, Georgia and Alabama where I have delivered a series of lectures on the evils of drinking and loose women. On this tour, I have been accompanied by my young friend and assistant, Clyde Lindstrome. Clyde, a young man of good family and excellent background, is a pathetic example of life ruined by excessive indulgence in liquor and loose women.

Clyde would appear with me at the lectures and sit on the platform, wheezing and staring at the audience through bleary, bloodshot eyes, sweating profusely, picking his nose, passing gas, and making obscene gestures, while I would point out what over-indulgence can do to a person.

This summer, unfortunately, Clyde died. A mutual friend has given me your name, and I wonder if you would be available to take Clyde's place on my campaign this winter.

Yours, in Faith

The Reverend Eletone Jones

Royko also loved to play games with sports columnist Ray Sons. These could be rather juvenile:

Dear Mr. Sons,

Mr. Royko told me to tell my story to you. I am a professional athlete. I cannot give my name at this time, because I want to be sure that you are sincere.

My problem is that I have only one nut.

Could you do a story about me, saying that I live with only one nut?

If you feel this story is of interest, please let me know by just dropping a line in your next column saying: Man with one nut— please contact me.

Hanke Gratteau remembered the time at the *Daily News* that Royko switched Ed Gilbreth's cigarettes. "He got a copy kid to steal one of Ed's cigarettes when Ed was away from his desk, and he very carefully emptied the tobacco and filled it with marijuana and got the copy kid to put it back in the pack. A little while later, Ed lights up and Mike strolls by his desk and starts sniffing the air. Ed doesn't know what's going on and Mike begins shouting, 'Oh my, God, Ed! What are you doing? You can't smoke marijuana at work. You'll get us all into trouble. Geez, Ed, what's wrong with you bringing marijuana into the newsroom?'

Rick Kogan said, "The amazing thing about his jokes is that they were so sophomoric. Here is this brilliant man, this great writer, and he loves to play practical jokes that kids do."

As Royko had written, it's hard to be a grown-up with the spirit of an adolescent. It was merely another of the contradictions that this seemingly moody, growling scold could get belly laughs by tricking or scaring friends and colleagues. But the man who loved to play jokes couldn't take them. No slight, intended or accidental, went unnoticed with Royko.

In March 1987, Eppie Lederer, aka Ann Landers, jumped from the *Chicago Sun-Times* to the *Tribune*. It was a great coup for the *Tribune*. If Royko was the most prestigious columnist in American newspapers, Ann Landers was the one with the biggest syndication in the world. Royko had always been friendly with Lederer, and he was happy to see her join the *Tribune*.

He and Judy were among the *Tribune* executives and top editors invited to a dinner welcoming Lederer. The host was *Tribune* president Charlie Brumback, who praised Lederer, noting that her column ran in more than a thousand newspapers. He also said her joining the *Tribune* was the most significant thing that had happened since he had been in charge of the newspaper.

Lois Wille, who also attended the dinner, said, "I remember being at the Casino Club. Charlie says, 'Ann Landers has 1,200 papers in her syndicate,' and then he added, 'Sorry, Mike.' Charlie was just clumsy. I think he thought he was kidding. I don't think anyone wondered if Mike was in trouble. But after we got home that night Mike called me; it must have been one in the morning. 'I've got to quit,' he said. 'What are you talking about?' I said. He replied, 'After what Brumback said up there, I've got to quit. He embarrassed me.' I told him I didn't think anything of it. We talked for a while, and I thought he would forget it by the next day."

But Royko was infuriated. The next day, he wrote Brumback a letter.

Mr. Charles Brumback:

First of all, I want to thank you for inviting me to your gala party for Eppie, although I regret that I attended. I didn't know that I was going to have to sit in a room with some of Chicago's most influential people, including the entire board of directors of the Tribune Co., and endure the embarrassment of your idiotic, insensitive clodlike remarks about me.

If you want to shoot off fireworks for Eppie, fine. If you want to hail her as the greatest thing since the invention of the wheel, that's fine, too. I like and admire Eppie.

But where in the hell do you get off praising Eppie at my expense? Was it really necessary to illustrate how successful her column is by saying that it is more successful than mine, that she has more readers than I have, that she is better known than I am, and that her switch to the Tribune was more significant than mine?

After your exercise in gracelessness, several genuinely confused people asked me what the heck that was all about? Was I in some kind of trouble?

When I got to my golf club the next day, a person who had been at the dinner wondered why you had elected to make a public fool of me.

I'm going to tell you something: If it wasn't for the fact that I respect people like Squires, Ciccone, Fuller and Wille—news people, not bookkeepers—I would have walked out of this joint this weekend. The fact is, I don't need this job.

But for the rest of the time I work here—which won't be one minute longer than my contract requires—I would appreciate it if you do not invite me to any more parties, speak to me, or have any correspondence with me. Nor do I want to be asked to do TV commercials for the Tribune or anything else that is in my contract. You and Eppie do them.

Sincerely,

Mike Royko [And you may call me Mr. Royko.]

P.S. As president, I know you have the authority to fire me. Make my day.

Later that year, Royko went into my office and said he was taking a few days off. "Now, don't say anything, nothing. My father died. I don't want anything written about it. Nothing."

Mike Royko Sr. died in a veterans home after ninety years of hard living, and hard drinking. He had been married four times.

"He had a love-hate relationship with father," Judy Royko said. "Ever since I knew Mike, not once did he contact his father. We went to the funeral, and Mike got the flag they give veterans' families. I don't know why Mike couldn't make up with him. He said his father was a difficult person and treated his mother badly. His father was a big drinker and had a terrible temper when he got drunk, and he must have done some things that really offended Mike."

Royko's father lived with him briefly at the house on Sioux Avenue, and the Royko children had vague memories of their grandfather.

"I remember," Rob Royko said, "that he used to take me for rides in his big Cadillac. I liked him. Mom thought he was a pain in the ass. He used to have a half pint in his jacket, and while we were driving around he'd take a swig. He said to me, 'I like David, but I like you better.' I guess I never ratted on him."

David Royko said, "He lived with us for a while during one of his divorces. I remember we took a great picture of him sitting in a chair holding one of our cats. Grandpa Royko was such a hard-ass; he had this smile on his face and it looked like he was ready to tear the cat's head off.

"To my mother, he was the bad guy. My mother blamed my father's alcoholism, his moodiness, all of his pain, on Grandpa Royko. She would tell me things that my dad told her about growing up. When Grandpa Royko came home whistling, that was a sign of trouble. He got physically abusive to my grandmother."

Royko saved a letter he received in 1969 from a woman who claimed his father had put something in her drink at his tavern and then raped her. As a result, she claimed, she had a child who was Royko's half-sister.

John Schackitano recalled a story Royko told about his father's propensity for women. "Mike told me he was working for his old man at one of the taverns when he was about fifteen, and he had a crush on a seventeen-year-old girl who was working as a waitress in the place. One day he goes into the back room and there's his dad screwing the seventeen-year-old and that was the end of his crush."

Royko's story reveals the strange mix of pride and anger he felt toward his father. In replying to an interviewer about his reputation for brawling, Royko had once said, "I'm not that tough. My dad just got beaten up in a fight a few weeks ago and he's in his sixties. He was a slugger for Joey Glimco's union."

Royko was always asked in interviews about his pugnaciousness, and his reply was a disclaimer, but there was always that sly twinkle in his words. "As for the brawls, I don't think I've been in more than about four real bar fights in eighteen years," he said in a 1981 interview. In another, he said, "I've only had two fights at the Billy Goat in twelve years and both of them were started by some rude assholes."

Royko excused his aggressiveness by comparing himself to others caught in the vice of celebrity: "I'm a victim of the Frank Sinatra syndrome. Whenver Frank Sinatra or Billy Martin (New York Yankee) goes somewhere, somebody tries to pick a fight. It's the same with me, only the reasons are different. People want to hit Sinatra or Martin to get their names in the paper. People want to slug me because I make them angry."

While Royko could repay rude strangers in kind, he was not by nature physically aggressive. Still, he seemed fascinated, awed, and envious of his father's crude toughness.

David Royko said, "I remember talks with my mom when we were driving somewhere. She told me that Dad felt like a failure in comparison to his father, that he didn't measure up to his father. That was after he had written *Boss* and after he had won a Pulitzer Prize and he still felt he wasn't as good as his father. She said Dad told her, 'I'm not a success. My father was a success.'"

16

In the summer of 1987, Royko became a father for the third time. He and Judy adopted a baby boy. They named him Sam, for Judy's father. Royko explained to friends that he was starting another family at age fifty-five because he wanted Judy to have a family in case anything happened to him.

"It was Mike's idea," Judy Royko recalled. "We were sitting at one of our favorite restaurants on Halsted not far from my apartment and he asked me if I had ever thought about having kids. This was before we were married. I was thirty-seven-years old and I said something like, 'If I do, I do. If I don't, I don't.' He said he would really like to have small kids again. We got married not long after that and having a child proved to be difficult so we began working at adopting. It was not easy."

David Royko recalled that his father had told Rob and him about the plans to adopt at Thanksgiving of 1986.

"It seemed like an out-of-the-blue-kind of thing, and I really didn't have a good sense of why he was doing it. I realized he was hoping he would get it right this time. He was very aware that he hadn't spent time with his first family, but it was unsettling," David Royko said.

David heard Sam before he saw him. "I'll never forget it. I was going to play a final club date with the band. We had set up in the afternoon and I had gone back to my apartment and I was just getting ready to walk out and go play the gig when the phone rings. I picked it up and it was dad and he said, 'David. Listen.' I could hear a crying baby. He said, 'That's your new little brother.'

"I said, 'Congratulations, what did you name him?' He said, 'Sam,' and I said, 'After Sianis or after our dog?' He said, 'Yeah, I thought you'd think I named him after the dog.'"

Royko's sister, Eleanor, also got a telephone call. "It was Mike. He said, 'You busy? Stop over sometime today. I've got something to show you.' So, Eddie and I drove over and we went in and there was a bassinette with the baby. Mike was all smiles. I looked at the baby and said, 'Who dressed him?' He said, 'I did,' and I said, 'That's why his pajamas are on backwards.'"

From that time on, Royko's life took another turn.

"He never wanted to leave Sam alone," Judy Royko said. "He never wanted to go out. He didn't like to go out much anyway, but after Sam he wouldn't even think of having a vacation without taking him."

David Royko said, "From then on, the two topics that he talked about almost exclusively were computers and the kid. That was it, not in that order. But the kid first for the first ten minutes, and then computers for the next two hours. He talked about every milestone of the baby's life and all the stuff he would do. I'll never forget how delighted he was when Sam was about two and pretending to be Solti in front of the television while Solti was conducting the Beethoven *Fifth*."

Royko would lay on the sofa with Sam on his stomach, mimicking whatever faces the child made. Soon, his office at the *Tribune* was filled with more than two-dozen snapshots of Sam as a baby and toddler.

Lois Wille remembered that Royko had been afraid he and Judy would not be able to adopt.

"He was afraid that because of his age they might not get a baby, and then he worried that there was a police file on his drunken-driver charges or the incident with the ketchup bottle. So he was really relieved when Sam came."

Lois Wille said, "He was so proud of Sam, how smart he was, and all the things he could do. He really talked about him a lot, more than his column."

Royko was also proud of his oldest son. David was working on his Ph.D. in clinical psychology at the Illinois School of Professional Psychology. He did his diagnostic studies at the Cook County jail, Lakeview Mental Health Center, and the University of Illinois Student Counseling Service. He received his Ph.D. in 1989 and joined the Cook County family services department, specializing in divorce counseling. In 1986, David married Karen Miller. The wedding took place at the cottage on Bohner's Lake.

Royko was especially proud of David's love of music, often boasting that David had the best music collection in Chicago and knew more about more kinds of music than anyone else in the city. David and Robbie had both played in bands since their teens.

"I started in high school playing the drums in a band, and I did it for ten or fifteen years. It was rough. To do a gig, you have to set up at three in the afternoon and by the time you're done playing it's two in the morning and you still have to load your equipment. By 1989 I was tired of it, and I wanted to write about music and it was bluegrass I wanted to write about," David said.

Royko was not an aficionado of the music his sons played.

"It was heavy metal, biker stuff. I remember once our band played at Roycemore High School, in the gym, and he came to that, but he wasn't very impressed by what we were playing and he didn't hide it."

Robbie had started playing guitar in his early teens. After he got kicked out of Roycemore High School, he went to Central YMCA where Royko had gotten his diploma. All his grades were incomplete, but he got all A's in jazz class. He was playing with a lot of bands. Sometimes he picked up $200 a day. Sometimes he played in the subway for spare change.

Rob was drifting from one job and one place to another. He lived for a while in Milwaukee, and then in Florida. In 1988, Rob's daughter, Carol, was born. Shortly after that, he and his wife separated. They were divorced in 1990. Rob moved for a short time to the cottage on Bohner's Lake.

Royko was thinking seriously of retirement, but he wanted to have enough money to take care of his various family responsibilities. He told an interviewer in 1988 that he was only continuing his column for the money. "I'm well paid. I'm not as well paid as any number of .260 hitters in baseball or as guys who spend a half-hour a day reading the news on television or pointing at weather maps. But I imagine I'm the highest paid person on this newspaper.

"But if a fairy godmother came along and said, 'Here is enough money so you can continue to live the way you've grown accustomed to living, which wouldn't require a vast fortune, and have the freedom from worry—I doubt if I'd ever write anything more than a postcard again. I don't have a burning desire to do any more writing. I do my job. And I do it as well as I can."

"I want to knock off the column before retirement age. I don't like seeing a great athlete when he's no longer got it, or a great politician hanging on after the voters have said no. I want to see myself coming back around the way Teddy Williams went—he hit a home run last time out. I'd like to do that."

Whatever Royko thought, his peers around the country felt he was a long way from burning out. A poll of newspapermen by the *Washington Journalism Review* named him the nation's best columnist in 1987, '88, and '89. That kind of recognition merely added to

Royko's diminished sense of conquest. He was like a general with no more wars to win, no battles left, no enemies still standing. But his intense competitive drive would not let him slow his grueling schedule. For nearly thirty years he had written five columns a week. There were never any Sundays for merely relaxing after golf. The only time Sunday was a free day was if he surrendered Saturday, or if he wrote on Friday, which was rare. But he refused to cut back.

He was beginning to worry about his health, especially with the new responsibilities of his family. He had always gone through fits of changing his lifestyle. He could quit drinking for weeks, or cut back as he did after his marriage to Judy. For a period he played handball, then tennis.

"But it never lasted," Hanke Gratteau said. "Something came up, some good story and he would cancel handball or tennis and that would be the end of it. Every year he would go to Dr. Quentin Young for his physical, and you knew he would be in a terrible mood because the news was always bad. Quit smoking, quit drinking, lose weight. But he couldn't. How could Mike Royko stop eating pork shanks? Most of what he liked wasn't good for him, but it was a combination of being in control and being Mike Royko."

Several *Tribune* editors encouraged him to write only four columns a week. He explained in baseball terms why he couldn't cut back: "I have a better batting average doing five a week. If one or two of them aren't so hot, and one is okay, but two of them are good, I'm still hitting .400. If I did three a week and two were bad I'd be down to .333."

It was more than that. He scoffed when people like William Safire and Dave Barry won Pulitzer Prizes. "How many columns a week do they write, two, one? I'm the only one who ever did five a week for this long."

Royko's preoccupation with his own health and lifestyle may have been in the back of his mind when he wrote his column after Mayor Harold Washington died of a massive heart attack on November 24,

1987. Washington had been reelected the previous April with little of the fear and acrimony that marked his 1983 surprise victory. And, as Royko had predicted with his famous "Don't worry, Uncle Chester, Harold Washington doesn't want to marry your sister" column, Chicago had generally calmed down and accepted Washington's jovial, if somewhat haphazard, manner of governing. Royko's first reaction to the news of Washington's death was to call his doctor, who also happened to be Washington's doctor. The conversation became a column:

"He ate junk foods. And when he ordered something to eat it was double everything and he seldom had enough rest. He wouldn't take the medication I prescribed for him," the doctor said.

Every deadly sign had been there in previous checkups. Blood pressure, a disaster. Cholesterol, a disaster. On again, off again, and on again with the cigarettes.

So Harold Washington, at sixty-five, died in much the way that he lived. On the one hand, brilliant and hard-driving. On the other, careless and indifferent.

He had the brains and ambition to overcome being a black man in a time of black subjugation in this country, and to become a lawyer, a state legislator and a congressman....

Yet, there was this other, puzzling, self-destructive side to his nature....

A man who could boast that he intended to live long enough to be mayor of Chicago for another twenty years. And then tell his driver to stop at a fast-food stand so he could wolf down the kind of greasy hamburgers that would make a doctor cringe.

So, I suppose that all we really know about Harold Washington is that he was human. And ever since humans got up on their hind legs, they've been their own worst enemies.

Washington's death triggered another of the wonderful political brawls that made Chicago famous. Alderman Edward Vrydolyak, the leader of the white majority in the council, had run against Washington in the spring election and lost. Vrdolyak, the gutsiest and perhaps smartest of the politicians who had been vying for power since the death of Mayor Daley, had been firmly in the camp of Daley's successor, Michael Bilandic, when Jane Byrne was elected. Within days, he was a close confidante of Byrne. But after Washington's first election, he was forced to fill the role of head of the opposition in the council. In the spring of 1987, he challenged Washington in the primary but lost. Weeks later, he threw up his hands in disgust and declared he was becoming a Republican.

When Washington suddenly died, the one man who had control of enough council members to claim the mayoralty had shot himself in the foot. The council—fifty Democrats—wasn't going to elect a Republican. But Vrdolyak and Alderman Edward Burke were doing everything possible to seat one of their allies. The council sessions lasted until past midnight, and the shouting and screaming of the blacks, Hispanics, and whites was carried on network television. It was the closing act of the century's most fascinating, perhaps most corrupt, but always entertaining big-city political circus. After three days, a black alderman, Eugene Sawyer, was appointed to fill the mayor's office until a special election would be held in 1989.

Nancy Ryan, who was Royko's legman at the time, remembered it as one of the most exciting periods of her stint with Royko: "The funniest time we had together was the two weeks after Washington died. Because prior to that, he [Royko] wasn't as preoccupied with Chicago politics. Plus, I think he was a little wary of writing about Washington. I think he felt like he had been burned during Washington's first campaign when a column he had written was interpreted as being racist. But after Washington died he was on the phone all day, and he wrote columns where the aldermen were saying dumb things, really to their own detriment, and it was like the old days for him. Then, there were the marathon nights at the council meetings. He went to city hall. He said it was the first time he had been there in years. City News Bureau put out a bulletin saying Mike Royko was at the council meeting.

"I got a taste of what it was like when he was writing exclusively about Chicago," Ryan said. "He had to have two columns every week that appealed to a national audience. It was amazing that he never had a backlog. He always wrote on deadline. So if even a really good Chicago story happened on a day he was writing for his syndicate, he had to ignore it. But he never complained. He was too much the practical businessman. And he really went over those syndicate numbers.

"He would spot some tiny newspaper that had dropped his column and he would get mad. I remember when the editor of the *Milwaukee Journal* called Mike and complained that he had been told he could no longer run the column. It was when the *Tribune* was trying to sell papers in Milwaukee and the *Tribune* said it was now in direct competition with the *Milwaukee Journal*. Mike said he got something like $8,000 a year from the *Milwaukee Journal* and he made a stink about it to the syndicate people and the column stayed in the *Journal*," Ryan said.

"But it was not only the money. He used it to get a sense of just how much appeal he had out there, and I think that if a lot of newspapers started dropping him he would have been terribly distraught and might have considered quitting. But during the years I was with him, 1987 and 1988, more and more newspapers were taking him and that made him feel great," Ryan said.

"But he always talked about retiring. He was very happy after Sam came in the summer of 1987, but it was sort of funny that what he talked about a lot was how happy he was that he could be a good provider. You would think that would be something he would take for granted, but it was a concern of his. He talked a lot about how hard it was the first years of his career when he didn't have much money.

"I was lucky that during the period I worked for him he was pretty content. He wasn't trying to quit drinking, which other people told me were terrible times. He wasn't drinking that much while I worked with him but he wasn't on the wagon. Of course, he was always gruff. If you asked him something he would act totally disinterested but then you just had to ask him one question about something topical or something he had written about, and he would just immediately enjoy the conversation. He'd do all the talking, of course, but that never bothered me. I just refused to allow myself to be intimidated by his gruffness. I just gave him a lot more slack in terms of his social graces than I would most people."

Ryan's favorite Royko column was one about a Sears personnel manager who had shown unusual insensitivity in the case of an older employee whose wife had just died. "I did all the reporting on that story and the next day the guy who was being such a jerk called and said, 'Well, congratulations, I just got fired.' Mike took an enormous amount of delight in that, and it was the first time I really felt the enormous impact that he had and that I had a role in it. I kind of had mixed feelings. The poor guy had lost his job, granted he was an asshole, but Mike never backed down on that kind of thing," Ryan said.

"I think a lot of reporters are kind of afraid of how they could actually affect people's lives, especially the non-politician, private citizen, and he [Royko] just reveled in it. He never ever really had second doubts or whatever. He was always pretty confident that the person who should have been screwed was," she said.

Ryan also had a least favorite column.

"He had just written this column about how he hated restaurants that didn't accept reservations, and he happened to call Harry Caray's to make reservations for lunch, and I heard this incredible loud slam on the phone. 'God damn it! I can't believe Harry Caray's doesn't take reservations,' he said. 'I just autographed a picture for him. I want you to go down there and get that picture back.' It was pretty early in my tenure, but you just don't say no to Mike," Ryan said.

"So I went over there after the lunch rush hour, and asked for the manager, and said, 'Hi, I'm Nancy Ryan. I work for Mike Royko, and he'd like his picture back, I tried to explain it to them and they kind of kept me preoccupied there, and then suddenly I smell this cologne, this cloud of cologne. I looked around and—I never realized Harry Caray was so little—and there he is saying, 'What's wrong? What's wrong? What's Mike's number, I'll call him.' I didn't give him Mike's direct line, I gave him the *Tribune*'s general number but he got through and I could hear the conversation, well, Harry's end of it. 'Com'on, Mike, when have I ever not given you anything you wanted, helped you out, or whatever?' I could just hear Mike's muffled yelling on the other end. He told me later he said to Harry, 'I've never asked for a fucking thing from you.' Somehow, it was dropped and Mike told me to forget it and leave the picture."

Ryan's tenure marked a sea change in Royko's relationship with his assistants. Ryan was a graduate of the University of Wisconsin. She was in her late twenties when Royko hired her from his usual recruiting pool, City News Bureau. She recalled that he usually talked with her about what he was planning to write, which he rarely did in his years at the *Daily News* or *Sun-Times*.

"Sometimes I had no idea what he was going to write, but most of the time he would talk to me unless it was a national story about Reagan or something on which I would have no input or no reporting to do." He also began to give Ryan career guidance, suggesting that despite her own misgivings she accept an assignment to the financial desk rather than become a general assignment reporter, which she wanted. Again, this was a much more personal approach than the way he treated Ellen Warren, Paul O'Connor, or Wade Nelson. And

Ryan and her successors rarely left the office. The era of sending Terry Shaffer off to "find gypsies" for a week was over.

Ryan did have the usual Royko work patterns to endure. "On Mondays, he was always moody and gruff. Usually on Tuesdays, too, but after that he got more cheerful. He was also considerate. I remember coming back from vacation once and he called me in and said, 'Are you depressed?' I asked him why would he ask that. He said, 'I'm always depressed at the end of my vacation. I think everyone is.' I never had a boss before or since who cared whether I was depressed after my vacation. And when my mother died—I wasn't working for him then—he called all around to find out where I was and what the arrangements were. He was very sweet. I think I had a real advantage because his previous relationship with his assistant hadn't gone too well."

When Royko came to the *Tribune* he had brought along Hanke Gratteau in her third stint with him and to save her from working for Rupert Murdoch. He also had a selfish motive, since the *Tribune* would take over Ribfest. Along with planning the Ribfest, Hanke planned her own wedding to Mark Brown, a reporter for the *Sun-Times*.

"Mike had graciously offered to let us hold our reception at Ridgemoor," Gratteau said. "And he insisted on buying the champagne. He went through an elaborate search for the best champagne for the money and of course Bob was in the wine business."

The wedding and Ribfest behind her, Gratteau was released to the city desk, where for several years she headed the *Tribune's* investigative team and eventually became deputy metropolitan editor. Her replacement was Lynn Terman, who stayed less than a year before returning to a job in Washington.

At that point, Royko was asked if he could choose his next legmen from among the many interns whom the *Tribune* would not be able to hire. Although the newspaper was making more money than ever, there were perpetual hiring freezes to reduce staff size. The young interns worked slavishly in the remote hope they would be hired full-time. But openings were rarely filled. One of the last people to move up from copy-boy to reporter was Paul Sullivan. Royko was asked to at least interview Sullivan. He agreed grudgingly.

Sullivan had been offered a job at the *Los Angeles Daily News*, which was then owned by the *Tribune*, but he didn't want to leave Chicago. He was drowning his sorrow at the Billy Goat one January night in 1985 when Royko sat down next to him.

"I hear they're fucking you," Royko said. Sullivan said he had a job offer in Los Angeles. "How would like to come to work for me?" Royko asked.

"Then," Sullivan said, "he got real drunk, and I didn't hear from him for three or four days and I didn't have enough nerve to go ask him. I had told everybody I was going to work for him so I just waited. When he finally got back to me he said I didn't have enough experience but he was going to take a chance with me. His last words were, 'Don't ever fuck me.'"

Sullivan did not have an easy apprenticeship. "Some days my job consisted of depositing his check at the bank in the Wrigley Building, going to the gyros place to get him an Italian beef with sweet peppers, then he'd close the drapes and take a nap, then wake up abruptly and tell me to get him some coffee, then bang out a column in about a half-hour. I would stick my head in every day around 5:30. Sometimes he'd wave me off, or just give me that Royko look."

When Sullivan's cousin called to ask if he could take off to go to the Cubs' opener in 1986, Royko answered the phone and told him, "I happen to know that Paul won't be able to make it."

Royko was also irritated that so many *Tribune* staffers—who all knew and liked Sullivan from his long stint as a copy clerk and his internship—were stopping at his desk to chat.

"He'd dial my extension and say, 'Get rid of her.'"

"When Jimmy Breslin won the Pulitzer in 1986, he [Royko] sent him a congratulatory telegram but when Breslin tried to call him back he wanted no part of him and wouldn't take the call," Sullivan recalled. "Royko said no one ever wins the Pulitzer for commentary twice, but he deserved to be the first. He always blew off ABC, CBS, NBC, all the time, unless it was *Nightline*. I got used to it but I couldn't believe it. He could have been much more famous and much richer, but he said it took up too much time."

In March 1986, Royko told the managing editor that Sullivan was due for a pay raise, but that he didn't think he deserved one. Then he said he was leaving on vacation. "You'll tell him, okay?" Royko said.

The editor called in Sullivan and informed him of Royko's appraisal of his work.

"I felt it was a personality conflict," said Sullivan. "For a while, Royko and I barely spoke to each other. There was no more small talk or calling me into the office to tell stories. I was drinking too much and so was he. It was like dueling hangovers some mornings."

Royko, who often said he could never be an editor because he was incapable of firing anyone, didn't know how to deliver bad news. He reverted to his old ways when he felt it was time for Sullivan to leave.

"He had a notice posted at City News that he was looking for a legman and the first I knew of it was an applicant dropped off her resume on my desk and asked me where I was going. I was fuming when Royko arrived at work and I went into his office and said, 'A woman dropped this off. She said you're looking for a legman.'

"He didn't look at me. He said, 'You're moving to sports. Talk to Ciccone.' I never considered a move to sports, so I was surprised and elated while still being pissed that he did it without telling me. That was the end of the longest two years of my life."

Sullivan prospered in sports, eventually becoming a *Tribune* baseball writer, where he covered both the Cubs and White Sox. At a gathering of his legmen years later, Royko said, "I had to treat Paulie different than the rest of you. I had to be a real asshole to him,

Marine-style. See, Paulie had been coddled by women editors all his life, and I had to be tough on him, toughen him up."

Sullivan said, "He never called me 'Paulie' when I worked for him so I felt that was a good sign. He said I was doing a good job covering the Sox....[Y]ou had to survive your term for Royko to have any respect for you, and if you went on to better things, he gets and deserves the credit. When I realized how much the screaming and abuse helped me to bear down and become a better reporter, I admitted to myself that he was the only reason I am where I am."

But the real problem was Sullivan had been jammed down Royko's throat. Royko was never forced to take a *Tribune* intern again.

After Sullivan, Royko hired Ryan from City News. He hired Ryan's successors without regard for the recommendations of the Metro editors, who wanted their own staff hired. This only increased the alienation and dissatisfaction with Royko from the rank-and-file, which later affected the politically correct criticisms of his columns by the staff.

Royko was always annoyed by the pressures of political correctness. His feelings became clear in April 1987, when he wrote a column about an interview with baseball executive Al Campanis of the Los Angeles Dodgers. Campanis told *Nightline*'s Ted Koppel that blacks were not being hired as managers or general managers because "they may not have some of the necessities." He was promptly fired from his job as vice president of personnel with the Dodgers. Royko did not defend Campanis's remark, but he was clearly sympathetic to the seventy-year-old and pointed out the irony that Campanis was on the show to celebrate the fortieth anniversary of Jackie Robinson breaking baseball's color line.

> Here we have an old guy who obviously revered Robinson as an athlete and a man. He had been Robinson's friend in the days when other players were shouting "nigger" from the dugout. But he

winds up being tarred as a racist and being fired because he went on a TV show to talk about what an extraordinary person Robinson was.

And he was dragged down by an English-born TV interrogator who didn't know Robinson and didn't see him play....I suppose Koppel was just doing his job. However, there is something that puzzles me. During his wrecking job on Campanis, Koppel disdainfully referred to something Campanis said as "garbage."

I've heard Koppel interview vicious international terrorists and cruel dictators, and not once have I heard him disdainfully tell them that their lies and self-justification were garbage. I guess it just shows what a dangerous guy old Al Campanis really is.

Several months later, Royko again was in a minority when he empathized with Jimmy "The Greek" Snyder who lost his CBS football analyst position after a television reporter stuck a microphone in his face and got him to say that "blacks are superior athletes because of selective breeding by slave owners and that if blacks become coaches, whites won't have any jobs in pro football."

Royko had once been in a poker game with Snyder. Royko recalled that the Greek never stopped talking during the poker game and by the end of night Royko had won and Snyder had lost:

It was clear that Jimmy couldn't talk and think at the same time. And this lack of brain-mouth coordination did him in....

Not knowing what hit him, Jimmy the Greek made a public apology, and to dramatize the depths of his remorse, he rushed to Jesse Jackson to seek forgiveness. Naturally, Jackson was gracious and compassionate, which was to be expected of a presidential candidate grabbing free network TV time....

This, of course, was the same Jackson who once referred to New York, which has a sizable Jewish population, as Hymietown. And if

that's not farce, I wasted many a Saturday afternoon watching the
Three Stooges....

What Snyder said about blacks was dumb. What Jackson said
about Jews was nasty. To even come close to matching Jackson, Snyder
would have had to have referred to Detroit or Oakland as Coontown.

These kinds of columns, while admitting that Campanis and Snyder had been stupid, were not the reaction some of the new generation of newsmen expected from the man who became famous
attacking the racism of southern rednecks and City Hall.

Royko wasn't the only *Tribune* columnist who would fall afoul of
the political correctness police.

In December 1995, the *New Yorker* published a profile of Ann Landers in which she had referred to Pope John Paul II as a "Polack." The
various Polish-American societies, of which Chicago had more than
any other place, were outraged. Callers and readers demanded Royko
defend his Polish heritage. As expected, he defended his friend:

So once and for all, let us get it straight. If you are truly Polish, you
are a Polack. Who says so? The Polish language says so. In Poland,
the word for someone who is Polish is Polack.

Thus, when Eppie, as we call her, described the Pope as a Polack,
she was 100 per cent correct....

So I would ask my fellow Polacks to calm down. This lady ain't
got a bigoted bone in her trim bod. And when life gets tough, you
couldn't ask for a better friend.

In later years it became de rigueur for columnists to attack political
correctness, but as usual Royko was far ahead of the pack.

Royko was also struggling with the *Tribune*'s various corporate
enterprises. It began with the Cubs. Although he believed Tribune
Company had done a far better job with the Cubs than the Wrigley

family, and although he supported the idea of a few night games, he was disgusted with the first one.

To begin with, Royko drove to Wrigley Field but learned that parking in the nearby lots for the historic evening cost fifty dollars. He decided to drive downtown to the *Tribune*, park his car, and take a cab, but there were no cabs available. When Royko went to the El, the platform was so crowded he couldn't squeeze on the train, so he finally caught a cab and got to the ballpark. He wasn't too thrilled by the spectacle:

> This is strictly a personal opinion, but lights in the Cubs Park cheapen the place. It looked glitzy, schlocky, more appropriate for a rock concert. Without the sky, the sun, the clouds, the park like setting is lost. It's no longer a fresh oasis in the city. It's just another glowing sports stadium. The uniqueness is gone.

It was hardly the ringing endorsement the corporate moguls at Tribune Company would have liked. But their opinion didn't matter to Royko. He wouldn't mince words for their sake. Royko never allowed peripheral ventures such as corporate interests or television to lessen to any degree the power and honesty of his columns. For example, his slamming of Koppel for the Campanis interview might have ended any future invitations to appear on the prestigious talk show. Royko had been interviewed by Koppel several times. There were few newspaper journalists in America who would have written anything that could jeopardize such invitations, but not Royko.

Royko was particularly irked when the *Tribune* bosses decided to hype the newspaper's annual Christmas holiday fund for the needy. It had been decided that every columnist in the newspaper from the sports pages to the auto writer to the gossip column should devote one column to soliciting donations for the Christmas fund and that an appropriate coupon would appear at the end of the column. Most columnists grumbled, but turned out some syrupy piece to please their bosses.

Royko was approached for his help in 1987. His eyes glared, his voice deepened, his chin sagged to his chest. "Yeah," he said. His column appeared the following week:

Those who believe in the great media conspiracy won't believe this, but in the 24 years I've been writing a column, no editor has ever told me what I should or shouldn't write about.

With one distasteful exception and this is it.

As some of you may have noticed, this paper runs a Christmas fund for the needy. So the editor suggests to all the columnists that we do one piece putting the arm on the readers for contributions

But for several valid reasons, I object to using an entire column for a purpose that could be interpreted as an act of kindness and charity, which makes my friends laugh at me.

First of all, almost all such columns are boring. By the time a reader gets to the third paragraph and realizes that he's being suckered into reading about the needy, the hopeless, the downtrodden and the destitute, he says, "Oh, man, I don't need this kind of downer."

And he promptly turns the page, seeking a story about murder, lust or political mischief—something to brighten his spirits....

But if you feel guilty for being prosperous, you can ease your middle-class conscience by using the form printed below to send a few bucks into the fund.

But please, don't tell anyone I asked you to do it. I have a reputation to uphold.

Mindful of the people who were tossing stock options and bonuses in his direction, Royko faithfully went along with the annual Christmas column, and each year he spoofed the entire idea. Perhaps unknowingly, he was already campaigning against what became a newspaper strategy of the 1990s, a fad called "civic journalism," in which newspapers would try to reach out to their communities; to participate in community affairs; to march in the parade instead of simply covering it. The remnants of this experiment are dinner parties for people who write letters to the editor with great frequency and newspaper-sponsored coffee klatches where readers are invited to give their opinions on what should be in the newspaper. These events are encouraged by modern corporate media companies who believe firmly in decision made by committee to spread the blame for ideas that do not work.

But the *Tribune* brass wasn't concerned with Royko's minor insurgencies. The newspaper was one of the top moneymakers in America. The *Chicago Sun-Times* circulation, which threatened the *Tribune* at the beginning of the 1980s, was in a free fall. Murdoch's foray into Chicago had proved a disaster. He imported editors from the *New York Post* to bring more of a sensational look to the *Sun-Times,* and that caused a 25,000 drop in daily circulation and a 44,000 hit on Sunday. The *Sun-Times's* once liberal editorial page—a deliberate contrast to the *Tribune*—took on a conservative slant that chilled the independent lakefront community, particularly Jewish readers who for the first time began to pick up the *Tribune.*

In the two years after Murdoch bought the paper, the *Sun-Times* was down by 50,000 daily. Murdoch was also engaged in buying up television stations around the country. Under Federal Communication Commission rules he would be forced to divest himself of some of his newspapers. In 1986, he sold the *Sun-Times* to an investment group headed by the man he had named publisher of the *Sun-Times,* Robert Page, and New York investment banker Leonard Shaykin. The price was $145 million. Murdoch had started the decline of a once proud newspaper and picked up $55 million in the bargain.

By 1989, the *Sun-Times* had fallen to nearly 500,000 in circulation both daily and Sunday, and Shaykin bought out Page's interests. By this time, the paper's staff and distribution facilities had shrunk. Its news pages had been reduced. The *Sun-Times* was no longer competitive with the *Tribune* in the growing suburban area, which appealed most to advertisers. Soon, the *Tribune* was boasting that it had more than 90 percent of the precious real estate advertising lineage. While the *Sun-Times* foreign, national and financial coverage was negligible compared with the *Tribune*, it remained competitive in sports and local coverage, especially the doings at City Hall where its reporters often came up with exclusives. But Royko's presence always gave the *Tribune* that extra dimension.

In 1989, he got involved with helping State's Attorney Richard M. Daley become mayor of Chicago. Eugene Sawyer's stewardship of the mayoralty after Washington's sudden death impressed no one, and Daley easily won the 1989 special primary to set up a three-way race in the general election against Alderman Timothy Evans and Edward Vrdolyak.

Royko's support of the younger Daley fascinated those who remembered how he had not only attacked the senior Daley but often lumped the Daley sons together as "Larry, Curley, and Moe."

But almost immediately after the elder Daley's death in 1976, Royko began defending Daley's son. More than anyone, Royko was responsible for Richard M. Daley's election to state's attorney in 1980, and he seemed determined to help him again. In January 1989, he defended Daley against those who argued that once elected the young candidate would want to be just like his father.

> Of course, he does. He'd be stupid not to. For that matter, just about anybody who wants to run Chicago would like to do it the way old man Daley did. By that I mean that the late mayor was boss. No ifs, ands or buts. He had complete power over city government, most of county government, the judiciary, Chicago's many state legislators, and the political machine that put them in office....

It might not be the democratic ideal. But from the perspective of the guy in charge, that kind of power makes the job a lot simpler.

It almost sounded as though Royko was overwhelmed with nostalgia for the autocracy he damned in *Boss*.

Lois Wille remembered in 1979 when the younger Daley made his first appearance at the *Sun-Times* editorial board, which was considering endorsements for the 1980 elections. "Mike would sit in sometimes just to hear candidates or meet someone he hadn't met before. When Richie came in, Mike was almost fatherly. He took his coat, asked him if he would like some coffee. He was really nice to him."

Undoubtedly, Phil Krone, Royko's consigliore for political matters, had a great deal to do with Royko's conversion, but from 1980 forward Royko steadfastly supported young Richie.

Black supporters of either Mayor Sawyer or Alderman Evans constantly attacked Daley on the grounds that returning Chicago's City Hall to a white candidate would be an insult to the African-American community and incite rioting.

Royko scoffed at such notions and for the first time actually declared seriously who he was going to vote for in an election.

Rev. Al Sampson has swung my vote. Next Tuesday morning, I'm going to vote for Richard M. Daley.

I don't remember ever hearing dumber political threats than those made by Sampson, a leading supporter of Mayor Eugene Sawyer.

Among his many ravings, Sampson has warned, "If Daley gets in City Hall, there will be protests in the streets. We will shut down this city."

That tells me that Sampson doesn't understand how an election works. The person with the most votes wins. If you don't get enough votes, you lose.

That appears simple and fair enough. But Sampson doesn't see it that way. He says:

"Daley has no right to that seat because it belongs to Harold Washington and Gene Sawyer is Harold Washington's man...."

If any of Daley's top supporters talked about shutting this town down if a black man is elected, or about hanging anyone who voted for a black man, or claimed that Daley had a divine right to the office because he was Richard J. Daley's son, I would vote for Sawyer. But they haven't said such things.

After Daley won the primary, Royko went after Vrdolyak, who had become the Republican nominee as a write-in candidate:

I wish Vrdolyak would say what's really in his heart. And that would be something like this:

"I have jumped in as a write-in candidate because I loathe Richie Daley. I envy and resent Richie Daley, and I will do anything I can to see Richie Daley go down to a humiliating defeat....I mean, you can't even imagine how the thought of Daley as mayor sticks in my craw."

Evans, running as a third-party candidate, said Daley was a racist because he did not hire enough black lawyers for his staff at the state's attorney's office. Royko retorted in a rather elaborate column citing the numbers of black lawyers graduating from Chicago law schools and tried to make the case that there weren't enough black lawyers interested in the relatively low paying prosecutor's office.

I can't look into Daley's soul or read his mind, so I have no idea how he feels about blacks. And even if he tried to explain how he felt I probably wouldn't understand him. But if Evans' supporters are going to try to tag him with a racist label, they'll have to come up with something more than those meaningless numbers.

Perhaps the most uncharacteristic thing Royko did in 1989 was to put on a black tie and go to Daley's inauguration. Whether he presumed he had now reached such celebrity status that he was considered a part of the Chicago establishment or whether he simply wanted to make amends for all those years of Daley onslaughts, his appearance surprised a number of people. Only a brief time later, Royko turned down a chance to have a "cheezborger" with George Bush or visit Bill and Hillary Clinton at the White House. It clearly wasn't the glamour of politics that attracted Royko. He simply liked the young Daley.

But as interesting as the restoration of the Daleys to the throne of Chicago might have been, it was upstaged by other events in Chicago. Politicians took a back seat to sport. In 1985, the Chicago Bears won a Super Bowl. It was their first title in twenty-two years, which was a miraculously brief period by Chicago standards. The Bears' fiery head coach, Mike Ditka, renowned for spitting gum at fans in San Francisco and being a champion television huckster, had become a city celebrity. Royko often wrote disparagingly of Ditka, but sometimes he defended the macho antics of the coach. When he wrote that the Bears were wrong for firing Ditka in 1992, the coach wrote him a note, which he kept.

"I'd like to have a beer with you sometime," Ditka wrote, surely the highest accolade Royko could get.

It had been a rough couple of decades for Chicago sports. The love affair that Royko and much of Chicago had with its sports teams was not mutual. The White Sox and the Cubs combined for a century of almost perfect failure; the Bears had a miserable record until their 1985 breakthrough; and the Stanley Cup, hockey's championship trophy, was such a rare visitor to Chicago that few people ever heard of it. The city was outmatched when it came to producing stars. New York had Ruth, Gehrig, DiMaggio, Mantle, and Mays. Even Royko's boyhood heroes, as idolized as they were in Chicago, were almost anonymous elsewhere. Phil Cavaretta and Andy Pafko weren't in the same league with stars from Boston like Ted Williams or St. Louis's Stan Musial.

But 1983 changed all that. Chicago, which had been without a bona fide supernova of sport for the entire century, now got the brightest one of all. Michael Jordan became Chicago to the world. He replaced Al Capone and the image of Chicago as the bloody home of Prohibition era gangsters.

Chicago was now the graceful, soaring, jump-shooting black man with a precious smile who sold Chevrolets, Nikes, and McDonald's hamburgers when he wasn't leading the league in scoring and endorsements.

Royko became a die-hard Bulls fan. One of his favorite fantasies almost took place in 1990. He called an editor and said, "There's no problem with my playing in a charity golf match, is there? It's not for *Tribune* charities, but some guy thinks they could raise a lot of money if they had a golf tournament between Chicago's three greatest Mikes—me, Jordan, and Ditka."

Weeks later Royko reported that the event had been canceled because Jordan couldn't fit it into his schedule. A short time later, Ditka lost his job, leaving Chicago with only two great Mikes.

After Daley's election and subsequent reelections in 1991 and 1995, politics held little interest for Chicago or Royko, which prompted some of the inaccurate criticisms that Royko had become more inter-ested in national affairs than his traditional targets in City Hall. Young Richard M. Daley managed to produce a few chuckles by fracturing his syntax in ways reminiscent of his father, but the old rogues were gone and the new mayor did not adopt his father's tendency to frighten political opponents in and around Chicago. The suburbs had grown in political power, and there was no more *Boss*. Politics in Chicago was missing an enemy for Royko to pick on. But he found them in other places. The Reagan White House was a fair target, and so was a scary political activist named Lyndon LaRouche. Royko was fond of calling LaRouche a former Communist and Stalinist thug.

In 1989, Royko was sued by LaRouche followers, whom Royko also labeled thugs. They got an injunction seeking to prevent Royko from writing any more mean things about them. The injunction was dismissed, and Royko cheerfully continued to bash LaRouche. But the fight got personal. Rob Royko entered the fray.

"I had been borrowing a lot of money from him [Royko] and he was pretty mad at me. He said, 'That's it, not another dime. I don't give a shit if you end up living under Wacker Drive,'" Rob Royko remembered. Lower Wacker Drive, curling along the banks of the Chicago River, had become a favorite place for the homeless in the 1980s. They dragged cardboard cartons to the loading docks that lined the street and slept there.

"Well, I was just about living on Lower Wacker," Rob said. "Actually, I had split with Kippie [his wife], and I needed $100 to pay rent. I was working at a job in the attorney general's office that Phil Krone had gotten me. That pissed my old man off because he didn't want me working in places where he thought I would embarrass him. On the other hand, he didn't want me unemployed. I called him and said I needed $100. 'I really need $80 for my rent and $20 to get to work for three or four days until payday.' He said no. We got in a huge argument over the phone.

"So I decided to get even. I called the LaRouche campaign headquarters and told them who I was and volunteered to join them. They went crazy. They put out a press release saying Mike Royko's kid was going to join them. I got a call to go over to the *Sun-Times* for an interview. They were eating it up. Before I got there, I heard from him [Royko]. I got the $100."

The Roykos moved again in 1989. They bought a house at Racine and Shubert on the near North Side. Royko continued his battle with the good life:

After the waiter took our orders, the blonde stared across the table at me and coldly said, "Why did you do that?"

Do what?

"You ordered veal shanks."

That's right. I love veal shanks. That's why I ordered it.

"But you know it isn't good for you. It is high in cholesterol."

I nodded and waved to the waiter. He came to the table and I told him I wanted to change my order.

"What will you have?" he asked.

I said I wanted the pork shank instead of the veal shank.

When he left, she said, "Why did you do that?"

Do what?

"You know perfectly well what. The pork shank has even higher cholesterol than the veal shank."

Then you should have left well enough alone when I ordered the veal shank.

"That's not the point. You shouldn't be eating any of that stuff. Pork shanks, veal shanks, red meats. They are all bad for you."

Royko also revealed his hope that he would inherit his father's genes.

To the best of my knowledge, my father never ate one can of tuna packed in water. No, the old man ate pork shanks, pork chops, slabs of beef smothered in gravy.

At age 80, he duked it out in a bar with two young punks who weren't a day over 65, and he came out with a split-decision draw.

He also gave up the Pall Malls he had been smoking for forty years and switched to low tar Carletons.

In 1989, Charlie Brumback, president of the *Chicago Tribune*, was named chairman of Tribune Company. He was succeeded by John

Madigan, who had joined the *Tribune* in 1975 from Salomon Brothers investment firm. Madigan was hired to plan the transition of the *Tribune* from a private to a publicly held company and had risen to executive vice president of the parent company.

James Squires, editor of the *Tribune*, and Madigan did not get along. Squires resigned in December 1989 and was succeeded by Jack Fuller, who had been the editor of the editorial page. Fuller had started his career at the *Daily News*. Royko knew him and liked him. Except for the Christmas-coupon column, Royko continued to write with no interference from his bosses.

In 1989, Royko brought out his sixth collection of columns, *Dr. Kookie, You're Right*. He received a $75,000 advance—twenty-five times what he had been given for *Boss*. *Dr. Kookie* became the replacement for Slats Grobnik, who only appeared in nostalgia columns about firecrackers and bowling. When Royko summoned him to discuss current events, Slats was now a mature man-in-the-street contemplating the rapid changes in American society, changes that Slats found disconcerting because he was becoming somewhat conservative, less certain, and more content. Slats's role as the glib punk who distrusted all forms of authority was not as essential for America of the 1990s.

Dr. Kookie, Royko explained, was the founder and prophet of the Church of Asylumism, which believed that the human race was descended from a race of people from a far-off planet who were sent to earth after their brains were addled by eating bad mushrooms. They wanted to become lawyers or form political parties, and they wanted to dictate to everyone how they should live. Foremost among these loons were the TV preachers who claimed God was telling them how people should live. Royko had a fair run of attacking Jerry Falwell, Jim Bakker, Jimmy Swaggart, Oral Roberts, and Pat Robertson.

In 1990, Royko was named for the third time as "the best newspaper writer in America" by the *Washington Journalism Review*. He told the magazine that he was planning to retire at age sixty and work on his golf game. "I could still be an Arnold Palmer. Or, I could be a cowboy. Or, I might run for mayor of Chicago. There's a lot of stuff I might do besides write a column."

Susan Kuczka, who succeeded Nancy Ryan as Royko's legman, said there was almost a daily discussion of whether Royko should retire. "He kept saying no one appreciated him anymore. He was trying hard not to drink a lot, and he was taking Valium. But once in a while, he'd tell me to run over to the Goat and get him a six-pack, and I would worry about him. But he still loved to go home. He loved to play with Sam and his Nintendo games. I think he wanted to spend as much time as possible with Sam because he knew he might not be around forever.

"I don't think he was that concerned with his health," Kuczka said. "He'd order skim milk and turkey sandwiches but drink ten cups of coffee. He'd talk about his high blood pressure or cholesterol, but he really didn't worry."

Although Royko was no longer hanging out in bars, he obviously missed what had been a large part of life. It showed when he mused to WJR that he might open his own piano bar and play songs by Cole Porter and Hoagy Carmichael. Actually, Royko needed another piano bar, because in August 1990 the Acorn on Oak closed.

I know a lot about bars. More, I'm sure than is good for my health....So I must take a moment to mourn the closing of one of the best gin mills in Chicago. Which is saying something. I doubt that any other city has a bar called "Stop and Drink." Here, a shot is a shot, a beer is a beer, and a corporate lawyer and a tuck pointer can belly up as equals.

The Acorn. Or more formally, the Acorn on Oak. Or, as some of us called it, Buddy's. It's closed down....

The first time I went there, many years ago, I was skeptical. I'd heard a lot of piano bar players banging out junk, second-rate Liberaces grabbing for the conventioneer's tip. Then Buddy smiled across the piano....

"Anything special you'd like to hear?" he said.

"Yeah, play 'Black and Blue.'"

He grinned. "Don't get many requests for that."

Then, wham, his fingers raced in a flourish from low E to high E, he leaned back, his eyes half-closed, a devilish grin and his piercing tenor froze the place.

And that was the beginning of a very fine friendship. During my bleakest years, before I met and married the blond, I spent many a night listening to Buddy. He made them less bleak.

The post-midnight gang feels homeless.

Royko also thought about retirement because it became harder and harder to focus on issues that he knew would catch the mood of most readers. One example was AIDS. He took on the AIDS question by ridiculing the media for fawning over Magic Johnson when the basketball star disclosed he had HIV. Slats Grobnik was back in the harness for that one:

I thought to be a hero you had to do something brave and not selfish and something that would help other people. When you cut through all the media hype, besides being a great basketball player, what did he do? He was out there hopping in the sack with one bimbo after another.

When the Chicago Transit Authority buses approved the use of an advertising poster with a man kissing a man, Royko noted, "The ad, we're told, was created by a group that wants to dispel myths about how people get AIDS. And kissing is one way you don't get it."

And I'm a bit puzzled by the statement that love doesn't cause AIDS.

Love isn't an issue at all, unless you define love as having anal sex with a stranger in a bathhouse.

It was after that column in 1990 that Royko's current leg person, Susan Kuczka, came to my office with a note in a sealed envelope: "I resign, effective immediately. Royko."

She told me, "He's really upset, you should go and talk to him."

It was about six o'clock and Royko was slumped in his chair when I walked into his office.

"I wrote my column for tomorrow, but that's the last one. I'm done," Royko said.

I asked if anything had happened at the paper to upset him.

"No, everything's fine. I just got off the phone with a woman whose son has AIDS."

I inwardly shuddered, since one of my responsibilities was to write letters to all the people who complained that Royko had hung up on them or cursed at them or said words they had never heard before.

"You usually don't have trouble with the crazies," I offered.

"She wasn't crazy. She was very polite. She said she understood exactly why I wrote about gays the way I did and she said she felt the same way until she learned her son was gay. Then she talked for a long time and about how her son hated being gay but couldn't do anything about it. She made me feel like shit. I can't write anymore. No matter what I write half the people get upset. Years ago, if I wrote a column saying the death penalty was too good for John Gacy, everybody would agree. If I wrote that column today, half the readers would start picketing the building. That's it. I'm done."

I called Royko at home and continued the conservation at two or three different intervals during the night. At midnight, Royko said he would be in the office the next day.

chapter 17

The role of American newspapers changed dramatically with the breakup of the Soviet Union and the end of the cold war. During the 1980s, the fear of an intercontinental ballistic showdown had waned in the wake of détente and perestroika, but the missiles were still in the silos, and Soviet encroachment in third-world countries continued to threaten America's claim as both the military and economic superpower. That ended with the fall of the Berlin Wall in 1989. Four decades of concern over the delicate balance between the Soviets and the West vanished. Television had not kept up with all the checkerboard movements of the Soviets in Eastern Europe, Africa, and Asia. Newspapers had been the chief source for this information. After 1989, Americans no longer had to rely on newspapers to inform them about the Soviet threat. East-West relations were suddenly of little value or concern to the American public.

Moreover, except for the brief flurry in the Persian Gulf, the U.S. president no longer seemed as crucial a figure as he had been during the cold war, when solemn young men hovered near him with the ominous black bag containing the instrument of nuclear war. Without the challenge of the Soviets, the president's henchmen also became more anonymous. The dramas surrounding John Foster Dulles, Dean Rusk, and Henry Kissinger had made them household names and even celebrities in the 1950s, 1960s, and 1970s. The most important cabinet members of the Clinton administration were those who dealt with the economy, and the average American never heard of them.

Washington news became stale unless it tweaked some emotional flash point, such as gays in the military or sex education in schools or the continued battles on abortion laws or flag burning. The day-to-day congressional séances and the routine briefings by various cabinet members were relegated more and more to the briefs columns and ignored entirely by television. In fairness to the American people, there were few monumental divisions in American politics. FDR's New Deal and LBJ's Great Society had created legislation that legally wiped out any vestige of discrimination and established programs to care for the poor, the elderly, and the disenfranchised. Certainly, discrimination continued to exist in society, and children and the elderly went hungry in some places, and health care, with its soaring costs, was an issue of concern for many. But there were no longer the great international, political, or social issues that galvanized the public's attention.

A new medium, the electronic delivery of news, wiped out the primary function of newspapers. Boris Yeltsin was elected president of Russia at 5 A.M. Eastern Standard Time, too late for any newspaper to break it in America. By the time the newspapers were able to report the story the following day, the entire world had moved on to something else. When Princess Diana was killed in a Paris car crash, Americans were riveted to their cable televisions, watching MSNBC, CNBC, and CNN. Hardly any East Coast newspaper was able to print a line about the tragic collision in their Sunday editions, and papers in the Midwest, such as the *Chicago Tribune*, rushed sparse stories into print, trailing far behind the detail reported on television.

The same thing happened in the O. J. Simpson murder trial. There was nothing newspapers could do to match the dramatic impact of television, which followed Simpson's white sports utility vehicle live as it moved down Southern California freeways. Breathless announcers reported Simpson had a gun on the car seat next to him. America might witness a celebrity suicide.

Most newspapers struggled to adjust to this new era of instant news and celebrity reporting. Their front pages no longer represented what editors thought were the most important stories, but what they felt would be the most interesting stories to their readers. Some of these stories were up-to-the-minute reports on the latest problem facing the baby boom generation; many more dealt with the celebrity or the crime that had been hyped by television on a slow day; others dealt with entertainment; sports figures moved to front pages.

Royko hated all this. He also hated overblown stories about Africa or Asia, the four- or five-part series stories, which major newspapers produce to provide interested readers with a glimpse of the world. Royko argued that the number of readers interested in such stories was minuscule and that such efforts were made only with an eye to winning a Pulitzer Prize. He scoffed at investigative stories, whose parameters had changed since he became a newspaperman. Many newspapers trumpeted stories as "investigations" when they were merely leaks from the prosecutor's office. Royko ranted when he saw a story in the *Tribune* inferring some politician or public figure had done something wrong and then read on the jump page the familiar, "Although there was no evidence of wrong doing or illegality…"

"Why in the hell are we reporting it if no one did anything wrong?" he often shouted. "We're just embarrassing a lot of people trying to win some goddamn prize."

In fact, a lot of readers were wondering about newspapers' motivations. Public opinion polls during the 1990s showed the media was considered less trustworthy than congressmen and about as trustworthy as used-car salesmen.

The new trends that showed up in the *Tribune* widened the gulf between Royko and the newspaper he had joined as a last resort. He began to spend more days at home, writing on his MacIntosh computer and sending stories into the office. One reason for his seclusion was that Lois Wille—his journalistic ideal, sounding board, conscience, and often the only person who didn't hastily agree with everything he said and did—retired.

Another reason to avoid the office was the ban on smoking in Tribune Tower. The ban applied to every office in every building owned by Tribune Company. It applied to the press box at Wrigley Field. It eventually applied to the entire ballpark. It applied to the small offices in the suburbs. It even applied to the sidewalks on Michigan Avenue in front of the Tribune Building, although these were owned by the city of Chicago. It did not apply to Mike Royko's office.

"Who's going to tell him?" was the reaction of everyone from the chairman of the company on down the chain of command.

Royko's office quickly became a haven for his former legmen and friends such as Hanke Gratteau, Rick Kogan, Nancy Ryan, Susan Kuczka, and others who smoked. Although he never pretended to go along with the ban, Royko obviously felt sheepish about being the sole violator. Whenever his editor walked in, Royko would nervously shovel things around his desktop to cover the ashtray, which seemed foolish since he knew that the *Tribune* bosses knew he smoked in his office, that no one was going to say anything, and that everyone in the building knew the smoking ban did not apply to his office. In a narrow way, that increased the rift between Royko and the staff, who perceived him as elitist and above the rules, which he was.

Still, the opportunity to smoke and listen to Royko provided some

new friends, such as science writer Peter Gorner. Royko liked to discuss his various ailments and treatments with Gorner, who was familiar with the latest medical theories and drugs for high blood pressure,
high cholesterol, and depression. But their friendship illustrated how
difficult it was for Royko and *Tribune* newsmen to form relationships.

Gorner had worked at City News Bureau through high school,
college, and graduate school. He had started there shortly after
Royko left for the *Daily News* in 1959. Gorner had been a top byline
writer at the *Tribune* for fifteen years when Royko joined the *Tribune*
in 1984. It would have seemed that they could be friends.

"I really didn't know him until 1993," Gorner said. "We were on an
elevator and he was talking to someone, which was rare, and I joined
in. After that, I began stopping at his office. He was sort of flattered
to know that I was someone who had read every column he had written. And we knew a lot of the same people, we had the City News
background, so we did a lot reminiscing."

Yet Gorner did not approach Royko at the *Tribune* for nearly ten
years. "It was something you didn't do," he said. "You had to be willing to break through that facade of toughness. He was very shy. But if
you did it, you could get an invitation to sit there and get three hours
of Royko free."

Janan Hanna, who followed Susan Kuczka as Royko's legman in
October 1990, said, "That was the best part of the job. Some days he
didn't want to talk, but other days he would say, 'Come in here,' and
he would go off for hours on stories that were fascinating. A lot of
times he was rehearsing his column. It was brilliant."

When Royko interviewed Hanna, he violated all the new laws
about privacy and discrimination. She recalled, "He asked my age,
which you weren't allowed to do, whether I was married, whether I
was dating anybody. I didn't care about that stuff. He asked about my
background and I told him I was a Palestinian. He asked a few more
questions and he finally said, 'Are you religious?'

"He could see I was somewhat taken aback and he shouted, 'Well, I have the right to know if you're going to pray outside my office five times a day.'"

Hanna also sent Royko into a fit of laughter when she told him her father sold carpets. "An Arab rug merchant!" he howled.

Then he apologized and hired her.

He was still a demanding editor, insisting that no telephone call be dismissed as worthless, that any caller might provide a column, that every detail be uncovered. Hanna's favorite recollection was the day Royko was watching CNN coverage of Queen Elizabeth's visit to Washington, D.C.

"She was supposed to visit some public housing and they interviewed a black woman she spoke with. The woman had made fried chicken thinking the queen would have lunch but she got stiffed. Royko watched this and said, 'Find that woman and ask her how she felt about being stiffed and what she would have asked the Queen.'

"So I did. It turned out the woman had to quit school in the third grade, walked three miles everyday to work as a maid, raised a bunch of children. She said she would have liked to ask, 'How's it feel being Queen?'

"The next day everyone had a story about the Queen's visit but Royko had this great column about the little old black lady who got stiffed. He had great instincts," Hanna said.

In the summer of 1992, Royko became suburban man. He and Judy moved to a three-story home in Winnetka, not far from Lake Michigan. Royko took over the third-floor room as an office. The room had a fireplace, views of the ravines that ran along the backyard of the house to Sheridan Road and the lake, and plenty of room for his books, a lifetime of columns, and all his computer gadgets. He put up a few pictures of himself with Mayor Richard M. Daley and threw a variety of his hats around the room. He began to spend more time writing from home. For one thing, it was a long commute from

Winnetka to the *Tribune*, and for another he enjoyed being with the family, which now included a little girl, Kate, whom the Roykos had adopted in 1992.

If the sixty-year-old Royko thought he would finally have a sugar-and-spice daughter to cuddle, he was mistaken. Kate was a bundle of energy who never wanted to sit still or lay quietly on his stomach the way Sam had as a baby.

His sister, Eleanor, recalled Royko saying, "After all these years I wanted a little girl I could cuddle and hug and here I get a kid who calls me 'Mike.'"

"One day," Judy Royko said, "we were taking the kids bowling, and they were in the back seat and fighting like kids will, and Mike turned around and said something like 'If you kids don't behave we will go back home.' I was driving and I looked in the rearview mirror and here was Kate—she was not even three—giving Mike the finger."

When Judy decided it was time to explain to Sam that he was adopted, Mike was so frightened the child's reaction would be negative that he refused to join the discussion and waited in another room of the house.

"He was trying so hard to be a good parent, and he worried that he wouldn't be there to see them grow up," Lois Wille said. "He called a lot just to talk about the kids and what they had been doing."

Some of Royko's friends assumed that he had moved from the city to please Judy and to ensure the children would go to good schools, but Royko was also distancing himself from much of what had been his past.

"It was hard to ever get him to go out," Judy Royko said. "It also was an effort to get him to be with people. More and more he chose not to make the effort. I would do it sometimes and then he would enjoy it but he wouldn't do it himself. He just wanted to be home and relax. His life was still so busy. He was still writing the column on weekends, on holidays."

When Hanna finished her two years with Royko in the fall of 1992, he hired Pam Cytrynbaum. He told her, "This is my contribution to Middle East peace. I just had a raghead and now I've got a Jew."

"He also asked me if I drank," Cytrynbaum said. "I was terrified. I had grown up reading Royko, this wild drinker. I was afraid if I told him I didn't drink he wouldn't want me. But I told the truth."

"Good, one of us has got to stay sober," Royko replied.

Actually, Royko was staying sober most of the time. He had gone through an out patient alcohol rehabilitation program that didn't completely take, but he only rarely stopped at the Goat for a drink and usually had a beer or two when he wound down at home.

He was also trying to quit smoking. "He had the patches for a while," Cytrynbaum said, "and he would go without cigarettes for a while, then he'd start again."

In 1994, Royko celebrated thirty years as a columnist. He was interviewed and feted. The *Tribune* held a luncheon at Harry Caray's Restaurant where old friends Lois Wille and Ann Landers (Eppie Lederer) joined the *Tribune* brass. His legmen surprised him with a party at his house where he entertained them as he had for years in his office with stories about them, how he hired them, what he remembered, and how proud he was of all of them.

Royko had reached a milestone that no one else had achieved. As he noted when he announced the previous year that he was cutting back to four columns, "At the *New York Times*, the *Washington Post* and other prestigious publications, the standard is two columns a week. If I had been at the *New York Times* the last twenty-eight years, I would have a backlog of columns taking me to year 2032, although I doubt that anyone will be able to read by then."

He was constantly thinking of retirement. "Almost every day," Judy Royko said, "the first thing he did at the computer was to figure out if he could afford to retire."

By this time, Royko was making in the neighborhood of $500,000 a year.

"He never really felt he was wealthy," his wife said.

Royko's never-ending worry over money was an understandable combination of youthful privation and ethnic insecurity that many members of his generation shared. There were always people who remembered the depression, or told tales of poverty in their hometowns in the Ukraine or Poland or any number of places where immigrants originated.

Above all, Royko wanted to take care of his family. And in 1994, family problems were building. David's wife, Karen, had given birth to twin boys, Jake and Ben, in July 1993. A year later, Ben was diagnosed with autism. Royko immediately told David and Karen he would help pay for any medical and therapeutic expenses the child required. That cost amounted to $30,000 a year.

Royko was still trying to come to terms with Rob, who had remarried and had two small boys, the oldest named Michael.

"He told me if I named a son after him he would cut me out of his will," Rob said. "I told him, 'You've already cut me out of the will.' I told Aunt Eleanor what he said and she said to ignore him, that he would be proud to have my son named after him. When I did, he didn't say anything."

Royko also missed many of his old buddies. He joined Sunset Ridge Country Club, a short drive from his home, because it was more convenient to take the children to the pool and most of the people Judy played tennis with lived in the area. But Sunset Ridge didn't have the same membership as Ridgemoor. There was old money and new money from investment banking, and law, and stocks and bonds. There weren't plumbers and cops and ward committeemen and aldermen and plasterers. Royko needed friends.

"He didn't want to walk into a club without any playing companions, and I was about the only friend he had who could afford a country club, so he said, hey, chump, you want to join Sunset Ridge?" Tim Weigel said. "Actually, he was very flattering when he recruited me. He conned me into joining with him. It was one of the best moves I ever made because I enjoy it thoroughly. We had a great time. His schedule was similar to mine and we wound up playing two, three mornings a week. Often, it was just him and me. Most of the time it was uproarious. It was like taking the bar out on to the golf course and having a few swings between the cigars."

Saturday, December 17, 1994, was Judy's birthday, and Royko went out to shop for a present. He hadn't had time earlier because Judy had just undergone surgery and had to remain in bed, while Royko attended to her and handled the children. "He went to the pro shop at Sunset Ridge," Judy Royko said, "and naturally the pro shop is right next to the bar so he had a glass of wine or two. He was driving home and talking to me from the phone in the car when the accident happened. I heard it."

Royko collided with another vehicle on Tower Road in Winnetka, hardly a mile from his home. Police were called, and Royko was arrested. He was charged with driving under the influence and resisting arrest. He posted 10 percent of his $1,000 bond and was released. "The few glasses of wine combined with all the medication he was taking and the stress he was under, the worry over my surgery, all had an effect," Judy Royko said.

The story was all over the Sunday television news. The *Tribune* printed a brief story on Monday.

Even Royko's celebrated ketchup incident of 1977 didn't get the media coverage of his latest arrest. Royko was not only a major celebrity, he was an institution. And drunken driving was no longer something that people got out of by slipping a Chicago cop a twenty-dollar bill. The campaign against drunken drivers had become a political issue, with everyone from the governor on down trying to pass more strict legislation. The lobbying group Mothers Against Drunk Driving (MADD) had become a powerful force in state legislatures. The *Chicago Tribune*, like most blue-chip companies, frowned on its employees drinking.

Royko was again embarrassed and depressed. He never liked Christmas, and this season would be one of his worst. He did not appear in the newspaper until after New Year's Day.

He relied on Slats to explain his latest troubles.

When we met for our traditional New Year's drink of Ovaltine, Slats Grobnik said: "Tell me about those pills. You buy them across the counter or does the doc have to write a prescription?"

"Pills? What pills?"

"Those Stupid Pills. I figure you been taking lately."

"Do you mind if we talk about something else?"

"How you got to be public enemy No. 1?"

"That's a slight exaggeration. But no, my attorney advised me against saying anything until we go to court."

"Oh, that ought to be fun with the cameras being shoved in your face and the TV reporters asking you how you feel, and if you regret being a jerk, and if you're ever going to do it again, and if you are thinking about hanging yourself, and how now you ain't got no credibility no more, and are you going to get in another line of work, and did your boss say you ought to go to the Betty Ford clinic, and how are you going to get around without wheels, and is your wife going to get a divorce, and how are you ever going to write something bad about someone else when now you are such a bum, and if you decide to hang yourself will you do it where they can get a shot for the 4 o'clock news, and..."

It was a skillful admission that Royko knew all the questions that people were asking and that he was asking them himself. It was also a case of the best defense being a good offense. He was trying to dilute the impact of what he knew would be a television circus when he appeared in court. The column concluded with a promise, to himself more than anyone else, that he was going to stop drinking. As Slats continued his harangue, Royko injected: "How about if we just have another Ovaltine?"

"You're no fun."

"Yes, and it's about time."

Royko had a hearing on January 23, 1995, and when he left the courtroom he was surrounded by television reporters who wanted him to explain what went on in the courtroom instead of listening to the proceedings themselves. He wrote a column calling them "ninnies." He wrote that the TV stations showed him and the cameras and the reporters moving through the parking garage but left out his asking the reporters why they were stupid.

Uncharacteristically, Royko made a mistake identifying which stations the reporters worked for and had to addend a correction to the next day's column. His now growing number of critics pointed to the error, saying Royko was losing his edge.

That might have been the end of it, but in June, a copy of the Winnetka police department arrest report was leaked to the media. Royko blamed State's Attorney Jack O'Malley's office for the leak, which proved more embarrassing than the arrest charges.

"What are you, Croatian? You fucking loser. What's your ethnicity, you fag?" Royko reportedly yelled at the police. "Why are you wearing those fag gloves?"

Royko was referring to the surgical gloves the police officers were wearing in order to protect themselves during physical contact with a person. The act of being treated as a criminal, especially one who might be infected with AIDS, would anger a sober person. Royko argued that it was the combination of his medicines—especially antidepressants—that exacerbated his reaction.

Gay and lesbian groups printed flyers with copies of Royko's remarks and distributed them on the North Side and in front of Tribune Tower. Gay rights groups urged the *Tribune* to fire Royko. The various radio and television stations broadcast excerpts of his remarks.

In July, the *Wall Street Journal* wrote a story saying that Royko had gone from "professional curmudgeon to crude bully."

"For the first time," Peter Gorner said, "he wasn't a hero to journalists. He was used to being a heroic figure, someone that every one admired. In the past, when he was arrested or got into some bar brawl, everyone winked. After all, that was what Mike Royko did. But there was no more wink-wink. Jack O'Malley wanted him to do jail time for the DUI."

O'Malley became the subject of several Royko columns, all of them critical, some inferring he botched the case of several brutal criminals. In 1996, O'Malley was defeated for reelection.

In June, Royko was fined $1,600, given two years probation, and had his driver's license suspended. Eddie Cronin, his sister Eleanor's husband, became his driver on the increasingly fewer visits he made to the office.

Howard Tyner, who had succeeded Jack Fuller as *Tribune* editor in 1993, said he often tried to get Royko to come into the office more. "He said he didn't have any friends here. He said he never felt at home here."

Royko, who always had trouble sleeping, was spending his nights going on-line, exchanging e-mails with other late-night friends such as Rick Soll, and entertaining himself in various chat rooms. He particularly enjoyed engaging right-wing conservatives. He gave himself screen names and baited liberals with his imitations.

"It was amazing," Gorner said. "People were getting two and three hours a night of free Royko."

One of the chat-room persons who always supported him was Trucker Babe, who was his legman, Pam Cytrynbaun. His favorite computer personality was named "USFirst."

"It's like Method acting," he told Cytrynbaum. "I created a character. I pictured a burly guy from the Southwest Side. Lithuanian. Thick neck, with a short, gray crew cut. He reads a lot, knows about politics. He was a low-level guard at a concentration camp, but hasn't been caught."

Royko also used the screen name "Mitch60," posting a message that said, "Don't trust this Perot fellow. His name sounds French, and you know we can't trust those Frenchies. Remember that movie, 'Casablanca?'"

Cytrynbaum said, "One of the real right-wingers wrote back: 'You're crazy or a mole for *Mother Jones* or the *Village Voice.*'" He had been caught.

Despite his fascination with the computer, Royko was less than agreeable when Tyner asked if he would go on-line to answer questions from *Tribune* readers. The *Tribune* was an early investor and believer in America Online and was attempting to create business synergies between its publishing, broadcast, and on-line subsidiaries.

"Royko did it," Tyner recalled, "with three standard replies. One was polite, one was not so polite and one was 'fuck you.'"

If his column had gotten too personal and introspective for some critics—which is what the *Wall Street Journal* said—it was also still remarkably astute and sometimes improbably prophetic.

Only two days after his withering attack on the television women covering his court appearance—which good friend Studs Terkel called an example of "winging the sparrow instead of the vulture"—Royko regained his perspective and wrote about the biggest story in the nation. He predicted that O. J. Simpson would walk out a free man.

Royko was always better than anyone at examining an issue from every imaginable position and often shared that tortured process with his readers in columns that were persuasive, insightful, and comical. He often mirrored exactly the process the readers were going through to reach similar conclusions.

When the story broke that ice-skater Nancy Kerrigan had been attacked by henchmen of her chief rival, Tonya Harding, Royko's column took readers through the process with him. At the start, he wrote, "That's it—it is a disgrace and she should be barred from competing in the Olympics."

> Then I remembered the basic principle of law and fairness—we are all innocent until proven guilty. And I quickly reversed myself and said: "Let her skate."

But when her ex-husband said he was part of the plot, Royko assumed that Tonya must have known about it. "Cast her into the darkness."

But, he then reasoned, there are many things husbands do that wives don't know about. "Let her have a chance for glory."

Tonya then admitted she knew that her crowd had bashed Kerrigan's knee.

"That clinches it....She is unworthy and should be banished."

But, Royko explained, he remembered her sad, deprived childhood with a mean, self-centered mother, living in dumps and spun again: "Let her skate."

Tonya, however, was an adult, and if she knew what had happened she should have told her coach. "But she kept lying. I wouldn't let her skate in my driveway," he concluded.

Back and forth he went, just as much of America and the U.S. Skating Committee was doing. Finally, he voted for Tonya, but not from any conclusion based on some lofty moral principle.

> What could be better show biz than the really mean and nasty girl competing against the really sweet and good girl?... Goodness and honesty being challenged by the forces of darkness and sleaze.
>
> My money will be on Tonya. Why should ice skating be any different from just about everything else in the world?"

In 1996, Royko made another effort to quit drinking, entering the Hazelden rehabilitation clinic in Minnesota. He told virtually no one. He called Tyner to say he needed time off for medical reasons and wanted an unpaid leave of absence. "He didn't say what he was doing but I knew and he knew I knew so we didn't have to say. I told him he would be paid no matter how much time he was off."

Judy Royko said, "He really tried hard with his drinking. But he never could bring himself to say, 'I am an alcoholic.'"

Peter Gorner recalled Royko telling him, "The doctor says I'm an alcoholic but I tell him he's wrong. I'm a drunk."

"He was under a lot of pressure from doctors," Gorner said. "He had one who was worrying about his heart, who wanted him to quit smoking; one who wanted him to quit drinking; one who wanted him to stop eating the stuff he loved. They were taking everything away."

Royko was even staying away from temptation. He sent an e-mail to former legman Janan Hanna, "I'm staying away from Ridgemoor and Sunset these days. I have found that when I finish a round of golf, my thirst becomes overwhelming.... Wish they invented a martini that was good for you."

Royko had a difficult moment—but a fine column—after he heard that Michael Jordan was going to play a round of golf at Sunset Ridge and promised Sam that he could try to get an autograph from the most celebrated athlete in the world.

The column headline read, "Oh, the humiliation a dad must endure."

Standing by the golf course pond with fishing rod in his hand, the man felt foolish and nervous....While he appeared to be spending some quiet time in the shade of an old oak tree, pulling out an occasional bluegill or small bass, he was really furtively watching and waiting for a certain person to walk by.

When that person appeared, the man would do something he had never done in his entire life and never thought he would do. He would ask a celebrity for an autograph. The prospect filled him with shame. He would have preferred to dive into the pond or run to his car and roar away with his pride intact. But he couldn't. The reason was standing next to him holding a fishing pole—his 8-year-old son....

Suddenly they were there, teeing off, then walking briskly down a patch near the pond, not ten feet from the man and his son.

The man blurted, "Good morning." Surely they'd pause and ask how they're biting.

Not a word. Not a glance. Eyes squinting, long strides, the Great One nodded once he was past them....

The man took a deep breath and said, "Michael?" The Great One stopped, turned and gave him a cold look....

"Uh, Michael, could I have a second of your time?"

Without hesitating, he said, "You've already had a second."

The man had an urge to back off and grab a fellow worm.

But the smiling, wide-eyed boy was looking up, as if at a god. So the man stammered: "Look, he hasn't worn anything else since you came back. He sleeps in that outfit. Could you maybe just initial...?"

A slight hesitation, then he stepped forward, took the marker and made a long squiggle on the jersey....With a slight smile, he said: "There you go, my little man."

Later, in the car, the man told his wife: "He said, 'There you go, my little man.'"

"Why, that's really nice," she said.

"Yeah. But I wonder if he was talking to me."

Royko understood better than anyone how difficult it was for a celebrity to preserve his privacy. He knew how much he detested having his own golf game interrupted by hand-shakers or even partners who wanted to talk about his column, so he understood he was crossing the boundary of "jerkism," but his devotion to Sam forced him to endure Jordan's cold stare. The incident was also another example of the ego pendulum swinging to its extensions. In a heady moment years earlier when he thought he would be one the three famous "Mikes" in a golf charity outing, Royko seemed comfortable to be classed in the same fame niche with Jordan. But after the autograph incident, he told a friend, with a mixture of disappointment and gratitude, "He had no idea who I was."

Still, Sam's joy made it worthwhile. And Royko was doing more things that he enjoyed. He and Judy and the children made several trips to Longboat Key in Florida, where he could fish, golf, and play with the children. In 1995 they bought a condominium, and the back door was fifty yards from the beach. As he had done with Weigel at Sunset Ridge, he began recruiting Dan Hurley to move to Longboat Key so he would have a golfing buddy.

But work life was often still strained. In January 1996, Royko was under fire again for an article he wrote about Maurica Taylor, who was being hauled into court as the defendant in a paternity suit. The charge was unusual, because Ms. Taylor was a woman.

Royko wrote that the entire bureaucratic screw-up was the fault of Ms. Taylor's mother, who decided to name her child Maurica:

> While I can understand the desire among many African-Americans to give their children names that are individualistic to provide a sense of uniqueness and identity (What, you don't like curbstone sociology?), sometimes it can lead to confusion and all sorts of problems.
>
> Some black names defy explanation. For example, one of the finest players in basketball is Anfernee Hardaway of the Orlando team.
>
> With due respect, what kind of a name is Anfernee?

The phones rang and the letters flowed. Royko called Lois Wille at her retirement home in Virginia for reassurance. He got none.

"I told him it was a dumb column. Why didn't he pick on all the people who name their little girls Taylor or Schuyler or Corey, and you can't tell if it's a girl or a boy?"

The following Tuesday Royko ran a letters column:

"This was not satire, not sarcasm and NOT funny! It wasn't even good writing. It was a complete departure from the key points of your article. But it was a grubby, mean, unnecessary slam that deserves some serious soul and head checking on the part of those who wrote, approved and printed it," one reader wrote. There was no sarcastic reply:

"Comment: You won't get an argument from me. It was a pretty bad column.

"Maybe not the worst I've ever written, but it's in the running. All I can say is that I truly wish I hadn't written it, and I feel worse about it than you do because my name is on it."

Another reader said, "My son is named after his grandfather, and he had a 'strange' name. But he's proud of it. My husband has a 'strange' name, but we all like it and are proud of it. We have pride when we choose African names. You need your butt kicked."

"Comment: No, I need my brain kicked. But in my case, maybe there's no difference. Let me close by saying that these letters represent not only the views of many black and some white readers. They also reflect the thinking of people who work on this paper—editors, reporters and others. I didn't intend the column to sound the way it turned out. Unfortunately, the need to babble as often as I do sometimes leads to unintended and unfortunate results. To those who were offended—readers and colleagues—I apologize. I'd slit my wrists, but the sight of blood makes me faint."

A month later came the column that Royko wrote about presidential candidate Patrick Buchanan's xenophobic preachings. It was satire, but Royko said he suspected that Buchanan even used the word "beaners," and it infuriated the Mexican-American community, led to picketing at Tribune Tower and wholesale criticism in the *Tribune* newsroom.

"He was very hurt by the whole thing," Peter Gorner said. "He really needed to be liked and now people were saying they didn't like him and the staff wasn't behind him. He wasn't their hero anymore.

"If I had seen the column before it was printed, I would have said something like, 'Sounds like you mean it?' meaning that some people might not get the satire. In one sense, he meant it, but he didn't mean to hurt anyone. But there wasn't anyone to question him. The people on the copy desk just checked his spelling, there wasn't anyone he respected who could say the column might offend some people.

"But I would have run it. It was Royko."

Lois Wille was quoted as saying, "Mike Royko hasn't changed. The times have changed."

In 1983, during the vituperative mayoral race between Harold Washington and Bernard Epton, Royko wrote that he was receiving letters accusing him of being a "nigger lover." His column went on to list racial epithets.

What do they say? Take your pick. Shine. Coon. Nigger. Jig. Spade. Boogy. Jungle Bunny. Buffalo. Cloud. Smoke. Darky. It's amazing now many imaginative racial epithets there are for blacks. They lead all other ethnic and racial groups by at least 20–1 in that regard.

Listing those epithets in 1983 was a means for Royko to damn discrimination and the white bigots who used them. A decade later, members of his own newsroom would rebel when he used far softer epithets in quoting the Los Angeles Police Department's radio conversations. The idea that Royko could or should change, which was voiced around the desks of the *Tribune*, was indicative of how the newspaper business had changed. Royko knew it and hated it.

When he heard that the *Tribune* had made another change in metro editors, he sent a message to Janan Hanna: "I remember the days when being city editor, as it used to be called, was almost a lifetime job. Of course, that was when editors were out-front take-charge hands-on ass-kicking chiefs, not administrators and group circle jerk leaders. Of course, that was when there were mostly grown-ups in the news room, not a bunch of ladder-climbing, moralizing, knee-jerk elitist, yogurt-eating, fruit juice drinking, back stabbing rats. Wish I was a serial killer."

In the summer of 1996 Royko was interviewed by Tower Productions in connection with the upcoming 150th anniversary of the *Chicago Tribune*. It had also been forty years since Royko started out as a rookie with Lerner Newspapers.

"There is a great deal about the newspaper business I miss," he said. "A great deal of the fun and the spirit has gone out of working for a newspaper. The major change is the kind of people who work on newspaper. There's been a change in the world view, community view of newspaper people. They are not in step with the general population, and there's the political correctness thing.

"There was more of a sense that a newspaper was family. If somebody got a story on page one, you'd say, 'Good story,' and not, 'Why not me?' You never heard the word career in the old days. It's not the world's biggest sin to have a couple of beers after work.

"Today, the average newspaper guy or woman went from high school to journalism school and had internships during the summer. They'd be better off if they had waited tables or worked on a garbage truck. You could fire a machine gun in the Tribune newsroom and not find one veteran of the Vietnam War. They don't view Clinton as a draft dodger. They were all draft dodgers. One guy bragged to me that his physician lied about his hearing. Newspaper people used to be native Chicagoans. We used to sit in the Billy Goat and play street corners. Name an intersection and someone had to name at least one business on that intersection. If you were stumped, you were embarrassed. Most newspaper people here today don't know anything about Chicago."

The 1996 Democratic National Convention was held in Chicago for the first time since 1968, and Royko was swamped with requests for interviews from all over the country. He refused them. He visited the convention arena only once. He did not go to the Republican convention in San Diego, just as he had skipped both conventions in 1992.

He was writing almost full-time from home.

In October, he sent Hanna an e-mail:

"I'm feeling pretty good. I switched to a new anti-depressant and it seems to work. I do my cardio workouts Mon Wed and Fri A.M. at Evanston Hospital. An hour of moderately vigorous exercise in pleasant, airy surroundings, with cheerful nurse-therapists who act as if my well being is the most important thing in their lives. There is one muscular little tart who leads me through a light weight lifting routine while giving me vamp looks. I'm thinking of having an affair with her, since she is qualified to revive me should my heart stop at a critical moment."

Royko underwent angioplasty procedure the day after Christmas and postponed his usual January trip to Florida until the spring.

He told friends that 1997 would be his final year. He told *Tribune* editor Tyner he planned to retire in September. He told Peter Gorner he was seeking financing to buy *North Shore* magazine, hire his friends, Jim Warren and Rick Kogan, and turn it into a real newsmagazine. He told Dan Hurley he was thinking of buying a castle in Ireland where they could both retire. A few weeks later he took the family to Longboat Key.

He filed his last column before leaving on March 21, 1997. The column dealt with subjects that Royko had written about extensively: the Chicago Cubs and Billy Goat and race.

It was time, Royko wrote, to stop blaming an old Greek and his goat for the curse he put on the Cubs in 1945 and to start pointing the finger at longtime owner P. K. Wrigley and the men he hired who refused to sign black ballplayers after Jackie Robinson broke the color line in 1947. It was not until 1953 that the Cubs signed black players. They were one of last teams in baseball to do it. It was racism, not the Goat, which haunted the Cubs.

On the morning of Wednesday, March 26, Royko was taking a shower when he felt a tremendous pain in his head. He told Judy, "It feels like someone is hitting my head with a ball peen hammer." He gulped aspirin and laid down for most of the day. The next day he felt better and he and Judy played golf. He shot the back nine in thirty-nine. On Friday, the family went bike riding, and on Saturday he booked a boat to go fishing on Sunday. Saturday night he fell asleep on the sofa watching the NCAA basketball tournament.

On Sunday, he told Judy he was feeling lousy. At five o'clock he told her to take him to the hospital in Sarasota.

"They did test after test all week," Judy Royko said. "On Friday, they did a spinal tap and said that he had suffered an aneurysm in the brain. They wanted to do surgery immediately and scheduled it for that afternoon."

Judy called Royko's cardiologist in Chicago, Dr. Joseph Messer, who had qualms about such a serious surgery being done in Sarasota. He made arrangements for specialists at the University of Florida Hospital in Gainesville to see Royko.

"We went in an ambulance that was doing ninety miles an hour along the shoulder to get by all the traffic. Still, it took three and a half hours," Judy said. At Gainesville, surgeons told Royko there was only a 50 percent chance that he would survive the operation and not be impaired in some way. "He accepted that and the next day he was in surgery for six and one-half hours. At 2 A.M., they called and told me everything had gone well," Judy Royko said.

In the meantime, *Tribune* editors tried to reach Hanke Gratteau to send her to Florida to help Judy with the children. She was not available, but Ellen Warren was. Warren flew to Florida. Judy had also called John Schackitano, who also immediately flew to Florida. Bob and Geri Royko had just returned to Madison from their Florida vacation when they learned of Royko's illness and flew back to Tampa, where they were met by David Royko. The three of them went to the condominium in Longboat Key, where Judy's sister, Connie, had arrived also to help with Sam and Kate.

Doctors told Royko that the aneurysm was unlikely to reoccur. At the same time they told him that three of the carotid arteries feeding blood to his brain were blocked and the fourth one was not in perfect shape. Royko remained in the University of Florida Hospital for nearly two weeks after the surgery, grumbling about wanting to go home. He had also been taken off his antidepressant medicine, which concerned Judy.

By the second week of April, Tribune Company executives were meeting at the Disneyworld Center in Orlando, and Tyner asked Tribune chairman John Madigan if the corporate plane could be used to take Royko back to Chicago.

"We picked him and Judy up in Gainesville and I thought he looked terrible. He slept most of the way. When we got to Chicago, we had a car waiting to take us to Rush Presbyterian Hospital. He perked up in the car and began talking about a column he was going to write in a few days about Tiger Woods. When we got to the hospital, he insisted on smoking a cigarette outside before he was admitted."

The *Tribune* had printed only a brief story saying Royko had suffered a minor stroke. Letters and e-mail containing good wishes arrived from all over the country.

After two days at Rush, Royko went home on Friday, April 18. "The first thing he did outside the hospital was to light a cigarette," Judy Royko said. Judy had employed twenty-four-hour nursing care and also insisted he put on nicotine patches and stop smoking. He had resumed taking his antidepressant medication before leaving Rush.

On Sunday, Judy thought it might be a good idea for him to talk to Jimmy Breslin, who had also gone through an aneurysm himself and had written about it. "I got Studs to track down a number and we finally got Breslin. He and Mike talked a long time and afterward Mike felt really good about it."

On the night of April 22, Royko climbed slowly to his third-floor office. Judy and the children were on the first floor. As he reached the doorway of his office he collapsed. Judy found him on the floor and called 911. An ambulance arrived at 9:15 P.M. and found him unconscious. He was rushed to Evanston Hospital, five miles south. He had suffered another aneurysm. The next day he was taken to Northwestern Memorial Hospital in downtown Chicago, where neurological specialists determined there was nothing they could do.

He died on April 29, 1997.

He had instructed Judy that he wanted a very private funeral. He was cremated and his ashes were buried in two places. One urn was placed in a mausoleum vault at Acacia Cemetery, where Carol was buried. Another was placed under a headstone beside his father at the same cemetery.

Epilogue

Mike Royko, who worried so much about his legacy, needn't have bothered. Chicago television stations reported his medical condition at each newscast during the final days of his life. When he died, all of them devoted a big chunk of their newscasts to pay him tribute. Camera crews descended on the Billy Goat, where Sam Sianis tearfully mourned his best friend and dozens of newsmen who rarely went there stopped by for a final drink in Royko's memory. The *Tribune* devoted an eight-page spread in its Sunday, May 4, edition to Royko, excerpting some of his best columns, providing a lengthy obituary, and printing tributes from around the nation. Television networks reported his death, and the *New York Times* gave him a lengthy and laudatory farewell. He would have enjoyed it.

When Pam Cytrynbaum went to work for Royko she began keeping a journal.

"Just wait'll I'm dead, kid," Royko told her. Cytrynbaum kept her promise and didn't open her journal to the public until after he was gone. She wrote a reminiscence of Royko for the *Chicago Tribune Magazine* in 1999. She remembered his words about retiring, which like so many of the millions of words he had written, were prophetic: "The day I go, nobody's gonna know it. It's not gonna be like Reston or any of those guys. I hate those memories columns. Crap. At the end of my last newspaper column it will say: 'Personal note: This is my last newspaper column. Time to go. Goodbye.' Or something like that. That's it."

Cytrynbaum asked, "Don't the readers deserve to get used to the idea, to say goodbye?"

"Jesus Christ, it's just 50 cents. You just go. One day, I'll be there, the next day I won't."

As with almost everyone who makes a mark, changes the way people think in some small or large measure, or makes them angry, happy, sad, or wiser, timing is everything. Royko was a product of his time. He was born into an enclave of American ethnic immigrants who were on the verge of shaking off the traditional discrimination and poverty that welcomed them to the United States. They were also clannish and proud of their heritage and willing to endure whatever it took to survive and thrive. Royko was proud of the way his family and friends fought through the depression and then fought and won a great war. His admiration for that generation was unrestrained. Their values—hard work, self-reliance, and the individual freedoms to be wise or foolish, poor or rich, happy or sad—became his values.

These were the themes he wrote about over and over. But he did not rejoice when that generation became the first members of the new American middle class and aped their predecessors in discriminating against the next wave of Americans trying to climb out of poverty. He was too young to have shared the experiences of the depression and World War II, but he witnessed firsthand the civil rights era that unalterably changed American culture in the second half of the century. He was in the perfect place to deride the anachronistic political processes that served only political leaders and not the people. He called Muhammad Ali a patriot when most of America thought him a traitor. He knew a stupid war when he saw it. He did not approve of the longhaired profane youths of the 1960s and 1970s, but he understood them. He was the first to label America's proliferation of guns as insanity. He watched as the new political reformers made all the same mistakes as the old political bosses. He was not afraid to take African-American political leaders to task for acting like Irish-American political bosses.

He could never tolerate arrogance in a million-dollar-a-year network anchor or a thirty-dollars-a-day bus driver. And he found arrogance everywhere in the spoiled and nouveau riche America of the eighties and nineties, when wealth meant celebrity and celebrity meant success.

H. L. Mencken brought his crusty social commentary and criticism to an America that was smug, almost Babbitt-like in its convictions about rugged individualism, free enterprise, fundamentalist religion, Anglo-Saxon values, and laissez-faire government. He appeared at a time when newspapers mirrored, as they always have, the mainstream sentiments of both political leadership and a public sense of well-being.

Mencken's influence waned during the 1930s, when the depression wiped away the collective American self-assurance and arrogance. Mencken was still there, but his favorite target, societal and governmental pomposity, had vanished. And the newspapers which were among Mencken's favorite targets no longer reeked of boosterism but reflected the somber times.

In the complacent 1950s, the survivors of the depression and the great triumph over fascism feasted on the economic and technological marvels of high wages, low interest, fast cars, cheap gas, electric kitchen gadgets, publicly financed higher education, television, and

Disneyland. That great sense of national satisfaction again was reflected in newspapers. Newspapers did not write about the great migration of blacks to northern cities, where their concentration would incubate centuries of bitterness and anger into a firebomb of racial unrest. The papers continued printing only what government leaders wanted them to say. Their publishers, editors, and most of their reporters saw themselves as little more than stenographers searching for presidents, governors, and mayors.

Mike Royko was in the vanguard of the newspapermen who in the 1960s revolted against the apathy of their predecessors. Since Royko was among the most talented, the most widely read, the most admired, and the most imitated, he was properly the heir to the Mencken tradition.

He was also a muckraker in the same vein as Upton Sinclair and Lincoln Steffens, rooting out malfeasance in political and bureaucratic cesspools rather than slaughter yards. He was a humorist equal to the best of James Thurber or Robert Benchley, and he far eclipsed their output.

Royko could be an echo or an aria. He sometimes wrote for everyman, sharing the frustrations, anger, and whims of blue-collar Americans. He was sometimes on stage alone, bellowing his sorrow or grief or outrage.

He was a storyteller who could twist an ending like O. Henry.

At his core, Royko was a newspaperman, a breed which he once described this way:

> If there's a big puddle in the street, a gentleman puts his coat down for a lady to walk on. But a newspaperman spits in the puddle, then goes to ask the sewer department why the puddle isn't draining.

He did all this with wit, indignation, satire, sensitivity, skepticism, anger, and love.

No other newspaperman of his time wrote for as long a period as Royko, and no one ever did it as often. He never had the kind of impact that a Walter Lippman had in the days when only newspapers provided information and thought. He was never read as widely as the gossip mavens in their heyday. But in his time, he was the best. And that time is past. Newspaper readership continues to decline, and the nation's senses are bombarded with opinions from talk shows and call-in shows and shows filled with journalists shouting at one another. It is doubtful that in the current or foreseeable future a single writer will be able to raise a public outcry, influence an electorate, ridicule the most powerful, and defend the weak the way Royko did so well for so long. He was not only the best. He was the last.

Royko would be delighted to know that a posthumous collection of his columns was a best-seller. One part of him would be astonished that he was ranked among the top twenty-five Chicagoans of the twentieth century in various publications that marked the new millennium. And part of him would also be disappointed that he wasn't ranked first.

After his death, one reader wrote the *Tribune* and said, "Mike Royko would have been a great carpenter. He always hit the nail on the head."

Royko would have replied: "Spare me the bullshit."

Index

PublicAffairs is a nonfiction publishing house founded in 1997. It is a tribute to the standards, values, and flair of three persons who have served as mentors to countless reporters, writers, editors, and book people of all kinds, including me.

I. F. Stone, proprietor of *I. F. Stone's Weekly*, combined a commitment to the First Amendment with entrepreneurial zeal and reporting skill and became one of the great independent journalists in American history. At the age of eighty, Izzy published *The Trial of Socrates*, which was a national bestseller. He wrote the book after he taught himself ancient Greek.

Benjamin C. Bradlee was for nearly thirty years the charismatic editorial leader of *The Washington Post*. It was Ben who gave the *Post* the range and courage to pursue such historic issues as Watergate. He supported his reporters with a tenacity that made them fearless and it is no accident that so many became authors of influential, bestselling books.

Robert L. Bernstein, the chief executive of Random House for more than a quarter century, guided one of the nation's premier publishing houses. Bob was personally responsible for many books of political dissent and argument that challenged tyranny around the globe. He is also the founder and longtime chair of Human Rights Watch, one of the most respected human rights organizations in the world.

———

For fifty years, the banner of Public Affairs Press was carried by its owner Morris B. Schnapper, who published Gandhi, Nasser, Toynbee, Truman and about 1,500 other authors. In 1983, Schnapper was described by *The Washington Post* as "a redoubtable gadfly." His legacy will endure in the books to come.

Peter Osnos, *Publisher*